T0137416

METHODS IN MOLECULAR BIOLOGY

Series Editor
John M. Walker
School of Life and Medical Sciences
University of Hertfordshire
Hatfield, Hertfordshire, AL10 9AB, UK

For further volumes:
http://www.springernature.com/series/7651

Noncanonical Amino Acids

Methods and Protocols

Edited by

Edward A. Lemke

*Structural and Computational Biology Unit & Cell Biology and Biophysics Unit,
EMBL, Heidelberg, Germany; Departments of Biology and Chemistry,
Pharmacy and Geosciences, Johannes Gutenberg-University, Mainz, Germany;
Institute of Molecular Biology (IMB), Mainz, Germany*

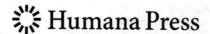 Humana Press

Editor
Edward A. Lemke
Structural and Computational Biology Unit &
Cell Biology and Biophysics Unit
EMBL
Heidelberg, Germany

Departments of Biology and Chemistry, Pharmacy and Geosciences
Johannes Gutenberg-University
Mainz, Germany

Institute of Molecular Biology (IMB)
Mainz, Germany

ISSN 1064-3745   ISSN 1940-6029   (electronic)
Methods in Molecular Biology
ISBN 978-1-4939-8525-8   ISBN 978-1-4939-7574-7   (eBook)
https://doi.org/10.1007/978-1-4939-7574-7

Printed on acid-free paper

This Humana Press imprint is published by Springer Nature
The registered company is Springer Science+Business Media, LLC
The registered company address is: 233 Spring Street, New York, NY 10013, U.S.A.

Preface

Over billions of years of evolution, nature has developed multiple types of organisms from just "20 + 2" naturally occurring amino acids. Over roughly the last three decades, mankind equipped itself with the tools for expanding the genetic code in ways that theoretically could enable us to extend this number to infinity. Only a few years ago, papers reporting on genetic code expansion typically started with an introduction reporting a list of things that the encoding of novel noncanonical amino acids (also known as "non-natural amino acids" or "unnatural amino acids") might enable scientists to achieve, such as the photochemical control of protein function, or the installation of epigenetic marks or chemical handles for further reactions. Today, it is probably fair to say that almost anything might be possible, taking into account that at least 200 different noncanonical amino acids with different functionalities have been encoded into a variety of organisms in a site-specific or proteome-wide manner.

In the 24 chapters constituting this book, we cover some of the most widely used protocols, providing detail and advice for users to get each method up and running for their chosen application. The protocols have been grouped into three parts, describing methods for protein production in the test tube, in prokaryotes, and in eukaryotes. However, these should not be seen as strict categories. Modern genetic code expansion systems are frequently orthogonal in many hosts, which means that they can be developed in one and then simply transferred to another. The choice of appropriate host enables different applications, ranging from cell-free to multicellular organism production of proteins having novel functionality. From this collection of protocols, readers will learn techniques that enable them to design new experiments and new areas of research to be created. Ideas that could simply not be thought of before.

I would like to thank all the authors for their extensive efforts in contributing to this book, and for taking the time to develop detailed protocols that also alert the reader to the many pitfalls that over the years they encountered and overcame as these methods matured.

I am particularly grateful to my research team who were involved extensively in helping me to compile this book.

Mainz, Germany   *Edward A. Lemke*

Contents

Contributors

MATHIS BAALMANN • *Institute of Pharmacy and Molecular Biotechnology, Ruprecht-Karls-Universität Heidelberg, Heidelberg, Germany*

ROBYN M. BARFIELD • *Catalent Pharma Solutions, Emeryville, CA, USA*

MARCEL BEST • *Institute of Pharmacy and Molecular Biotechnology, Ruprecht-Karls-Universität Heidelberg, Heidelberg, Germany*

ROBERT J. BLIZZARD • *Department of Biochemistry and Biophysics, Oregon State University, Corvallis, OR, USA*

SIMONE BRAND • *Chemical Genomics Centre of the Max Planck Society, Dortmund, Germany*

NEDILJKO BUDISA • *Biocatalysis Group, Department of Chemistry, Berlin Institute of Technology/TU Berlin, Berlin, Germany*

ABHISHEK CHATTERJEE • *Department of Chemistry, Boston College, Chestnut Hill, MA, USA*

PENG R. CHEN • *Peking-Tsinghua Center for Life Sciences, Beijing, China; Beijing National Laboratory for Molecular Sciences, Synthetic and Functional Biomolecules Center, College of Chemistry and Molecular Engineering, Peking University, Beijing, China*

IRENE COIN • *Institute of Biochemistry, University of Leipzig, Leipzig, Germany*

LLOYD DAVIS • *Centre for Integrative Physiology, University of Edinburgh, Edinburgh, UK*

MARK B. VAN ELDIJK • *Division of Chemistry and Chemical Engineering, California Institute of Technology, Pasadena, CA, USA*

SIMON J. ELSÄSSER • *Science for Life Laboratory, Division of Translational Medicine and Chemical Biology, Department of Medical Biochemistry and Biophysics, Karolinska Institutet, Stockholm, Sweden*

SARAH B. ERICKSON • *Department of Chemistry, Boston College, Chestnut Hill, MA, USA*

GEMMA ESTRADA GIRONA • *Structural and Computational Biology Unit, European Molecular Biology Laboratory, Heidelberg, Germany; Cell Biology and Biophysics Unit, European Molecular Biology Laboratory, Heidelberg, Germany*

TRUE E. GIBSON • *Department of Biochemistry and Biophysics, Oregon State University, Corvallis, OR, USA*

YUKI GOTO • *Department of Chemistry, Graduate School of Science, The University of Tokyo, Tokyo, Japan*

SEBASTIAN GREISS • *Centre for Integrative Physiology, University of Edinburgh, Edinburgh, UK*

CHRISTIAN P.R. HACKENBERGER • *Department of Chemical-Biology, Leibniz-Forschungsinstitut für Molekulare Pharmakologie (FMP), Berlin, Germany; Department of Chemistry, Humboldt Universität zu Berlin, Berlin, Germany*

JINGXUAN HE • *Department of Chemistry and Chemical Biology, University of New Mexico, Albuquerque, NM, USA*

JONAS HELMA • *Department of Biology II, Center for Integrated Protein Science Munich, Ludwig Maximilians Universität München, Planegg-Martinsried, München, Germany*

JAN C.M. VAN HEST • *Institute for Molecules and Materials, Radboud University Nijmegen, Nijmegen, The Netherlands*

CHRISTIAN HOFFMANN • *Georg August University Göttingen, Göttingen, Germany; Accurion GmbH, Göttingen, Germany*

YOSHIHIKO IWANE • *Department of Chemistry, Graduate School of Science, The University of Tokyo, Tokyo, Japan*

JI-YONG KANG • *Department of Neuroscience, School of Medicine, Tufts University, Boston, MA, USA*

TAKAYUKI KATOH • *Department of Chemistry, Graduate School of Science, The University of Tokyo, Tokyo, Japan*

DAICHI KAWAGUCHI • *Graduate School of Pharmaceutical Sciences, The University of Tokyo, Tokyo, Japan*

PÉTER KELE • *Chemical Biology Research Group, Research Centre for Natural Sciences, Institute of Organic Chemistry, Hungarian Academy of Sciences, Budapest, Hungary*

RACHEL E. KELEMEN • *Department of Chemistry, Boston College, Chestnut Hill, MA, USA*

CHRISTINE KOEHLER • *Structural and Computational Biology Unit, EMBL, Heidelberg, Germany; Cell Biology and Biophysics Unit, EMBL, Heidelberg, Germany; Department of Biology, Johannes Gutenberg-University Mainz, Mainz, Germany; Institute of Molecular Biology (IMB), Mainz, Germany*

ESZTER KOZMA • *Chemical Biology Research Group, Research Centre for Natural Sciences, Institute of Organic Chemistry, Hungarian Academy of Sciences, Budapest, Hungary*

MICHAEL LAMMERS • *Institute for Genetics and Cologne Excellence Cluster on Cellular Stress Responses in Aging-Associated Diseases (CECAD), University of Cologne, Cologne, Germany*

EDWARD A. LEMKE • *Structural and Computational Biology Unit & Cell Biology and Biophysics Unit, EMBL, Heidelberg, Germany; Departments of Biology and Chemistry, Pharmacy and Geosciences, Johannes Gutenberg-University, Mainz, Germany; Institute of Molecular Biology (IMB), Mainz, Germany*

HEINRICH LEONHARDT • *Department of Biology II, Center for Integrated Protein Science Munich, Ludwig Maximilians Universität München, Planegg-Martinsried, München, Germany*

XIANG LI • *Department of Biomedical Engineering, University of California, Irvine, CA, USA*

SHIXIAN LIN • *Beijing National Laboratory for Molecular Sciences, Synthetic and Functional Biomolecules Center, College of Chemistry and Molecular Engineering, Peking University, Beijing, China*

CHANG C. LIU • *Department of Biomedical Engineering, University of California, Irvine, CA, USA; Department of Chemistry, University of California, Irvine, CA, USA; Department of Molecular Biology and Biochemistry, University of California, Irvine, CA, USA*

WENSHE R. LIU • *Department of Chemistry, Texas A&M University, College Station, TX, USA*

YING MA • *Biocatalysis Group, Department of Chemistry, Berlin Institute of Technology/ TU Berlin, Berlin, Germany*

RYAN A. MEHL • *Department of Biochemistry and Biophysics, Oregon State University, Corvallis, OR, USA*

CHARLES E. MELANÇON III • *Department of Chemistry and Chemical Biology, University of New Mexico, Albuquerque, NM, USA; Department of Biology, Center for Biomedical Engineering, University of New Mexico, Albuquerque, NM, USA*

HEINZ NEUMANN • *Georg August University Göttingen, Göttingen, Germany; Max-Planck-Institute of Molecular Physiology, Dortmund, Germany*

PETRA NEUMANN-STAUBITZ • *Georg August University Göttingen, Göttingen, Germany; Max-Planck-Institute of Molecular Physiology, Dortmund, Germany*

IVANA NIKIĆ-SPIEGEL • *Werner Reichardt Centre for Integrative Neuroscience (CIN), The Eberhard Karls University of Tübingen, Tübingen, Germany*

GIULIA PACI • *Structural and Computational Biology Unit, European Molecular Biology Laboratory, Heidelberg, Germany; Cell Biology and Biophysics Unit, European Molecular Biology Laboratory, Heidelberg, Germany*

DAVID RABUKA • *Catalent Pharma Solutions, Emeryville, CA, USA*

MARTINO L. DI SALVO • *Dipartimento di Scienze Biochimiche "A. Rossi Fanelli", Sapienza Università di Roma, Rome, Italy*

MORITZ J. SCHMIDT • *University of Konstanz, Constance, Germany*

DOMINIK SCHUMACHER • *Department of Chemical-Biology, Leibniz-Forschungsinstitut für Molekulare Pharmakologie (FMP), Berlin, Germany; Department of Chemistry, Humboldt Universität zu Berlin, Berlin, Germany*

LISA SEIDEL • *Institute of Biochemistry, University of Leipzig, Leipzig, Germany*

EIKO SEKI • *RIKEN Structural Biology Laboratory, Yokohama, Japan*

HIROAKI SUGA • *Department of Chemistry, Graduate School of Science, The University of Tokyo, Tokyo, Japan; JST-CREST, Tokyo, Japan*

DANIEL SUMMERER • *University of Konstanz, Constance, Germany; Department of Chemistry and Chemical Biology, TU Dortmund University, Dortmund, Germany*

JEFFERY M. THARP • *Department of Chemistry, Texas A&M University, College Station, TX, USA*

LEI WANG • *Department of Pharmaceutical Chemistry, University of California, San Francisco, CA, USA; Cardiovascular Research Institute, University of California, San Francisco, CA, USA*

RICHARD WOMBACHER • *Institute of Pharmacy and Molecular Biotechnology, Ruprecht-Karls-Universität Heidelberg, Heidelberg, Germany*

YAO-WEN WU • *Chemical Genomics Centre of the Max Planck Society, Dortmund, Germany; Max-Planck-Institute of Molecular Physiology, Dortmund, Germany*

TATSUO YANAGISAWA • *RIKEN Structural Biology Laboratory, Yokohama, Japan*

SHIGEYUKI YOKOYAMA • *RIKEN Structural Biology Laboratory, Yokohama, Japan*

HUANGTAO ZHENG • *Peking-Tsinghua Center for Life Sciences, Beijing, China; Academy for Advanced Interdisciplinary Studies, Peking University, Beijing, China*

Part I

Engineering in the Test Tube

Chapter 1

Leveraging Formylglycine-Generating Enzyme for Production of Site-Specifically Modified Bioconjugates

Robyn M. Barfield and David Rabuka

Abstract

Enzymatic modification of proteins can generate uniquely reactive chemical functionality, enabling site-specific reactions on the protein surface. Formylglycine-generating enzyme (FGE) is one enzyme that can be exploited in this fashion. FGE binds its consensus sequence (CXPXR, known as the "aldehyde-tag") and converts the cysteine to a formylglycine (fGly). fGly-containing proteins contain a bioorthogonal aldehyde on their surface that can be modified selectively in the presence of the 20 canonical amino acids. Here, we describe protocols for the generation of a site-specifically modified protein, an antibody-drug conjugate (ADC), using aldehyde-tagging protocols and aldehyde-reactive conjugation chemistry.

Key words Aldehyde tag, Antibody-drug conjugate (ADC), Formylglycine (fGly), Formylglycine-generating enzyme (FGE), Site-specific, SMARTag™

1 Introduction

The generation of site-specifically modified bioconjugates has gained widespread interest over the past several years due to their defined molecular structures, expanded utility, simplified analytics, and potential benefits as therapeutics [1]. Site-specific bioconjugation reactions require bioorthogonal chemistries, or chemistries that are directed to a single chemical entity not found in naturally occurring polypeptides [2]. The aldehyde tag technology incorporates a chemoenzymatic solution to site-specificity. A five amino acid consensus sequence (CXPXR, also referred to as the "aldehyde tag") is genetically encoded into the protein of interest at the desired location within the peptide backbone [3, 4]. During protein expression, the cysteine within the consensus sequence is oxidized to a formylglycine (fGly) residue by formylglycine-generating enzyme (FGE; Fig. 1). FGE resides in the endoplasmic reticulum (ER) and co-translationally modifies the cysteine. fGly bears an aldehyde functional group that serves as a chemical handle for subsequent conjugation. The aldehyde is an ideal site for chemistries

Edward A. Lemke (ed.), *Noncanonical Amino Acids: Methods and Protocols*, Methods in Molecular Biology, vol. 1728, https://doi.org/10.1007/978-1-4939-7574-7_1, © Springer Science+Business Media, LLC 2018

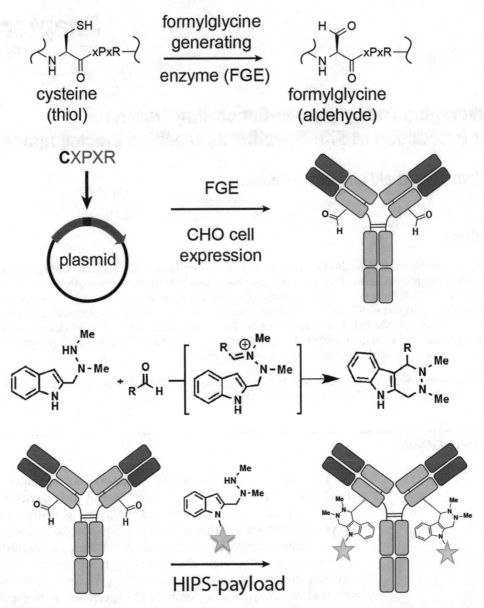

Fig. 1 Generation of a SMARTag™ bioconjugate. (**a**) FGE oxidizes cysteine to fGly (within the context of its consensus sequence). (**b**) A site-specific aldehyde is engineered into an antibody backbone by cloning the minimal FGE consensus sequence/aldehyde tag (CXPXR) into one or multiple sites within the coding region of the antibody using simple molecular biology techniques. The antibody is produced in a cell line stably or transiently overexpressing human FGE. FGE converts the cysteine within the aldehyde to an fGly, and an antibody containing a reactive aldehyde is produced. (**c**) Overview of the aldehyde-reactive hydrazino-*iso*-Pictet-Spengler (HIPS) ligation. (**d**) Aldehyde-tagged mAbs purified from cell media using standard techniques can be directly conjugated to a HIPS-payload to generate a site-specific bioconjugate

that are tuned to only react with carbonyls, found nowhere else in a native polypeptide. We have developed a unique aldehyde-specific conjugation chemistry, known as the Hydrazino-*iso*-Pictet-Spengler (HIPS) reaction. This reaction can be performed near neutral pH and results in the formation of a stable C–C bond between the protein being modified and the chemical HIPS payload (Fig. 1) [5, 6]. The conjugates formed using HIPS chemistry are remarkably stable both in solution and in vivo, making HIPS conjugation ideal for development of bioconjugates to be used in a therapeutic setting.

The aldehyde-tag technology can be used to generate a variety of different bioconjugate classes [6–11]. Here, we detail use of this technology for the generation of antibody-drug conjugates (ADCs). ADCs combine the target specificity and long half-life of an antibody with the high potency of a synthetic small molecule to offer a distinct therapeutic advantage over either molecule on its own [12]. It is becoming increasingly clear that site-specific conjugation offers significant advantages over conventional lysine and cysteine conjugation chemistries for ADCs. Site-specific payload placement allows for control over both the drug-to-antibody ratio (DAR) and the conjugation site, both of which play an important role in governing the pharmacokinetics (PK), disposition, and efficacy of the ADC [6, 13, 14]. This chapter describes the production and analysis of site-specific ADCs using aldehyde tag technology, also referred to as SMARTag™ technology throughout. The first step involves genetically engineering the aldehyde tag into either the heavy or light chain constant region of a monoclonal antibody (mAb). Next, heavy and light chain expression vectors are co-transfected into mammalian cells along with a human FGE expression vector. The secreted aldehyde tag containing antibody is then purified using standard techniques and conjugated under mild conditions to an aldehyde-reactive HIPS-Maytansine payload-linker. The ADC is then purified and the integrity and DAR of the final ADC is analyzed.

2 Materials

2.1 Aldehyde-Tag Incorporation Using Site-Directed Mutagenesis

1. IgG1 heavy chain expression vector (5 ng/μL) (*see* **Note 1**).
2. Forward and reverse primers with desired aldehyde tag sequence insertions and substitutions (100 ng/μL) (*see* **Note 2**).
3. dNTP mix (100 mM).
4. 10× PfuUltra reaction buffer.
5. PfuUltra high-fidelity DNA polymerase (2.5 U/μL).
6. DpnI (10 U/μL).
7. Supercompetent *E. coli* cells, such as XL1-Blue.
8. LB medium.

9. LB-agar plates containing selection antibiotic appropriate for heavy chain or light chain expression construct.

10. LB medium containing selection antibiotic appropriate for heavy chain or light chain expression construct.

2.2 SMARTag™ mAb Production

1. ExpiCHO-S cells (ThermoFisher).

2. ExpiCHO™ expression medium (ThermoFisher).

3. Human FGE expression vector [15] (*see* **Note 3**).

4. IgG1 heavy chain expression vector (containing aldehyde tag encoding sequence at chosen site inserted in Subheading 2.1).

5. IgG1 light chain expression vector.

6. 100 mM $CuSO_4$ [15].

7. ExpiFectamine™ CHO transfection reagent (ThermoFisher).

8. OptiPRO™ SFM complexation reagent (ThermoFisher).

9. ExpiFectamine™ CHO enhancer reagent (ThermoFisher).

10. ExpiCHO™ Feed (ThermoFisher).

11. 5 mL Mab Select SuRe column (GE Healthsciences).

12. FPLC Binding and Wash Buffer: 20 mM NaCitrate, pH 7.2, 150 mM NaCl.

13. FPLC Elution Buffer: 20 mM NaAcetate, pH 3.5, 50 mM NaCl.

14. Neutralization Buffer: 1 M triethanolamine, pH 7.4.

15. mAb storage buffer: 20 mM NaCitrate, pH 5.5, 50 mM NaCl.

16. TFF Pellicon 3 Cassette with Ultracel PLCTK membrane, 88 cm2, 30 kDa MWCO.

17. 15 mL 30 kDa MWCO centrifugal filter.

2.3 fGly Quantification

1. mAb from Subheading 2.2.

2. 1 M NH_4HCO_3, pH 8.8.

3. 0.5 M Dithiothreitol (DTT).

4. 1 µg/µL trypsin.

5. 0.5 M iodoacetimide.

6. 3 M methoxylamine.

7. 1 M HCl.

8. 1 M NaCitrate, pH 5.5.

9. Jupiter™ 150 × 1.0-mm C18 column (Phenomenex).

10. Skyline software (University of Washington).

2.4 Conjugation Reaction and Conjugate

1. 500 mM NaCitrate, pH 5.5.

2. 20 mM NaCitrate, pH 5.5, 50 mM NaCl.

3. mAb from Subheading 2.2.

4. 10% Triton X-100.

5. 100 mM HIPS-Glu-2PEG-Maytansine drug-linker dissolved in N, N-Dimethylacetamide (DMA) [16] (*see* **Note 4**).

6. TFF Pellicon 3 Cassette with Ultraecl PLCTK membrane, 88 cm2, 30 kDa MWCO.

7. 5 mL HiTrap Phenyl HP FPLC Hydrophobic Interaction Chromatography (HIC) column (GE Health Sciences).

8. FPLC sample diluent: 2 M $(NH_4)_2 SO_4$, 25 mM $NaPO_4$, pH 6.8.

9. FPLC buffer A: 1 M $(NH_4)_2 SO_4$, 25 mM $NaPO_4$, pH 6.8.

10. FPLC buffer B: 25% isopropanol, 18.75 mM $NaPO_4$, pH 6.8.

11. 15 mL 30 kD MWCO centrifugal filter.

2.5 Conjugate
Analysis

1. TSK gel Butyl-NPR HIC column, 4.6 mm × 35 mm (Tosoh).

2. HIC buffer A: 1.5 M $(NH_4)_2 SO_4$, 25 mM $NaPO_4$, pH 6.8.

3. HIC buffer B: 25% isopropanol, 18.75 mM $NaPO_4$, pH 6.8.

4. ChemStation software (Agilent Technologies).

5. 0.5 mM DTT.

6. 1× PBS.

7. PLRP-S HPLC column, 1000 Å, 2.1 × 50 mm, 8 μm (Agilent).

8. PLRP buffer A: 0.1% trifluoroacetic acid in H_2O.

9. PLRP buffer B: 0.1% trifluoroacetic acid in CH_3CN.

10. G3000 SW XL SEC column, 7.8 × 300 mm (Tosoh).

11. SEC buffer: 300 mM NaCl, 25 mM $NaPO_4$, pH 6.8.

3 Methods

3.1 Aldehyde-Tag
Incorporation Using
Site-Directed
Mutagenesis

1. In a PCR tube, combine 5 μL 10× reaction buffer, 2 μL heavy chain expression vector, 1.25 μL forward primer, 1.25 μL reverse primer, 1 μL dNTP mix, and 39.5 μL ddH$_2$O. Then add 1 μL PfuUltra polymerase.

2. Incubate the sample in a thermal cycler for 2 min at 95 °C followed by 18 cycles of 30 s at 95 °C, 1 min at 52 °C, and 6 min at 68 °C.

3. Chill the sample on ice for 2 min, add 1 μL DpnI, mix, pulse-spin, and incubate at 37 °C for 1 h.

4. Add 1 μL of the DpnI-treated sample to 50 μL competent cells that have been thawed on ice. Gently mix and incubate on ice for 30 min.

5. Heat the transformation reaction at 42 °C for 45 s and then incubate on ice for 2 min.

6. Add 500 μL LB medium and shake in a 37 °C incubator for 30 min.

7. Plate 250 μL on two different LB-agar plates with antibiotic, and incubate overnight at 37 °C.

8. Inoculate single colonies in LB-medium containing appropriate selection antibiotic and shake in a 37 °C incubator for 8–16 h.

9. Isolate plasmid DNA using a Qiagen miniprep kit. Sequence the plasmid DNA to confirm insertion of the aldehyde-tag encoding sequence and to confirm there are no undesired mutations present.

3.2 SMARTag™ mAb Production

1. Subculture ExpiCHO-S™ cells at 37 °C in an 8% CO_2 incubator to a density of 4–6 × 10^6 viable cells/mL at 125 rpm (*see* **Note 5**).

2. Day—1: Dilute the cells to 3–4 × 10^6 and grow overnight at 37 °C in an 8% CO_2 incubator to a density of 4–6 × 10^6 viable cells/mL at 125 rpm.

3. Day 0: Dilute the cells to 6 × 10^6 viable cells/mL with pre-warmed ExpiCHO™ expression medium to a final volume of 200 mL in a 1 L flask. Add $CuSO_4$ to a final concentration of 100 μM [15].

4. Dilute 60 μg IgG1 heavy chain (containing aldehyde tag encoding sequence), 90 μg IgG1 light chain, and 50 μg FGE expression constructs with 7.9 mL cold OptiPRO™ SFM complexation reagent in a 50 mL Falcon tube. Invert the tube to mix.

5. Combine 640 μL ExpiFectamine™ CHO Reagent with 7.4 mL OptiPRO™ medium in a 50 mL Falcon tube. Invert the tube to mix.

6. Add the ExpiFectamine/OptiPRO™ mixture to the DNA/OptiPRO mixture, invert the tube to mix, and incubate at room temperature for 1–5 min.

7. Slowly transfer the DNA solution to a flask from **step 3** in Subheading 3.2 (*see* **Note 6**).

8. Incubate the cells at 37 °C overnight in an 8% CO_2 incubator with shaking at 125 rpm.

9. Day 1: Add 1.2 mL ExpiFectamine™ CHO enhancer and 32 mL ExpiCHO™ feed to the flask and transfer the flask to **32 °C** in a **5%** CO_2 incubator with shaking at 125 rpm (*see* **Note 7**).

10. Day 5: Add 32 mL ExpiCHO™ feed to the flask and continue incubating at 32 °C until Day 10.

11. Day 10: Transfer cultures to 50 mL Falcon tubes and spin for 15 min at 4000 × *g*. Filter supernatant through a 0.45 μm membrane.

12. Load the supernatant at 2 mL/min onto a 5 mL Mab Select SuRe chromatography column equilibrated in binding buffer. Wash with 10 column volumes (CVs) of wash buffer, and elute with 10 CVs of elution buffer into collection vials containing 10% neutralization buffer by volume (*see* **Note 8**).

13. Formulate the mAb in 20 mM NaCitrate, pH 5.5, 50 mM NaCl by diafiltration using a 30 kDa TFF cassette and exchanging for eight diavolumes (*see* **Note 9**).

14. Collect mAb from TFF and concentrate in a 15 mL 30 kDa MWCO Amicon centrifugal filter to ~18 mg/mL (*see* **Note 10**).

3.3 fGly Quantification

1. To 20 μg of mAb from Subheading 2.2, add 1 μL NH_4HCO_3, 1 μL DTT, and water to a final volume of 20 μL. Vortex and pulse-spin sample. Incubate the sample at 37 °C for 15–30 min (*see* **Note 11**).

2. Remove the sample from 37 °C. Add 1.5 μL HCl and vortex.

3. Add 1 μL NH_4HCO_3 and vortex (*see* **Note 12**).

4. Add 1 μL trypsin and 3 μL iodoacetimide. Vortex and pulse-spin sample. Incubate at 37 °C for 60 min in the dark (*see* **Note 13**).

5. Remove the sample from 37 °C. Add 5 μL NaCitrate and 1.3 μL methoxylamine. Vortex and pulse-spin sample. Incubate at 37 °C overnight.

6. Analyze the processed sample by LC-MRM/MS using a 150 × 1.0-mm C18 column enclosed in a column heater set to 65 °C. Calculate LC-MRM/MS transition masses and integrate the data using Skyline [17] (*see* **Note 14**; Fig. 2).

3.4 Conjugation Reaction and Conjugate Purification

1. To 15 mg SMARTag™ mAb, add 100 μL 500 mM NaCitrate, pH 5.5, 25 μL 10% TX-100, 25 μL 100 mM payload-linker, and 20 mM NaCitrate, pH 5.5, 50 mM NaCl to 15 mg of SMARTag™ mAb to a final volume of 1000 μL (*see* **Note 15**). Incubate the reaction mixture for 18–24 h at 37 °C (*see* **Note 16**).

2. Remove unconjugated payload-linker by diafiltration using a 30 kDa TFF cassette and exchanging for 12 diavolumes into 20 mM NaCitrate, pH 5.5, 50 mM NaCl (*see* **Note 17**).

3. Collect ADC from TFF and add FPLC sample diluent to the ADC sample to a final concentration of 0.8 M (*see* **Note 18**).

4. Load the diluted ADC at 4 mL/min onto a 5 mL HiTrap Phenyl HP HIC chromatography column equilibrated in Buffer A, then wash with two column volumes of Buffer A. An isocratic hold at 31% B for 7 CVs is used to elute any DAR 0 material. This is followed by a 7 CV linear gradient of 41–95% B to elute DAR 1 and DAR 2 species and then an isocratic hold at 95% B to elute any remaining DAR 2 species. Fractions

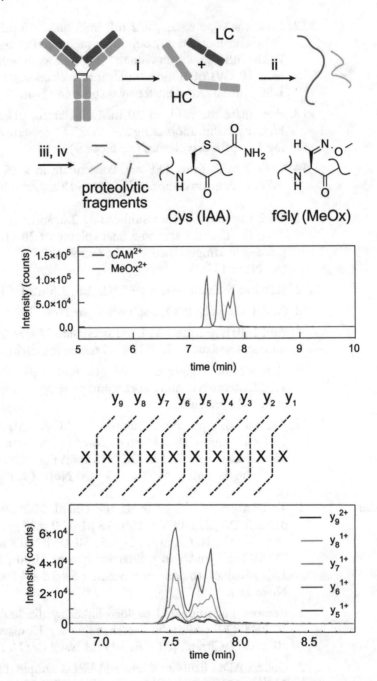

Fig. 2 LC-MRM/MS method to quantify fGly on an antibody backbone. (**a**) Monoclonal antibody is disassembled and digested, and the Cys and fGly functional groups are trapped as the carboxyamidomethyl (CAM) and methyl oxime (MeOx) species, respectively, for characterization, through the following steps: (i) DTT, pH $\geq$ 7.5, 37 °C, 15 min; (ii) HCl, then NH_4HCO_3; (iii) trypsin, iodoacetamide, pH $\geq$ 7.5, 37 °C, 1 h; (iv) methoxylamine, pH 5.5, 16 h. (**b**) Each aldehyde tag location is characterized by a unique tryptic peptide containing the installed tag sequence. Targeted MRM-MS is used to quantify the peptides modified as CAM or MeOx. The most intense precursor/product ion transitions are chosen for peak area integration. (**c**) Ion chromatogram of the CAM- and MeOx-containing peptides. The MeOx-containing peptide separates into the diastereomers present as a result of oxime formation. This research was originally published in The Journal of Biological Chemistry © the American Society for Biochemistry and Molecular Biology [18]

are collected into vials containing 33% 20 mM NaCitrate, pH 5.5, 50 mM NaCl by volume (*see* **Note 19**).

5. Pool fractions to give desired final DAR.

6. Formulate the mAb in 20 mM NaCitrate, pH 5.5, 50 mM NaCl by diafiltration using a 30 kDa TFF cassette and exchanging for eight diavolumes (*see* **Note 20**).

7. Collect ADC from TFF and concentrate in a 15 mL 30 kDa MWCO Amicon centrifugal filter to desired final concentration. Sterilize using a 0.2 µM filter.

3.5 Conjugate Analysis

1. HIC analysis for quantifying DAR: Inject 20 µg of ADC, diluted 1:1 in HIC Buffer A, onto a butyl HIC column equilibrated with 88% Buffer A and 12% Buffer B (*see* **Note 21**).

2. Elute at 1.0 mL/min with a liner gradient from 12 to 42% B for 5.5 min, followed by a liner gradient from 42 to 70% B for 2 min, followed by a linear gradient from 70 to 100% B for 1 min and a 1 min hold at 100% B. Monitor the protein at 280 nm UV absorbance.

3. Reverse phase analysis for quantifying DAR: Incubate 20 µg of ADC, diluted in PBS to 90 µL plus 10 µL DTT at 37 °C for 30 min.

4. Load the sample onto a PLRP column equilibrated with 75% PLRP buffer A and 25% PLRP buffer B.

5. Elute at 2.0 mL/min with a linear gradient from 25 to 50% B for 8 min, followed by a linear gradient of 50 to 95% B for 0.1 min and a 1.9 min hold at 95% B. Monitor the protein at 280 nm UV absorbance.

6. SEC analysis for quantifying % monomer: Inject 20 µg of ADC onto an SEC column equilibrated with SEC buffer.

7. Elute at 1.0 mL/min for 20 min with an isocratic hold at 100% buffer A. Monitor the protein at 280 nm UV absorbance.

8. Use ChemStation software to quantify ratios of different DAR species for HIC and PLRP and to quantify ADC monomer, high molecular weight, and low molecular weight species for SEC (*see* **Note 22**; Fig. 3).

4 Notes

1. Numerous locations have been identified along the heavy and light chain constant regions where the aldehyde tag sequence can be installed without perturbing the native IgG1 structure while also allowing for efficient conjugation of payload. Drake et al. discusses several of these sites [6]. While the minimal FGE recognition sequence across species is CXPXR, for human FGE we have found the optimal sequence to be LCTPSR and therefore include this six residue sequence. Here, a location in

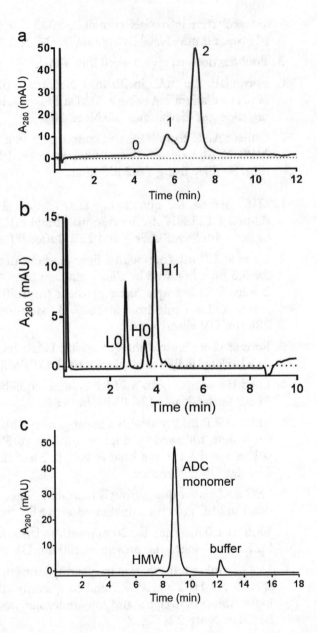

Fig. 3 Analysis of a representative SMARTag™ ADC with a maytansine payload conjugated to the heavy chain. (**a**) Hydrophobic interaction chromatography shows separation between unconjugated mAb (0), mAb conjugated to one payload (1), and mAb conjugated to two payloads (2). The separation between the different species allows for accurate quantification of DAR as well as conjugate enrichment (on a HIC preparative column), if desired. (**b**) Reverse phase chromatography of a DTT-reduced ADC is an orthogonal method to quantify DAR. This method shows separation between unconjugated light chain (L0), unconjugated heavy chain (H0), and heavy chain conjugated to 1 payload (H1). (**c**) Size exclusion chromatography demonstrates a highly monomeric SMARTag™ ADC with only a very small proportion of high molecular weight species (HMW)

the heavy chain CH1 region is highlighted and therefore site-directed mutagenesis of the heavy chain expression vector is described.

2. Depending on the exact location and surrounding residues of the aldehyde tag sequence, LCTPSR can be introduced as a complete insertion or as a combination of insertions and substitutions. For the particular CH1 location described, primers are used to introduce LCTPSR such that the heavy chain constant region sequence is changed from SWNSGALTSGVHTFPA to SWNSGALCTPSRGVHTFPA.

3. As an alternative to transfecting three separate vectors (expressing human FGE, IgG1 heavy chain, and IgG1 light chain), the human FGE DNA sequence can be incorporated into the light chain expression vector under a separate EF1 promoter. In this case, only two expression vectors (IgG1 heavy chain and IgG1 light chain/human FGE) are transfected at a ratio of 2 heavy chain:3 light chain. Conversion of cysteine to fGly is more efficient under these conditions, likely due to a greater transfection efficiency of FGE into cells where the heavy and light chains are being expressed.

4. The SMARTag™ system is compatible with a variety of different payloads, including cytotoxins (such as maytansine), nucleic acids, and fluorophores. Depending on the payload used for conjugation, different solvents may be optimal for dissolving the payload. It is important to note that even trace amounts of formaldehyde and acetone can react with the indole on the HIPS portion of the deprotected payload-linker. Use of high-quality solvents to dissolve the payload is important to mitigate this risk. Avoid using solvents such as methanol and other alcohols when handling unreacted HIPS reagents as they can be contaminated with aldehydes or carbonyls that can react with the HIPS reagent. In addition, any glassware that comes in contact with the HIPS reagent should never come in contact with acetone.

5. A transient expression system for generating SMARTag™ mAbs (where heavy chain, light chain, and FGE constructs are transfected) is outlined here for quicker production of aldehyde-tagged mAbs. SMARTag™ mAbs can also be generated using cell lines stably expressing heavy chain, light chain, and FGE. Historically, we have demonstrated that the GPEx stable transduction system works very efficiently with SMARTag™ methodology [15].

6. During **steps 7–12** in Subheading 3.2, care should be taken to avoid exposure of the cell culture medium to white light. We have observed that fGly, in combination with cell culture media and light, can degrade over time. It is best to work in red light

and/or cover flasks during cell growth and purification. Once the fGly-containing mAb is eluted from the Mab Select SuRe resin, this degradation of fGly no longer takes place and the fGly is stable in the solution.

7. The drop in temperature from 37 °C to 32 °C as well as the feeding protocol help improve titers of the mAb.

8. Depending on the scale and levels of mAb expression, the mAb can be purified in batch instead of on an FPLC. Although Tris buffer is commonly used for the neutralization step, avoid using Tris because of its potential reactivity with the aldehyde on the fGly.

9. 20 mM NaCitrate, pH 5.5, 50 mM NaCl works well as both a storage buffer and conjugation buffer for many SMARTag™ mAbs. Therefore, formulating the mAb in this buffer minimizes mAb manipulation before conjugation.

10. Concentrating mAbs to 18 mg/mL supports conjugation reactions at 15 mg/mL. In order to minimize the payload:mAb equivalents used in the conjugation reaction, SMARTag™ mAbs are conjugated at 15 mg/mL. Small amounts of high molecular weight (HMW) species may form during concentration. This HMW species will dissociate upon conjugation, generating highly monomeric final product.

11. The pH of the sample should be ≥7.5 in order for DTT to work effectively. It is important to confirm pH especially when using mAb with a low concentration. In this case, the contribution of the mAb buffer to the total reaction volume is more substantial.

12. It is important for the pH of the sample to be ≥7.5 for efficient digestion by trypsin.

13. Iodoacetimide should be made up fresh with H_2O. Iodoacetimide is light sensitive; keep the samples in the dark during the iodoacetimide reaction.

14. Skyline is a free, open source software program developed at the University of Washington (https://skyline.gs.washington.edu/labkey/project/home/software/Skyline/begin.view). Skyline enables rapid method development and data analysis of LC-MRM/MS-based studies and other proteomics applications. The MRM transitions include the following modifications of the LCTPSR aldehyde-tag containing peptide: cysteine, cysteine modified with iodoacetamide (carboxyacetamidomethyl-Cys; CAM), fGly, fGly hydrate, fGly methyl oxime, and glycine. The Skyline predicted peptides are analyzed by LC-MRM/MS and the most significant precursor/product ion pairs are combined to determine relative abundance of each component. We do not detect unreacted Cys, fGly, or fGly hydrate above background.

15. The optimal pH range for conjugation of HIPS payloads to SMARTag™ mAbs is 5.0–6.0. Alternative buffers can be used but conjugation efficiency should be tested. Depending on the solubility of the HIPS-linker-payload to be used for conjugation, different amounts and combinations of organic solvents, surfactants, and excipients can be tested for optimal conjugation.

16. The protocol outlined here calls for 2.5 mM of HIPS-linker-payload and a reaction time of 16–24 h. Alternatively, 1.7 mM or 0.85 mM HIPS-linker-payload can be used for a reaction time of 48 or 72 h, respectively. The molarity of the HIPS payload is more important than the number of payload:mAb equivalents. Rotation of the conjugation mixture is optional depending on reaction volume and payload. Conjugation yields are typically in the range of 95% relative to the amount of fGly. The fGly can be present as the fGly hydrate. The presence of fGly hydrate does not impact conjugation efficiency as it is in equilibrium with the aldehyde.

17. Diafiltration is the most scalable method for the removal of free payload. For small-scale reactions and for reactions where the residual payload concentration is not a concern, alternative methods can be used (i.e., desalting and size exclusion chromatography).

18. The FPLC HIC chromatography step is optional. If the DAR after the free payload removal step meets specification, the HIC chromatography step to enrich for higher DAR species can be eliminated. In that case, after the collection from the TFF, concentrate the ADC to the desired final concentration and sterilize using a 0.2 μM filter.

19. The conjugation process outlined in **step 1** in Subheading 3.4 is scalable for 0.1 mg up to at least 1000 mg. The HIC purification process outlined in **step 4** in Subheading 3.4 may require slight alterations (in the gradient, for instance) as the process is scaled up. In addition, when using a different payload-linker, the choice of HIC resin, gradient, and buffers will need to be adjusted to achieve optimal separation of different DAR species. For hydrophobic payload-linkers, such as the maytansine payload described here, separation by HIC chromatography works well. For hydrophilic payloads, separation of different DAR species by HIC chromatography might not be possible.

20. 20 mM NaCitrate, pH 5.5, 50 mM NaCl works well as a storage buffer for many SMARTag™ ADCs. Depending on the exact mAb/payload combination, other formulations could be ideal. SMARTag™ ADCs can be stored for at least 1 month at 4 °C and at least 1 year at −80 °C.

21. HIC and PLRP methods to determine DAR rely on the increased hydrophobicity imparted by conjugation of hydrophobic payloads, such as maytansine. These methods may not

separate differentially conjugated species for more hydrophilic payloads. In that case, DAR can be estimated by looking at the relative abundance of conjugated and unconjugated species by mass spectrometry.

22. ChemStation is a software package to control Agilent liquid chromatography systems. If using another HPLC system, the appropriate software can be substituted.

References

1. Drake PM, Rabuka D (2015) An emerging playbook for antibody-drug conjugates: lessons from the laboratory and clinic suggest a strategy for improving efficacy and safety. Curr Opin Chem Biol 28:174–180

2. Patterson DM, Nazarova LA, Prescher JA (2014) Finding the right (bioorthogonal) chemistry. ACS Chem Biol 9:592–605

3. Appel MJ, Bertozzi CR (2015) Formylglycine, a post-translationally generated residue with unique catalytic capabilities and biotechnology applications. ACS Chem Biol 10:72–84

4. Wu P, Shui W, Carlson BL, Hu N, Rabuka D, Lee J, Bertozzi CR (2009) Site-specific chemical modification of recombinant proteins produced in mammalian cells by using the genetically encoded aldehyde tag. Proc Natl Acad Sci U S A 106:3000–3005

5. Agarwal P, Kudirka R, Albers AE, Barfield RM, de Hart GW, Drake PM, Jones LC, Rabuka D (2013) Hydrazino-pictet-spengler ligation as a biocompatible method for the generation of stable protein conjugates. Bioconjug Chem 24:846–851

6. Drake PM, Albers AE, Baker J, Bañas S, Barfield RM et al (2014) Aldehyde tag coupled with HIPS chemistry enables the production of ADCs conjugated site-specifically to different antibody regions with distinct in vivo efficacy and PK outcomes. Bioconjug Chem 25:1331–1341

7. Liang SI, McFarland JM, Rabuka D, Gartner ZJ (2014) A modular approach for assembling aldehyde-tagged proteins on DNA scaffolds. J Am Chem Soc 136:10850–10853

8. Liu J, Hanne J, Britton BM, Shoffner M, Albers AE et al (2015) An efficient site-specific method for irreversible covalent labeling of proteins with a fluorophore. Sci Rep 19:16883

9. Hudak JE, HH Y, Bertozzi CR (2011) Protein glycoengineering enabled by the versatile synthesis of aminooxy glycans and the genetically encoded aldehyde tag. J Am Chem Soc 133:16127–16135

10. Hudak JE, Barfield RM, de Hart GW, Grob P, Nogales E et al (2012) Synthesis of heterobifunctional protein fusions using copper-free click chemistry and the aldehyde tag. Angew Chem Int Ed Engl 51:4161–4165

11. Carrico IS, Carlson BL, Bertozzi CR (2007) Introducing genetically encoded aldehydes into proteins. Nat Chem Biol 3:321–322

12. Sievers EL, Senter PD (2013) Antibody-drug conjugates in cancer therapy. Annu Rev Med 64:15–29

13. Junutula JR, Raab H, Clark S, Bhakta S, Leipold DD et al (2008) Site-specific conjugation of a cytotoxic drug to an antibody improves the therapeutic index. Nat Biotechnol 26:925–932

14. Strop P, Liu SH, Dorywalska M, Delaria K, Dushin RG (2013) Location matters: site of conjugation modulates stability and pharmacokinetics of antibody drug conjugates. Chem Biol 20:161–167

15. York D, Baker J, Holder PG, Jones LC, Drake PM et al (2015) Generating aldehyde-tagged antibodies with high titers and high formylglycine yields by supplementing culture media with copper(II). BMC Biotechnol 16:23

16. Albers AE, Garofalo AW, Drake PM, Kudirka R, de Hart GW et al (2014) Exploring the effects of linker composition on site-specifically modified antibody-drug conjugates. Eur J Med Chem 88:3–9

17. MacLean B, Tomazela DM, Shulman N, Chambers M, Finney GL, Frewen B, Kern R, Tabb DL, Liebler DC, MacCoss MJ (2010) Skyline: an open source document editor for creating and analyzing targeted proteomics experiments. Bioinformatics 26:966–968

18. Holder PG, Jones LC, Drake PM, Barfield RM, Bañas S, de Hart GW, Baker J, Rabuka D (2015) Reconstitution of formylglycine-generating enzyme with copper(II) for aldehyde tag conversion. J Biol Chem 290:15730–15745

Chapter 2

Artificial Division of Codon Boxes for Expansion of the Amino Acid Repertoire of Ribosomal Polypeptide Synthesis

Yoshihiko Iwane, Takayuki Katoh, Yuki Goto, and Hiroaki Suga

Abstract

In ribosomal polypeptide synthesis, the 61 sense codons redundantly code for the 20 proteinogenic amino acids. The genetic code contains eight family codon boxes consisting of synonymous codons that redundantly code for the same amino acid. Here, we describe the protocol of a recently published method to artificially divide such family codon boxes and encode multiple nonproteinogenic amino acids in addition to the 20 proteinogenic ones in a reprogrammed genetic code. To achieve this, an in vitro translation system reconstituted with 32 in vitro transcribed $tRNA_{SNN}$'s (S = C or G; N = U, C, A or G) was first developed, where the 32 tRNA transcripts can be charged with 20 proteinogenic amino acids by aminoacyl-tRNA synthetases in situ and orthogonally decode the corresponding 31 NNS sense codons as well as the AUG initiation codon. When some redundant $tRNA_{GNN}$'s are replaced with $tRNA_{GNN}$'s precharged with nonproteinogenic amino acids by means of flexizymes, the nonproteinogenic and proteinogenic aminoacyl-tRNAs can decode the NNC and NNG codons in the same family codon box independently. In this protocol, we describe expression of model peptides, including a macrocyclic peptide containing three kinds of N-methyl-amino acids reassigned to the vacant codons generated by the method of artificial division of codon boxes.

Key words Ribosome, Translation, Manipulation of genetic code, Nonproteinogenic amino acid, Nonstandard peptide, Codon box, tRNA, Anticodon, Flexizyme

1 Introduction

Ribosomal peptide synthesis enables us to construct peptide libraries with extraordinary diverse sequences as high as 10^{12} from the mRNA libraries containing a randomized region within their sequences. Particularly, a custom-made Flexible In-vitro Translation (FIT) system enables us to express "nonstandard" macrocyclic peptide libraries [1–7], in which the genetic code reprogramming method [8, 9] is employed to assign nonproteinogenic amino acids (Naa's) to certain sense codons. For example, a FIT system assigning N-chloroacetyl amino acids and N-methyl-amino acids to the

Edward A. Lemke (ed.), *Noncanonical Amino Acids: Methods and Protocols*, Methods in Molecular Biology, vol. 1728,
https://doi.org/10.1007/978-1-4939-7574-7_2, © Springer Science+Business Media, LLC 2018

initiation and elongation codons, respectively, can express thioether-macrocyclic *N*-methyl-peptide libraries [10]. Moreover, further integration of such macrocyclic peptide libraries with an in vitro display method [11, 12], such as RaPID (Random non-standard Peptides Integrated Discovery) system [13], allows us to rapidly screen potent binders against a designated target of proteins, leading to development of useful probes for biology as well as drugs for therapeutics. For example, we have discovered macrocyclic peptide probes for co-crystallization, which improved the resolution by stabilizing the protein tertiary structure [14]. Also, many potent and selective inhibitors have been discovered against a variety of proteins with therapeutic potentials [10, 15–17].

The above FIT system is based on a sense codon-suppression approach involving; (1) the preparation of a reconstituted in vitro translation system [18] where certain proteinogenic amino acids (Paa's) and their cognate aminoacyl-tRNA synthetases (AARSs) are omitted to create vacant codons, and (2) the addition of pre-charged Naa-tRNAs prepared by flexizymes (flexible tRNA-acylation ribozymes) [1] to the translation system for the reassignment of the Naa's to the vacant codons. Despite the fact that this approach has been successful to isolate potent macrocyclic nonstandard peptides, some Paa's must be sacrificed for creating the vacant codons in order to use Naa's for the incorporation. Considering that there are as many as 61 sense codons in the genetic code, one may think that there should be extra codons to accommodate additional repertoire of Naa's. But such an engineering has been difficult because all of the 61 sense codons are decoded by 45 kinds of native *E. coli* tRNAs [19] via "expanded" wobble base pairing mechanisms [20–22]. For example, the four codons constituting the valine codon box (GUN; N = U, C, A, or G) are decoded by two kinds of tRNAs possessing GAC and cmo⁵UAC (cmo⁵U: uridine 5-oxyacetic acid) anticodons, which are acylated with valine by valyl-tRNA synthetase (Fig. 1a). Due to the genetic code redundancy, the omission of some Paa's has been required to create vacant codons. If we could manipulate the decoding system and reduce the redundancy, more Naa's could be utilized without sacrificing any of the 20 Paa's. As such a method, we have recently developed a method of "artificial division of codon boxes" [23], which encodes more than two Naa's with all 20 Paa's retained in a reprogrammed genetic code. This "proto-col" discusses this new approach to reprogram the genetic code.

1.1 The Methodology of Artificial Division of Codon Boxes

To achieve the goal, we reconstituted an in vitro translation system with 32 in vitro transcribed tRNAs possessing SNN (S = G or C) anticodons. These 32 tRNA transcripts can be charged with 20 Paa's by endogenous AARSs and orthogonally decode the corresponding 31 NNS elongation codons as well as the AUG initiation codon (Fig. 1b). Among the 32 tRNAs, we can omit some

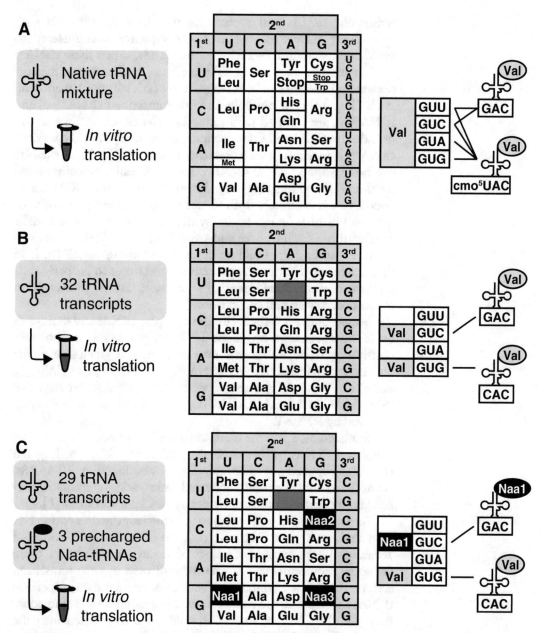

Fig. 1 Schematic representation of the artificial division of codon boxes. (**a**) The genetic code of a reconstituted translation system containing *E. coli* native tRNA mixture. The GUN codons in valine codon box are decoded by two kinds of native tRNA^{Val}s as illustrated. (**b**) A reprogrammed genetic code where 32 in vitro transcribed tRNAs decodes 31 NNS (S = C or G) elongation codons along with the AUG initiation codon. The UAG stop codon is vacant since RF1 is omitted from our translation system. The UAA and UGA codons, on the other hand, designate termination by means of RF2. (**c**) A reprogrammed genetic code containing 23 building blocks by means of artificial division of three codon boxes. The three nonproteinogenic amino acids (Naa's) are assigned to the black background codons by replacing the redundant tRNA$^{Val}_{GAC}$, tRNA$^{Arg}_{GCG}$, and tRNA$^{Gly}_{GCC}$ with the three Naa-tRNA$^{EnAsn}_{GNN}$'s prepared by flexizyme-mediated aminoacylation

redundant tRNAs and replace them with tRNA$^{AsnE2}_{SNN}$'s with appropriate anticodons, whose body sequence is engineered to avoid aminoacylation by endogenous AARSs. When these tRNA$^{AsnE2}_{SNN}$'s are precharged with different Naa's by flexizymes, the corresponding codons can be reassigned to Naa's without abandoning Paa's (Fig. 1c). For example, in vitro transcribed tRNA$^{Val}_{GAC}$ and tRNA$^{Val}_{CAC}$ are charged with valine by valyl-tRNA synthetase and decode the GUC and GUG codons, respectively (Fig. 1b). When tRNA$^{Val}_{GAC}$ is replaced with precharged Naa-tRNA$^{AsnE2}_{GAC}$, the Naa can be reassigned to the GUC codon while valine remains coded by the GUG codon (Fig. 1c). In other words, the GUN family codon box is artificially divided and reprogrammed to accommodate two distinct amino acids. By this strategy we have artificially divided three codon boxes and reassigned three distinct Naa's to them, expanding the building block repertoire up to 23 [23]. In this article, we describe the procedures and notes to properly conduct the reprogrammed translation reaction. The methods section is composed of several subsections.

Subheading 3.1 is the methods to prepare tRNAs and flexizymes by in vitro transcription. Since T7 RNA polymerase cannot efficiently synthesize RNAs that do not contain a guanosine at the 5′-terminus [24], four tRNAs (tRNA$^{Gln}_{CUG}$, tRNA$^{Pro}_{CGG}$, tRNA$^{Pro}_{GGG}$, and tRNA$^{Trp}_{GCA}$) are synthesized with a 5′-terminal leader sequence (5′-GGAACGCGCGACUCUAAU-3′), which is subsequently removed by RNase P.

Subheading 3.2 is the method to synthesize nonproteinogenic aminoacyl-tRNAs (Naa-tRNAs). tRNA$^{AsnE2}_{SNN}$'s, orthogonal tRNAs engineered not to be recognized by endogenous 20 AARSs, are acylated with chemically activated amino acids by means of artificial aminoacylation ribozymes flexizymes (eFx and dFx) [1].

Subheading 3.3 is the method to artificially divide a codon box with a minimal set of tRNAs. In order to utilize Naa efficiently and accurately by the method of artificial division of codon boxes, there are some requirements: (1) The reconstituted in vitro translation system should not contain native tRNA contaminant, especially tRNAs that decode the codons to be designated for Naa introduction. (2) The Naa of interest should be incorporated into the nascent polypeptide with sufficient efficiency that is comparable to Paa's. (3) The NNC and NNG codons in the artificially divided codon box should be orthogonally decoded by Naa-tRNA$_{GNN}$ and proteinogenic aminoacyl-tRNA$_{CNN}$, respectively, without cross reading. In order to verify these points, we recommend testing the artificial division using a simplified translation system containing a minimal set of tRNA transcripts and a single artificially divided codon box. As an example, Fig. 2 represents verification of artificial division of the GGN glycine codon box. While the control reaction using native tRNA yielded the peptide containing Gly at the sixth position (P1-Gly) (Fig. 2b, lane 1), the P1-Gly was not expressed

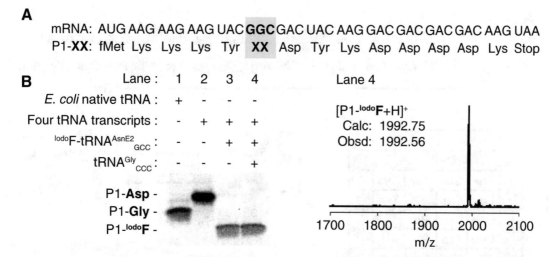

Fig. 2 Confirmation of artificial division of the GGN glycine codon box with a minimal set of tRNA transcripts. (**a**) Sequences of mRNA and the corresponding peptide. P1-XX denotes the peptide containing XX at the sixth position. (**b**) Evaluation of translation efficiency and accuracy by autoradiography after tricine-SDS-PAGE and MALDI-TOF-MS. Four tRNA transcripts represent 5 μM each tRNA$^{Ini}_{CAU}$, tRNA$^{Asp}_{GUC}$, tRNA$^{Tyr}_{GUA}$, and 30 μM tRNA$^{Lys}_{CUU}$. Figure materials are reproduced from a previous article [23]

in a negative control that did not contain tRNA transcript corresponding to the GGC codon (Fig. 2b, lane 2). This meant that the reconstituted translation system was not contaminated with native tRNA$^{Gly}_{GCC}$, and instead Asp-tRNA$^{Asp}_{GUC}$ misread the GGC codon resulting in the P1-Asp expression. When *p*-iodophenylalanyl-tRNA$^{AsnE2}_{GCC}$ (^{Iodo}F-tRNA$^{AsnE2}_{GCC}$) was added, the P1-^{Iodo}F peptide was expressed with 60% efficiency compared to the native tRNA control (Fig. 2b, lane 3 vs. 1), suggesting that the incorporation efficiency was sufficient. Addition of tRNA$^{Gly}_{CCC}$ transcript did not disturb the expression of P1-^{Iodo}F peptide (Fig. 2b, lane 4 and Fig. 2c), showing that the GGC codon was accurately decoded by ^{Iodo}F-tRNA$^{AsnE2}_{GCC}$ without being cross read by Gly-tRNA$^{Gly}_{CCC}$. Expression level of the model peptide is evaluated by tricine-SDS-PAGE [25, 26] followed by autoradiographic detection of [^{14}C]-Asp in the C-terminal FLAG peptide (DYKDDDDK). Decoding accuracy of the codons in the artificially divided codon boxes is evaluated by MALDI-TOF-MS (Matrix-Assisted Laser Desorption Ionization-Time of Flight Mass Spectrometry) after purification of the peptide product by ANTI-FLAG M2 agarose. The purification step greatly helps detection of the peptide in the MS measurement.

Subheading 3.4 is the method to simultaneously divide three codon boxes and synthesize a macrocyclic *N*-methyl-peptide (Fig. 3). The macrocyclic *N*-methyl-peptide "CM$_{11}$–1" was previously developed as an E6AP inhibitor [10]. The method of artificial division of codon boxes was applied to synthesize CM$_{11}$–1

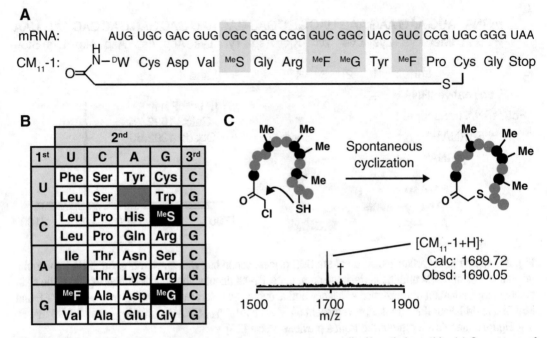

Fig. 3 Artificial division of three codon boxes to express a macrocyclic *N*-methyl-peptide. (**a**) Sequences of mRNA and the corresponding peptide. (**b**) A reprogrammed genetic code containing three *N*-methyl-amino acids in artificially divided codon boxes. (**c**) MALDI-TOF-MS of the peptide product. The macrocyclic structure is formed by spontaneous cyclization reaction between ClAc^DW and the downstream Cys residue at the 13th position [10]. † indicates a peak corresponding to the linear peptide containing the unreacted *N*-terminal chloroacetyl group. Figure materials are reproduced from a previous article [23]

under the reprogrammed genetic code containing 22 building blocks [23]. In this system, ClAc^DW (*N*-chloroacetyl-D-tryptophan) is used as the initiator amino acid by replacing Met with precharged ClAc^DW-tRNA^Ini_CAU. Three kinds of *N*-methyl-amino acids (^MeF (*N*-methylphenylalanine), ^MeS (*N*-methylserine), and ^MeG (*N*-methylglycine)) are assigned to the GUC, CGC, and GGC codons in artificially divided codon boxes, respectively.

2   Materials

All materials should be prepared in an RNase-free manner. Use RNase-free H_2O, tubes, pipette tips, and gloves. Solution containing proteins and RNA should be stored in low binding tubes such as BM4006 (BM Equipment).

2.1 Preparation of tRNAs by In Vitro Transcription

1. DNA templates coding for tRNAs: Prepare them as reported previously [23]. Briefly, the templates are synthesized by primer extension followed by two-step PCR using appropriate primers (*see* Tables 1 and 2 for primer information) as follows: 1 µM each designated forward and reverse primers are mixed

Table 1
List of oligonucleotides sequences

Name	Sequence
O-1	5′-GGCGTAATACGACTCACTATAG-3′
O-2	5′-GTAATACGACTCACTATAGGAACGCGCGACTCTAAT-3′
O-3	5′-GTAATACGACTCACTATAGGCGGGGTGGAGCAGCCTGGTAGCTCGTCGG-3′
O-4	5′-GAACCGACGATCTTCGGGTTATGAGCCCGACGAGCTACCAGGCT-3′
O-5	5′-TGGTTGCGGGGGCCGGATTTGAACCGACGATCTTCGGG-3′
O-6	5′-TGGTTGCGGGGGCCGGATTT-3′
O-7	5′-GTAATACGACTCACTATAGGGCTATAGCTCAGCTGGGAGAGCGCCTGC-3′
O-8	5′-GAACCGCAGACCTCCTGCGTGCGAAGCAGGGCTCTCCCAGCT-3′
O-9	5′-TGGTGGAGCTATGCGGGATCGAACCGCAGACCTCCTGC-3′
O-10	5′-TGGTGGAGCTATGCGGGATC-3′
O-11	5′-GTAATACGACTCACTATAGGGCTATAGCTCAGCTGGGAGAGCGCCTGC-3′
O-12	5′-GAACCGCAGACCTCCTGCATGCCATGCAGGGCGCTCTCCCAGCT-3′
O-13	5′-TGGTGGAGCTATGCGGGATCGAACCGCACCTCCTGC-3′
O-14	5′-TGGTGGAGCTATGCGGGATC-3′
O-15	5′-GTAATACGACTCACTATAGGCCCGTAGCTCAGCTGGATAGAGCGCTGC-3′
O-16	5′-CGAACCTGAGACCTCTGCCTCCGGAGGGCAGCGCTCTATCCAGCT-3′
O-17	5′-TGGCGCGCCCGACAGGATTCGAACCTGAGACCTCTGC-3′
O-18	5′-TGGCGCGCCCGACAGGATTC-3′
O-19	5′-GTAATACGACTCACTATAGCGCCCGTAGCTCAGCTGGATAGAGCGCTGC-3′

(continued)

Table 1
(continued)

Name	Sequence
O-20	5'-CGAACCTGAGACCTCTGCCTTAGGAGGGCAGCGGCTCTATCCAGCT-3'
O-21	5'-TGGCGCGCCCGACAGGATTCGAACCTGAGACCTCTGC-3'
O-22	5'-TGGCGCGCCCGACAGGATTC-3'
O-23	5'-GTAATACGACTCACTATAGCGCCGTAGCTCAGCTGGATAGAGCGCTGC-3'
O-24	5'-CGAACCTGAGACCTCTGCCTTCGCAGGGCAGCGGCTCTATCCAGCT-3'
O-25	5'-TGGCGCGCCOGACAGGATTCGAACCTGAGACCTCTGC-3'
O-26	5'-TGGCGCGCCCGACAGGATTC-3'
O-27	5'-GTAATACGACTCACTATAGCCTCTGTAGTTCAGTCGGTAGAACGGCGGA-3'
O-28	5'-GAACCAGTGACATACGGATTAACAGTCCGCCGTTCTACCGACT-3'
O-29	5'-TGGCGCCTCTGACTGACTCGAACCAGTGACATACGGA-3'
O-30	5'-TGGCGCCTCTGACTGACTC-3'
O-31	5'-GTAATACGACTCACTATAGGGAGGGTAGTTCAGTCGGTTAGAATACCTG-3'
O-32	5'-GAACCCGGACCCCTGCGTACAGGCAGGTATTCTAACCGACT-3'
O-33	5'-TGGCGGAACGGACGGGACTCGAACCCGGACCCCTGC-3'
O-34	5'-TGGCGGAACGGACGGGACTC-3'
O-35	5'-GTAATACGACTCACTATAGGCGCGTTAACAAAGCGGTTATGTA-3'
O-36	5'-CGCGAACCGGACTAGACGGATTTGCAGTCCGCTACATAACCGCTT-3'
O-37	5'-TGGAGGGCGGTTCCGGAGTCGAACCGGACTAGACGGA-3'
O-38	5'-TGGAGGGCGGTTCCGGAGTC-3'
O-39	5'-GGAACGCGCGACTCTAATTGGGTATCGCCAAGCGGTAAGGCACCGGA-3'

O-40   5′-GAACCTCGGAATGCCGGAATCAGAGTCCGGTGCCTTACCGCTT-3′

O-41   5′-TGGCTGGGGTACGAGGATTCGAACCTCGGAATGCCGGA-3′

O-42   5′-TGGCTGGGGTACGAGGATTC-3′

O-43   5′-GTAATACGACTCACTATAGTCCCCTTCGTCTAGAGGCCAGGACACCGC-3′

O-44   5′-CGAACCCCTGTTACCGCCGTGAGAGGGCGGTGTCCTGGGCCTCT-3′

O-45   5′-TGGCGTCCCCTAGGGATTCGAACCCCTGTTACCGCC-3′

O-46   5′-TGGCGTCCCCTAGGGATTC-3′

O-47   5′-GTAATACGACTCACTATAGCGGCGTAGTTCAATGGTAGAACGAGAGC-3′

O-48   5′-CGAACCCTCGTATAGAGCTTGGGAAGCTCTCGTTCTACCATTG-3′

O-49   5′-TGGAGCGGGCGAAGGGAATCGAACCCTCGTATAGAGC-3′

O-50   5′-TGGAGCGGGCGAAGGGAATC-3′

O-51   5′-GTAATACGACTCACTATAGCGGGCGTAGTTCAATGGTAGAACGAGAGC-3′

O-52   5′-CGAACCCTCGTATAGAGCTTGCAAGCTCTCGTTCTACCATTG-3′

O-53   5′-TGGAGCGGGCGAAGGGAATCGAACCCTCGTATAGAGC-3′

O-54   5′-TGGAGCGGGCGAAGGGAATC-3′

O-55   5′-GTAATACGACTCACTATAGTGGCTATAGCTCAGTTGGTAGAGCCCTGG-3′

O-56   5′-GAACCCACGACAACTGGAATCACAATCCAGGGCTCTACCAACTG-3′

O-57   5′-TGGGGTGGCTAATGGGATTCGAACCCACGACAACTGGA-3′

O-58   5′-TGGGGTGGCTAATGGGATTC-3′

O-59   5′-GTAATACGACTCACTATAGGGCTTGTAGCTCAGGTGGTTAGAGCGCACC-3′

O-60   5′-GAACCACCGACCTCACCCTTATCAGGGGTGCGCTCTAACCACCT-3′

O-61   5′-TGGTGGGCCTGAGTGACTTGAACCACCGACCTCACCC-3′

(continued)

Table 1
(continued)

Name	Sequence
O-62	5'-TGGTGGGCCTGAGTGGACTT-3'
O-63	5'-GTAATACGACTCACTATAGGCGAAGGTGGCGGAATTGGTAGACGCGCTAG-3'
O-64	5'-GTCCGTAAGGACACTAACTTTTGAAGCTAGCGGCGTCTACCAATT-3'
O-65	5'-TGGTGCGAGGGGGGGACTTGAACCCCACGTCCGTAAGGACACTAAC-3'
O-66	5'-TGGTGCGAGGGGGGGACTT-3'
O-67	5'-GTAATACGACTCACTATAGCGAAGGTGGCGGAATTGGTAGACGCGCTAG-3'
O-68	5'-GTCCGTAAGGACACTAACACCTGAAGCTAGCGGGTCTACCAATT-3'
O-69	5'-TGGTGCGAGGGGGGGACTTGAACCCCACGTCCGTAAGGACACTAAC-3'
O-70	5'-TGGTGCGAGGGGGGGACTT-3'
O-71	5'-GTAATACGACTCACTATAGCGAAGGTGGCGGAATTGGTAGACGCGCTAG-3'
O-72	5'-GTCCGTAAGGACACTAACACCTCAAGCTAGCGGGTCTACCAATT-3'
O-73	5'-TGGTGCGAGGGGGGGACTTGAACCCCACGTCCGTAAGGACACTAAC-3'
O-74	5'-TGGTGCGAGGGGGGGACTT-3'
O-75	5'-GTAATACGACTCACTATAGGGTCGTTAGCTCAGTTGGTAGAGCAGTTGA-3'
O-76	5'-GAACCTGCGACCAATTGATTAAGAGTCAACTGCTCTACCAACT-3'
O-77	5'-TGGTGGGTCGTGCAGGATTCGAACCTGCGACCAATTGA-3'
O-78	5'-TGGTGGGTCGTGCAGGATTC-3'
O-79	5'-GTAATACGACTCACTATAGGCTACGTAGCTCAGTTGGTTAGAGCACATC-3'
O-80	5'-GAACCTGTGACCCATCATTATGAGTGATGTGCTCAACCAACT-3'
O-81	5'-TGGTGGCTACGACGGGATTCGAACCTGTGACCCATCA-3'

O-82	5′-TGGTGGCTACGACGGGATTC-3′
O-83	5′-GTAATACGACTCACTATAGCCCGGATAGCTCAGTCGGTAGAGCAGGGGA-3′
O-84	5′-GAACCAAGGACACGGGGATTTCAGTCCCTGCTCTACCGACT-3′
O-85	5′-TGGTGCCCGGACTCGGAATCGAACCAAGGACACGGGGA-3′
O-86	5′-TGGTGCCCGGACTCGGAATC-3′
O-87	5′-GGAACGCGGCGACTCTAATCGGTGATTGGCGCAGCCTGGTAGCGCACTTC-3′
O-88	5′-GAACCTCCGACCCCTTGGTCCCGAACGAAGTGCGCTACCAGGCT-3′
O-89	5′-TGGTCGGTGATAGAGGATTCGAACCTCCGACCCCTTCG-3′
O-90	5′-TGGTCGGTGATAGAGGATTC-3′
O-91	5′-GGAACGCGGCGACTCTAATCGGTGATTGGCGCAGCCTGGTAGCGCACTTC-3′
O-92	5′-GAACCTCCGACCCCTTCGACCCCATGAAGTGCGCTACCAGGCT-3′
O-93	5′-TGGTCGGTGATAGAGGATTCGAACCTCCGACCCCTTCG-3′
O-94	5′-TGGTCGGTGATAGAGGATTC-3′
O-95	5′-GTAATACGACTCACTATAGGAGAGATGCCGGAGCGGCTGAACGGACCGG-3′
O-96	5′-AGAGTTGCCCTACTCCGGTTTCGAGACGCCGGTCCGTTCAGCCGCT-3′
O-97	5′-TGGCGGGAGAGAGGGGGATTTGAACCCCGGTAGAGTTGCCCTACTCCG-3′
O-98	5′-TGGCGGGAGAGAGGGGGATTT-3′
O-99	5′-GTAATACGACTCACTATAGGAGAGATGCCGGAGCGGCTGAACGGACCGG-3′
O-100	5′-AGAGTTGCCCTACTCCGGTTTAGCAGACGCCGGTCCGTTCAGCCGCT-3′
O-101	5′-TGGCGGGAGAGAGGGGGATTTGAACCCCGGTAGAGTTGCCCTACTCCG-3′
O-102	5′-TGGCGGGAGAGAGGGGGATTT-3′
O-103	5′-GTAATACGACTCACTATAGGAGAGATGCCGGAGCGGCTGAACGGACCGG-3′

(continued)

Table 1
(continued)

Name	Sequence
O-104	5'-AGAGTTGCCCTACTCCGGTTTTCCAGACCGGTCCGTTCAGCCGCT-3'
O-105	5'-TGGCGGAGAGAGGGGATTTGAACCCCGGTAGAGTTGCCCTACTCCG-3'
O-106	5'-TGGCGGAGAGAGGGGATTT-3'
O-107	5'-GTAATACGACTCACTATAGCCGATATAGCTCAGTTGGTAGAGCAGCGCA-3'
O-108	5'-GAACCTACGACCTTCGCATTACGAATGCGCTGCTCTACCAACT-3'
O-109	5'-TCGTGCCGATAATAGGAGTCGAACCTACGACCTTCGCA-3'
O-110	5'-TGGTGCCGATAATAGGAGTC-3'
O-111	5'-GTAATACGACTCACTATAGCCGATATAGCTCAGTTGGTAGAGCAGCGCA-3'
O-112	5'-GAACCTACGACCTTCGCACTACCAATGCGCTGCTCTACCAACT-3'
O-113	5'-TGGTGCCGATAATAGGAGTCGAACCTACGACCTTCGCA-3'
O-114	5'-TGGTGCCGATAATAGGAGTC-3'
O-115	5'-GGAACGCGGCGACTCTAATAGGGGCGTAGTTCAATTGGTAGAGCACCGGT-3'
O-116	5'-GAACTCCCAACACCCGGTTTTGGAGACCGGTGCTCTACCAATT-3'
O-117	5'-TGGCAGGGGCGGAGAGACTCGAACTCCCAACACCCGGT-3'
O-118	5'-TGGCAGGGGCGGAGAGACTC-3'
O-119	5'-GTAATACGACTCACTATAGGTGGGGTTCCCGAGCGGCCAAAGGGAGCAG-3'
O-120	5'-GAAGTCTGTGACGGCAGATTTACAGTCTGCTCCCTTTGCCGCT-3'
O-121	5'-TGGTGGTGGGGGAAGGATTCGAACCTTCGAAGTCTGTGACGGCAGA-3'
O-122	5'-TGGTGGTGGGGGAAGGATTC-3'
O-123	5'-GTAATACGACTCACTATAGGGTGATTAGCTCAGCTGGGAGAGCACCTCC-3'

O-124   5'-GAACCGCGACCCCCTCCATGTCAAGGAGGTGTCTCCCAGCT-3'

O-125   5'-TGGTGGGTGATGACGGGATCGAACCGCCGACCCCTCC-3'

O-126   5'-TGGTGGGTGATGACGGGATC-3'

O-127   5'-GTAATACGACTCACTATAGGGTGATTAGCTCAGCTGGGAGAGCACCTCC-3'

O-128   5'-GAACCGCGACCCCCTCCGTGTGAGGGAGGTGTCTCCCAGCT-3'

O-129   5'-TGGTGGGTGATGACGGGATCGAACCGCCGACCCCCTCC-3'

O-130   5'-TGGTGGGTGATGACGGGATC-3'

O-131   5'-GTAATACGACTCACTATAGGCTCTGTAGTTCAGTCGGTAGAACGGCGGA-3'

O-132   5'-GAACCAGTGACATACGGAATGTCAATCCGCCGTTCTACCGACT-3'

O-133   5'-TGGCGGCTCTGACTGGAACCAGTGACATACGGA-3'

O-134   5'-TGGCGGCTCTGACTGGACTC-3'

O-135   5'-GAACCAGTGACATACGGATTGGCAATCCGCCGTTCTACCGACT-3'

O-136   5'-GAACCAGTGACATACGGATCCGCCAGTCCGCCGTTCTACCGACT-3'

O-137   5'-GTAATACGACTCACTATAGGGATCGAAAGATTTCCGC-3'

O-138   5'-ACCTAACGCTAATCCCCTTTCGGGGCCGCGGAAATCTTTCGATCC-3'

O-139   5'-ACCTAACGCTAATCCCCT-3'

O-140   5'-ACCTAACGCCATGTACCCTTTCGGGGATGCGGAAATCTTTCGATCC-3'

O-141   5'-ACCTAACGCCATGTACCCT-3'

O-142   5'-TAATACGACTCACTATAGGGCTTTAATAAGGAGAAAAACATGAAGAAGA-3'

O-143   5'-GTCGTCGTCCTTGTAGTCGCCGTACTTCTTCTTCATGTTTTTCTC-3'

O-144   5'-CGAAGCTTACTTGTCGTCGTCGTCCTTGTAGTC-3'

O-145   5'-TAATACGACTCACTATAGGCTTTAATAAGGAGAAAAACATGTGCGACGTGCGCGGGCG-3'

O-146   5'-CGAAGCTTACCCGCACGGGACGTAGCCGCCCGCCGCCACGTCG-3'

O-147   5'-CGAAGCTTACCCGCACG-3'

Table 2
List of oligonucleotides used for the preparation of DNA templates coding tRNAs and flexizymes

Name	Anticodon	Extension		First PCR		Second PCR	
		Forward primer	Reverse primer	Forward primer	Reverse primer	Forward primer	Reverse primer
tRNA Ini	CAU	O-3	O-4	O-1	O-5	O-1	O-6
tRNA Ala	CGC	O-7	O-8	O-1	O-9	O-1	O-10
tRNA Ala	GGC	O-11	O-12	O-1	O-13	O-1	O-14
tRNA Arg	CCG	O-15	O-16	O-1	O-17	O-1	O-18
tRNA Arg	CCU	O-19	O-20	O-1	O-21	O-1	O-22
tRNA Arg	GCG	O-23	O-24	O-1	O-25	O-1	O-26
tRNA Asn	GUU	O-27	O-28	O-1	O-29	O-1	O-30
tRNA Asp	GUC	O-31	O-32	O-1	O-33	O-1	O-34
tRNA Cys	GCA	O-35	O-36	O-1	O-37	O-1	O-38
tRNA Gln	CUG	O-39	O-40	O-2	O-41	O-2	O-42
tRNA Glu	CUC	O-43	O-44	O-1	O-45	O-1	O-46
tRNA Gly	CCC	O-47	O-48	O-1	O-49	O-1	O-50
tRNA Gly	GCC	O-51	O-52	O-1	O-53	O-1	O-54
tRNA His	GUG	O-55	O-56	O-1	O-57	O-1	O-58
tRNA Ile	GAU	O-59	O-60	O-1	O-61	O-1	O-62
tRNA Leu	CAA	O-63	O-64	O-1	O-65	O-1	O-66
tRNA Leu	CAG	O-67	O-68	O-1	O-69	O-1	O-70
tRNA Leu	GAG	O-71	O-72	O-1	O-73	O-1	O-74
tRNA Lys	CUU	O-75	O-76	O-1	O-77	O-1	O-78
tRNA Met	CAU	O-79	O-80	O-1	O-81	O-1	O-82
tRNA Phe	GAA	O-83	O-84	O-1	O-85	O-1	O-86
tRNA Pro	CGG	O-87	O-88	O-2	O-89	O-2	O-90
tRNA Pro	GGG	O-91	O-92	O-2	O-93	O-2	O-94
tRNA Ser	CGA	O-95	O-96	O-1	O-97	O-1	O-98
tRNA Ser	GCU	O-99	O-100	O-1	O-101	O-1	O-102
tRNA Ser	GGA	O-103	O-104	O-1	O-105	O-1	O-106
tRNA Thr	CGU	O-107	O-108	O-1	O-109	O-1	O-110
tRNA Thr	GGU	O-111	O-112	O-1	O-113	O-1	O-114

(continued)

Table 2
(continued)

Name	Anticodon	Extension		First PCR		Second PCR	
		Forward primer	Reverse primer	Forward primer	Reverse primer	Forward primer	Reverse primer
tRNA Trp	CCA	O-115	O-116	O-2	O-117	O-2	O-118
tRNA Tyr	GUA	O-119	O-120	O-1	O-121	O-1	O-122
tRNA Val	GAC	O-123	O-124	O-1	O-125	O-1	O-126
tRNA Val	CAC	O-127	O-128	O-1	O-129	O-1	O-130
tRNA AsnE2	GAC	O-131	O-132	O-1	O-133	O-1	O-134
tRNA AsnE2	GCC	O-131	O-135	O-1	O-133	O-1	O-134
tRNA AsnE2	GCG	O-131	O-136	O-1	O-133	O-1	O-134
eFx	–	O-137	O-138	O-1	O-139	–	–
dFx	–	O-137	O-140	O-1	O-141	–	–

in 10 μL of the PCR mixture (10 mM Tris–HCl (pH 9.0), 50 mM KCl, 2.5 mM MgCl$_2$, 0.25 mM each dNTPs, 0.1% (v/v) Triton X-100, and 45 nM *Taq* DNA polymerase). Primer extension is conducted by denaturing (95 °C for 1 min) followed by 5 cycles of annealing (50 °C for 1 min) and extending (72 °C for 1 min). The resulting 10 μL of reaction mixture is 20-fold diluted with 190 μL of the PCR mixture containing 0.5 μM each designated forward and reverse primers, and the first PCR is conducted by 5 cycles of denaturing (95 °C for 40 s), annealing (50 °C for 40 s), and extending (72 °C for 40 s). 1 μL of the resulting 1st PCR mixture is 200-fold diluted with 199 μL of the PCR mixture containing 0.5 μM each designated forward and reverse primers, and the second PCR is conducted by 12 cycles of denaturing (95 °C for 40 s), annealing (50 °C for 40 s), and extending (72 °C for 40 s). Amplification of the second PCR product is confirmed by 3% agarose gel electrophoresis and ethidium bromide staining. The resulting DNA is purified by phenol/chloroform extraction and ethanol precipitation, and then dissolved in 20 μL of water. Store them at −20 °C (stable at least for a year).

2. DNA templates coding for enhanced flexizyme (eFx) and dinitro-flexizyme (dFx): Prepare them as reported previously [7]. Briefly, the DNA templates coding for eFx and dFx are

synthesized by primer extension followed by PCR using appropriate primers (*see* Tables 1 and 2 for primer information) as follows: 1 μM each designated forward and reverse primers are mixed in 100 μL of the PCR mixture (10 mM Tris–HCl (pH 9.0), 50 mM KCl, 2.5 mM $MgCl_2$, 0.25 mM each dNTPs, 0.1% (v/v) Triton X-100, and 45 nM *Taq* DNA polymerase). Primer extension is conducted by denaturing (95 °C for 1 min) followed by 5 cycles of annealing (50 °C for 1 min) and extending (72 °C for 1 min). 1 μL of the resulting reaction mixture is 200-fold diluted with 199 μL of the PCR mixture containing 0.5 μM each designated forward and reverse primers, and PCR is conducted by 12 cycles of denaturing (95 °C for 40 s), annealing (50 °C for 40 s), and extending (72 °C for 40 s). Amplification of the PCR product is confirmed by 3% agarose gel electrophoresis and ethidium bromide staining. The resulting DNA is purified by phenol/chloroform extraction and ethanol precipitation, and then dissolved in 20 μL of water.

3. 10× T7 buffer: 400 mM Tris–HCl (pH 8.0), 10 mM spermidine, 0.1% (v/v) Triton X-100. Store it at −20 °C (stable at least for a year).

4. 25 mM each NTPs: 25 mM ATP, 25 mM GTP, 25 mM CTP, 25 mM UTP. Store it at −20 °C (stable at least for a year).

5. RQ1 RNase-Free DNase: a commercially available product (Promega, M6101).

6. 2× Denaturing PAGE loading buffer: 8 M urea, 2 mM Tris, 2 mM EDTA-Na_2, 0.01% (w/v) bromophenol blue (BPB). Store it at room temperature (stable at least for a year).

7. 5× TBE: 445 mM Tris, 445 mM boric acid, 10 mM EDTA (pH 8.0). Store it at room temperature (stable at least for a month).

8. Denaturing 8% polyacrylamide gel: 8% (w/v) acrylamide/bisacrylamide (19:1), 6 M urea, 44.5 mM Tris, 44.5 mM boric acid, 1 mM EDTA (pH 8.0), 0.1% (w/v) ammonium persulfate (APS), 0.075% (v/v) N,N,N′,N′-tetramethylethylenediamine (TEMED). Pour the solution between assembled gel plates (BIO CRAFT BE-140) and let it stand at room temperature for 30 min to allow polymerization.

9. RNA component of RNase P (M1 RNA): Prepare M1 RNA by in vitro transcription as described previously [27]. Briefly, M1 RNA is synthesized by in vitro transcription with T7 RNA polymerase from a DNA template coding M1 RNA gene (*rnpb*) (a generous gift from the laboratory of Dr. Michael E. Harris). The transcription product is purified by denaturing PAGE. Store it at −20 °C (stable at least for a year).

10. Protein component of RNase P (C5 protein): Prepare C5 protein as described previously [28]. C5 protein is expressed as a fusion with chitin-binding domain and purified by affinity chromatography with subsequent removal of the affinity tag by intein cleavage. This system relies on an Impact (NEB) expression system with a (pTYB3)-derived plasmid coding C5 protein gene (*pnpa*) (pMHRNPAP; a generous gift from the laboratory of Dr. Michael E. Harris). *E. coli* 2566 (NEB) is transformed with pMHRNPAP. 1 L of LB broth is inoculated and grown at 37 °C to an A_{600} of 0.5–0.8. Expression is then induced by the addition of IPTG to 0.3 mM, and incubation is continued at 25 °C for an additional 6 h. The cell is harvested by centrifugation at 5000 × g for 10 min, and the pellet is frozen at −80 °C. To assist in lysis, the cell pellet is subjected to three freeze-thaw cycles (−80 °C to room temperature). The cell pellet is resuspended in 25 mL of Buffer A (20 mM Tris–HCl (pH 8.0), 2 M NaCl, 0.1 mM EDTA, and 0.1% (v/v) reduced Triton X-100) in which a cOmplete protease inhibitor cocktail tablet (Roche) is dissolved. Cells are lysed by sonication on ice, and the sonicate is clarified by centrifugation at 20,000 × g for 1 h at 4 °C. A column involving a 40 mL bed of chitin beads (NEB) is equilibrated with ten column volumes of Buffer A at 1 mL/min. The lysate is loaded onto the equilibrated column at 0.5 mL/min and washed with 20 column volumes of Buffer A (1 mL/min) followed by two volumes of Buffer B (20 mM Tris–HCl (pH 8.0), 500 mM NaCl, 0.1 mM EDTA, 0.01% (v/v) reduced Triton X-100). The column is flashed by gravity flow with three column volumes of Buffer B containing 50 mM fresh DTT. After the flow stops, the column is incubated at 4 °C for 16 h. The tag-free C5 protein is eluted from the column with five column volumes of Buffer B containing 5 mM fresh DTT (0.5 mL/min) and fractionated. Fractions containing C5 protein are identified by SDS-PAGE and combined. The solution is dialyzed against 1 L of Buffer B containing 5 mM DTT overnight at 4 °C. The solution is further dialyzed against a buffer containing 5 mM HEPES-KOH (pH 7.6), 10 mM KOAc, 0.1 mM DTT overnight at 4 °C. Concentration of the resulting C5 protein is identified using the Bradford assay (Bio-Rad). Store it at −80 °C (stable at least for a year).

2.2 Synthesis of Nonproteinogenic Aminoacyl-tRNAs

1. Chemically activated amino acids: Synthesize amino acid substrates activated with appropriate active groups (Table 3; *see* **Note 1**) as described previously [1, 2, 4]. General procedure for the synthesis of N^{α}-methyl-aa-CME substrates (N^{α}-methyl-Phe-CME): A mixture of N^{α}-Boc-N^{α}-methyl-Phe (69.8 mg, 0.25 mmol), triethylamine (27.8 mg, 0.275 mmol), and chloroacetonitrile (0.1 mL) in 0.1 mL of *N*,*N*-dimethylformamide is stirred at room temperature for 12 h. After the reaction,

Table 3
Appropriate combinations of active groups, flexizymes, and reaction time for each acylation substrate

Activated Naa	Flexizyme	Reaction time (h)
  IodoF-CME	eFx	2
  ClAcDW-CME	eFx	2
  MeF-CME	eFx	6
  MeS-DBE	dFx	6
  MeG-DBE	dFx	2

diethylether (9 mL) is added and the solution is washed with 1 M HCl (3 mL × 3), sat. NaHCO$_3$ (3 mL × 3), and brine (5 mL ×1). The organic layer is dried over MgSO$_4$ and concentrated under reduced pressure. The crude residue is dissolved in 1.0 mL of dichloromethane and mixed with 2 mL of 4 M HCl/ethylacetate and incubated for 20 min at room temperature. The residue is washed with diethylether (3 mL) three times to obtain N^α-methyl-Phe-CME. For the synthesis of N^α-methyl-aa-DBE substrates (N^α-methyl-Gly-DBE), a mixture of N^α-Boc-N^α-methyl-Gly (47.3 mg, 0.25 mmol), triethylamine (27.8 mg, 0.275 mmol), and 3,5-dinitrobenzylchloride (59.6 mg, 0.275 mmol) in 0.1 mL of N,N-dimethylformamide is stirred at room temperature for 12 h, followed by the same purification and deprotection described above. To a solid of amino acid substrate add DMSO-d$_6$ to make a 200 mM stock solution. Store it at −20 °C (generally stable at least for a year). To make 25 mM working solution, mix 10 μL of the 200 mM stock solution with 70 μL of DMSO. Store them at −20 °C (generally stable at least for a year).

2.3 Artificial Division of a Codon Box with a Minimal Set of tRNAs

2.3.1 In Vitro Translation

1. 10× Four tRNA transcripts mixture: 50 μM each tRNA$^{Ini}_{CAU}$, tRNA$^{Asp}_{GUC}$, tRNA$^{Tyr}_{GUA}$, and 300 μM tRNA$^{Lys}_{CUU}$. Store it at −20 °C (stable at least for a year).

2. 10× TS (FD+) solution: 500 mM HEPES-KOH (pH 7.6), 1 M KOAc, 130 mM Mg(OAc)$_2$, 20 mM ATP, 20 mM GTP, 10 mM CTP, 10 mM UTP, 200 mM creatine phosphate, 1 mM 10-formyl-5,6,7,8-tetrahydrofolic acid, 20 mM spermidine, 10 mM DTT (see **Note 2**). Store the aliquots at −80 °C (stable at least for a year).

3. 10× 19 amino acids mixture: 2 mM each 19 Paa's except for Asp. Store it at −20 °C (stable at least for a year).

4. 2.5 μM DNA template coding for a model peptide: Prepare it as reported previously [23]. Briefly, the template is synthesized by primer extension followed by PCR using appropriate primers as follows: 1 μM each forward and reverse primers (O-142 and O-143, see Table 1 for their sequences) are mixed in 100 μL of the PCR mixture (10 mM Tris–HCl (pH 9.0), 50 mM KCl, 2.5 mM MgCl$_2$, 0.25 mM each dNTPs, 0.1% (v/v) Triton X-100, and 45 nM *Taq* DNA polymerase). Primer extension is conducted by denaturing (95 °C for 1 min) followed by 5 cycles of annealing (50 °C for 1 min) and extending (72 °C for 1 min). 5 μL of the resulting reaction mixture is 200-fold diluted with 995 μL of the PCR mixture containing 0.5 μM each forward and reverse primers (O-1 and O-144, see Table 1 for their sequences), and PCR is conducted by 12 cycles of denaturing (95 °C for 40 s), annealing (50 °C for 40 s), and extending (72 °C for 40 s). Amplification of the PCR product

is confirmed by 3% agarose gel electrophoresis and ethidium bromide staining. The resulting DNA is purified by phenol/chloroform extraction and ethanol precipitation, and then dissolved in 10 μL of water. The concentration is measured by 8% native PAGE and ethidium bromide staining with 100 bp Quick-Load DNA Ladders (New England BioLabs) as reference. Store them at −20 °C (stable at least for a year).

5. 10× AARS solution: 7.3 μM AlaRS, 0.3 μM ArgRS, 3.8 μM AsnRS, 1.3 μM AspRS, 0.2 μM CysRS, 0.6 μM GlnRS, 2.3 μM GluRS, 0.9 μM GlyRS, 0.2 μM HisRS, 4.0 μM IleRS, 0.4 μM LeuRS, 11 μM LysRS, 0.3 μM MetRS, 6.8 μM PheRS, 1.6 μM ProRS, 0.4 μM SerRS, 0.9 μM ThrRS, 0.3 μM TrpRS, 0.2 μM TyrRS, and 0.2 μM ValRS. Store the aliquots at −80 °C (stable at least for a year; *see* **Note 3**).

6. 10× RP solution: 3 mM $Mg(OAc)_2$, 12 μM ribosome, 6 μM MTF, 27 μM IF1, 4 μM IF2, 15 μM IF3, 2.6 μM EF-G, 100 μM EF-Tu, 6.6 μM EF-Ts, 2.5 μM RF2, 1.7 μM RF3, 5 μM RRF, 40 μg/mL creatine kinase, 30 μg/mL myokinase, 1 μM inorganic pyrophosphatase, 1 μM nucleotide diphosphate kinase, and 1 μM T7 RNA polymerase. Store the aliquots at −80 °C (stable at least for a year; *see* **Note 3**).

7. 10× native tRNA: 15 mg/mL of *E. coli* total tRNA (Roche, 109550).

2.3.2 Tricine-SDS-PAGE and Autoradiography for Quantification of Expressed Peptides

1. 2× Tricine-SDS-PAGE loading buffer: 900 mM Tris–HCl (pH 8.45), 8% (w/v) SDS, 30% (v/v) glycerol, 0.01% (w/v) xylene cyanol. Store it at room temperature (stable at least for a year).

2. Tricine-SDS-polyacrylamide gel: This gel is composed of two layers in assembled gel plates (Bio-Rad Mini-PROTEAN 3). The lower separation gel (approximately 5.5 cm height) is composed of 15% (w/v) acrylamide/bisacrylamide (19:1), 1 M Tris–HCl (pH 8.45), 0.1% (w/v) SDS, 10.4% (v/v) glycerol, 0.1% (w/v) APS, 0.066% (v/v) TEMED. The upper stacking gel (approximately 1 cm height) is composed of 4% (w/v) acrylamide/bisacrylamide (29:1), 750 mM Tris–HCl (pH 8.45), 0.075% (w/v) SDS, 0.2% (w/v) APS, 0.1% (v/v) TEMED.

3. Tricine-SDS-PAGE cathode buffer: 100 mM Tris–HCl (pH 8.45), 100 mM tricine, 0.1% (w/v) SDS. Store it at room temperature (stable at least for a month).

4. Tricine-SDS-PAGE anode buffer: 200 mM Tris–HCl (pH 8.9). Store it at room temperature (stable at least for a month).

2.3.3 MALDI-TOF-MS to Evaluate the Translation Accuracy

1. ANTI-FLAG M2 Affinity Gel: a commercially available product (Sigma, A2220).

2. 2× TBS: 100 mM Tris–HCl (pH 7.6), 300 mM NaCl. Store it at room temperature (stable at least for a year).

3. Solid phase extraction tip (SPE C-tip): a commercially available product (Nikkyo Technos, T300-C18-4).

4. C-tip elusion solution: 80% (v/v) acetonitrile, 0.5% (v/v) acetic acid. Store it at room temperature (stable at least for a year).

5. C-tip wash solution: 4% (v/v) acetonitrile, 0.5% (v/v) acetic acid. Store it at room temperature (stable at least for a year).

2.4 Simultaneous Division of Three Codon Boxes and Synthesis of a Macrocyclic N-Methyl-Peptide

1. 10× tRNA transcripts mixture: 300 μM tRNA$^{Lys}_{GAU}$, 300 μM tRNA$^{Ile}_{CUU}$, 300 μM tRNA$^{Glu}_{CUC}$, 50 μM tRNA$^{Asp}_{GUC}$, 50 μM tRNA$^{Tyr}_{GUA}$, 10 μM tRNA$^{Phe}_{GAA}$, 10 μM tRNA$^{Leu}_{CAA}$, 10 μM tRNA$^{Leu}_{GAG}$, 10 μM tRNA$^{Leu}_{CAG}$, 10 μM tRNA$^{Met}_{CAU}$, 10 μM tRNA$^{Val}_{CAC}$, 10 μM tRNA$^{Ser}_{GGA}$, 10 μM tRNA$^{Ser}_{CGA}$, 10 μM tRNA$^{Pro}_{GGG}$, 10 μM tRNA$^{Pro}_{CGG}$, 10 μM tRNA$^{Thr}_{GGU}$, 10 μM tRNA$^{Thr}_{CGU}$, 10 μM tRNA$^{Ala}_{GGC}$, 10 μM tRNA$^{Ala}_{CGC}$, 10 μM tRNA$^{His}_{GUG}$, 10 μM tRNA$^{Gln}_{CUG}$, 10 μM tRNA$^{Asn}_{GUU}$, 10 μM tRNA$^{Cys}_{GCA}$, 10 μM tRNA$^{Trp}_{CCA}$, 10 μM tRNA$^{Arg}_{CCG}$, 10 μM tRNA$^{Ser}_{GCU}$, 10 μM tRNA$^{Arg}_{CCU}$, and 10 μM tRNA$^{Gly}_{CCC}$. Procedures to prepare this mixture are described in Subheading 3.4 (**steps 1–12**).

2. 10× TS (FD-) solution: 500 mM HEPES-KOH (pH 7.6), 1 M KOAc, 130 mM Mg(OAc)$_2$, 20 mM ATP, 20 mM GTP, 10 mM CTP, 10 mM UTP, 200 mM creatine phosphate, 20 mM spermidine, 10 mM DTT (*see* **Note 4**). Store the aliquots at −80 °C (stable at least for a year).

3. 10× TCEP: 10 mM tris(2-carboxyethyl)phosphine hydrochloride (TCEP-HCl), 500 mM HEPES-KOH (pH 7.6). Store the aliquots at −20 °C (stable at least for a month).

4. 10× 18 amino acids mixture: 2 mM each 18 Paa's except for Asp and Met. Store it at −20 °C (stable at least for a year).

5. 2.5 μM DNA template coding for a model peptide: Prepare it in the same way with **item 4** in Subheading 2.3.1 except that different forward and reverse primers are used: extension is conducted with O-145 and O-146 (*see* Table 1 for their sequences), while PCR is conducted with O-1 and O-147.

6. The other materials are the same as **items 5** and **6** in Subheading 2.3.1.

3 Methods

All the steps should be conducted in an RNase-free manner. Use RNase-free H$_2$O, tubes, pipette tips, and gloves. Solution containing proteins and RNA should be handled in low binding tubes such as BM4006 (BM Equipment).

3.1 Preparation of tRNAs by In Vitro Transcription

3.1.1 Preparation of tRNAs Containing a Guanosine at the 5′-Terminus

1. Mix 75.8 μL of H_2O, 20 μL of DNA template coding for a tRNA, 20 μL of 10× T7 buffer, 20 μL of 100 mM DTT, 18 μL of 250 mM $MgCl_2$, 30 μL of 25 mM each NTPs, 2.25 μL of 2 M KOH, 10 μL of 100 mM GMP, and 4 μL of 6 μM T7 RNA polymerase.

2. Incubate it at 37 °C overnight.

3. Add 4 μL of 100 mM $MnCl_2$ and 1 μL of 1 U/μL RQ1 RNase-Free DNase.

4. Incubate it at 37 °C for 30 min.

5. Add 30 μL of 500 mM EDTA (pH 8.0), 20 μL of 3 M NaCl, and 210 μL of isopropanol.

6. Centrifuge it at $15,300 \times g$ for 5 min at 25 °C.

7. Remove the supernatant.

8. Wash the precipitate with 100 μL of 70% (v/v) ethanol.

9. Centrifuge it at $15,300 \times g$ for 3 min at 25 °C.

10. Remove the supernatant.

11. Dry the precipitate at room temperature for 10 min.

12. Dissolve the precipitate in 20 μL of H_2O.

13. Add 20 μL of 2× denaturing PAGE loading buffer into the solution.

14. Heat the solution at 95 °C for 3 min.

15. Apply the solution into a well of denaturing 8% polyacrylamide gel assembled in PAGE apparatus (BIO CRAFT BE-140).

16. Run PAGE with 230 V for 1.5 h at room temperature using 1× TBE as running buffer.

17. Remove the glass plate and sandwich the gel between two sheets of plastic wrap and put in on a TLC plate coated with a fluorescent indicator (Merck Millipore, 1.05715.0001).

18. Visualize the shadow of tRNA using a UV lamp and draw the band edge on the plastic wrap.

19. Cut the gel and collect it into a 5 mL tube.

20. Crush the gel using a micropipette tip.

21. Add 4 mL of 0.3 M NaCl into the tube.

22. Set the tube to a mixing rotator and mix it by gentle inversion at room temperature for 1 h.

23. Centrifuge the tube at $15,300 \times g$ for 5 min at 25 °C.

24. Collect the supernatant and filtrate it using a 0.45 μm syringe filter (Merck Millipore, SLLHH25NB).

25. Add twice the volume of ethanol.

26. Centrifuge it at $15,300 \times g$ for 15 min at 25 °C.

27. Remove the supernatant.

28. Wash the precipitate with 100 μL of 70% (v/v) ethanol.

29. Centrifuge it at $15{,}300 \times g$ for 3 min at 25 °C.

30. Remove the supernatant.

31. Dry the precipitate at room temperature for 10 min.

32. Dissolve the precipitate in 10 μL of H_2O.

33. Measure the concentration by A_{260} of the tRNA solution. Usually, tenfold diluted solution is appropriate for 1 mm optical path length measurement.

34. Store the solution at −20 °C (stable at least for a year).

3.1.2 Preparation of tRNAs That Do Not Have a Guanosine at the 5′-Terminus (tRNA$^{Gln}_{CUG}$, tRNA$^{Pro}_{CGG}$, tRNA$^{Pro}_{GGG}$, and tRNA$^{Trp}_{GCA}$)

1. Transcription and DNase reaction are conducted as **steps 1–4** in the Subheading 3.1.1.

2. Mix 1.67 μL of 300 μM M1 RNA and 4.66 μL of H_2O in a new test tube.

3. Heat the tube at 95 °C for 2 min.

4. Put it at room temperature for 5 min, and then put it on ice.

5. Add 1.0 μL of 10× T7 buffer, 1.0 μL of 100 mM DTT, and 1.67 μL of 300 μM C5 protein into the solution.

6. Add 1.0 μL of the RNase P solution prepared in **step 5** and 1.34 μL of 3 M $MgCl_2$ into the tRNA solution prepared in **step 1**.

7. Incubate it at 37 °C overnight.

8. Add 30 μL of 500 mM EDTA (pH 8.0), 20 μL of 3 M NaCl, and 210 μL of isopropanol.

9. Isopropanol precipitation, PAGE purification, ethanol precipitation, and concentration measurement are conducted as **steps 6–33** in Subheading 3.1.1.

10. Store the solution at −20 °C (stable at least for a year).

3.1.3 Preparation of Flexizymes

1. Mix 69 μL of H_2O, 20 μL of DNA template coding for a flexizyme (eFx or dFx), 20 μL of 10× T7 buffer, 20 μL of 100 mM DTT, 24 μL of 250 mM $MgCl_2$, 40 μL of 25 mM each NTPs, 3 μL of 2 M KOH, and 4 μL of 6 μM T7 RNA polymerase.

2. Incubate it at 37 °C overnight.

3. DNase reaction, isopropanol precipitation, PAGE purification, ethanol precipitation, and concentration measurement are conducted as **steps 3–34** in Subheading 3.1.1 except that 12% denaturing polyacrylamide gel is used instead of 8%.

3.2 Synthesis of Nonproteinogenic Aminoacyl-tRNAs

1. Mix 18 μL of H_2O, 6 μL of 500 mM HEPES-KOH (pH 7.5), 6 μL of 250 μM flexizyme of choice (Table 3; *see* **Note 1**), and 6 μL of 250 μM tRNA (*see* **Note 5**).

2. Heat the solution at 95 °C for 2 min.

3. Incubate it at room temperature for 5 min.

4. Add 12 μL of 3 M MgCl$_2$.

5. Incubate it at room temperature for 5 min.

6. Put the tube on ice.

7. Add 12 μL of 25 mM chemically activated amino acid substrate in DMSO (Table 3).

8. Incubate it on ice for several hours (depending on the amino acid substrate; Table 3; *see* **Note 6**) to conduct aminoacylation of the tRNA.

9. Add 240 μL of 0.3 M NaOAc (pH 5.2) to quench the reaction.

10. Add 600 μL of ethanol.

11. Centrifuge the solution at 15,300 × g for 15 min at 25 °C.

12. Remove the supernatant.

13. Wash the precipitate with 150 μL of 70% ethanol containing 0.1 M NaOAc (pH 5.2) and vortex it for 5 s.

14. Centrifuge the solution at 15,300 × g for 10 min at 25 °C.

15. Remove the supernatant.

16. Repeat **steps 13–15** one more time.

17. Wash the precipitate with 150 μL of 70% ethanol.

18. Centrifuge the solution at 15,300 × g for 3 min at 25 °C.

19. Remove the supernatant.

20. Dry the precipitate at room temperature for 10 min.

21. Store it at −80 °C. It is generally stable at least for 1 day.

3.3 Artificial Division of a Codon Box with a Minimal Set of tRNAs

3.3.1 In Vitro Translation

1. Prepare master mix as follows: mix 2.5 μL of H$_2$O, 2.5 μL of 10× four tRNA transcripts mixture (tRNA$^{Ini}_{CAU}$, tRNA$^{Asp}_{GUC}$, tRNA$^{Tyr}_{GUA}$, and tRNA$^{Lys}_{CUU}$), 2.5 μL of 10× TS (FD+) solution, 2.5 μL of 10× 19 amino acids mixture, 2.5 μL of 500 μM [^{14}C]-Asp or [^{12}C]-Asp (*see* **Note 7**), 2.5 μL of 2.5 μM DNA template coding for a model peptide, 2.5 μL of 10× AARS solution, and 2.5 μL of 10× RP solution.

2. Dissolve 1500 pmol tRNA$^{AsnE2}_{GCC}$ precharged with ^{Iodo}F in 1.5 μL of 1 mM NaOAc (pH 5.2) to make a 1000 μM solution (*see* **Note 8**).

3. Prepare the lane 1 sample by mixing 4.0 μL of the master mix, 0.5 μL of 1 mM NaOAc (pH 5.2), and 0.5 μL of 10× native tRNA.

4. Prepare the lane 2 sample by mixing 4.0 μL of the master mix, 0.5 μL of 1 mM NaOAc (pH 5.2), and 0.5 μL of H$_2$O.

5. Prepare the lane 3 sample by mixing 4.0 μL of the master mix, 0.5 μL of 1000 μM tRNA$^{\text{AsnE2}}_{\text{GCC}}$ precharged with ^{Iodo}F, and 0.5 μL of H_2O.

6. Prepare the lane 4 sample by mixing 4.0 μL of the master mix, 0.5 μL of 1000 μM tRNA$^{\text{AsnE2}}_{\text{GCC}}$ precharged with ^{Iodo}F, and 0.5 μL of 50 μM tRNA$^{\text{Gly}}_{\text{CCC}}$ transcript.

7. Incubate the solutions prepared in **steps 3–6** at 37 °C for 10 min (*see* **Note 9**), then put them on ice to stop the reaction.

8. Samples are analyzed using the following protocol by tricine-SDS-AGE (Subheading 3.3.2) and MALDI-TOF-MS (Subheading 3.3.3).

3.3.2 Tricine-SDS-PAGE and Autoradiography for Quantification of Expressed Peptides

1. Add 5 μL of 2× tricine-SDS-PAGE loading buffer into each of 5 μL translation reaction mixtures.

2. Standard solution is also prepared by mixing 5 μL of 32 μM [^{14}C]-Asp and 5 μL of 2× tricine-SDS-PAGE loading buffer.

3. Heat the solutions prepared in **steps 1** and **2** at 95 °C for 5 min.

4. Apply 5.0 μL each of the solutions into wells of Tricine-SDS-polyacrylamide gel assembled in PAGE apparatus (Bio-Rad Mini-PROTEAN 3).

5. Run PAGE with 150 V for 40 min at room temperature using tricine-SDS-PAGE cathode and anode buffers.

6. Remove the glass plate and sandwich the gel between a thick paper and plastic wrap.

7. Set the gel in a gel dryer (Bio-Rad, model 583), and dry it at 80 °C for 30 min under vacuum.

8. Set the gel in an imaging plate cassette (Fujifilm, BAS-MS 2025) and put it at 4 °C overnight.

9. Scan the imaging plate by a Typhoon FLA 7000 (GE Healthcare).

10. Measure the band intensities of the expressed peptides in the translation reaction samples and 80 pmol of [^{14}C]-Asp in the standard sample. Note that the [^{14}C]-Asp moves faster than the peptide during tricine-SDS-PAGE. Amount of the expressed peptides can be calculated based on the relative band intensities.

3.3.3 MALDI-TOF-MS to Evaluate the Translation Accuracy

1. Suspend the ANTI-FLAG M2 affinity gel by pipetting, and transfer 5.0 μL of the gel suspension to a test tube.

2. Centrifuge it by a tabletop centrifuge for 10 s.

3. Remove the supernatant.

4. Wash the gel with 15 μL of 100 mM Gly-HCl (pH 3.5).

5. Centrifuge it by a tabletop centrifuge for 10 s.

6. Remove the supernatant.

7. Wash the gel with 25 μL of 1× TBS.

8. Centrifuge it by a tabletop centrifuge for 10 s.

9. Remove the supernatant.

10. Repeat **steps 7–9** one more time.

11. Add 5.0 μL of 2× TBS into the 5.0 μL translation product.

12. Transfer this solution into the tube prepared in **step 10**.

13. Set the tube to a mixing rotator and mix it by gentle inversion at room temperature for 1 h.

14. Centrifuge it by a tabletop centrifuge for 10 s.

15. Remove the supernatant.

16. Wash the gel with 25 μL of 1× TBS.

17. Centrifuge it by a tabletop centrifuge for 10 s.

18. Remove the supernatant.

19. Add 15 μL of 0.2% (v/v) trifluoroacetic acid.

20. Incubate it at room temperature for 5 min, while the peptides are eluted from the gel. Do not leave it for longer time to prevent oxidation of the peptides.

21. Set an SPE C-tip [29] in a tabletop centrifuge.

22. Add 15 μL of C-tip elusion solution into the SPE C-tip.

23. Centrifuge it until the solution goes through the resin.

24. Add 15 μL of C-tip wash solution into the SPE C-tip.

25. Centrifuge it until the solution goes through the resin.

26. Centrifuge the tube prepared in **step 20** by a tabletop centrifuge for 10 s.

27. Transfer the supernatant into the SPE C-tip prepared in **step 25**.

28. Centrifuge the SPE C-tip until the solution goes through the resin.

29. Add 15 μL of C-tip wash solution into the SPE C-tip.

30. Centrifuge it until the solution goes through the resin.

31. Repeat **steps 29** and **30** one more time.

32. Prepare 10 μL of C-tip elusion solution saturated with the matrix α-cyano-4-hydroxycinnamic acid (Bruker Daltonics).

33. Mix the 10 μL of the saturated solution with 10 μL of C-tip elusion solution to make the 50% saturated matrix solution.

34. Add 1 μL of the 50% saturated matrix solution into the SPE C-tip prepared in **step 31**.

35. Swing the SPE C-tip by hand to make the solution contact the resin.

36. Elute the solution onto a target plate (Bruker Daltonics, MTP 384 target plate ground steel) using a syringe.

37. Wait until the solution evaporates and the matrix is crystalized.

38. Set the plate in a MALDI-TOF-MS instrument (Bruker Daltonics, ultrafleXtreme).

39. Externally calibrate the mass spectrometer using peptide calibration standard II and/or protein calibration standard I (Bruker Daltonics).

40. Measure MALDI-TOF-MS of the peptide product.

3.4 Simultaneous Division of Three Codon Boxes and Synthesis of a Macrocyclic N-Methyl-Peptide

1. Mix 3000 pmol of $tRNA^{Lys}_{GAU}$, 3000 pmol of $tRNA^{Ile}_{CUU}$, 3000 pmol of $tRNA^{Glu}_{CUC}$, 500 pmol of $tRNA^{Asp}_{GUC}$, 500 pmol of $tRNA^{Tyr}_{GUA}$, and 100 pmol each of the other tRNAs (*see* **Notes 10** and **11**).

2. Add 10% volume of 3 M NaCl and 220% volume of ethanol.

3. Centrifuge it at $15,300 \times g$ for 15 min at 25 °C.

4. Remove the supernatant.

5. Wash the precipitate with 100 µL of 70% (v/v) ethanol.

6. Centrifuge it at $15,300 \times g$ for 3 min at 25 °C.

7. Remove the supernatant.

8. Repeat **steps 5–7** one more time.

9. Dry the precipitate at room temperature for 20 min.

10. Dissolve the precipitate in 5 µL of H_2O.

11. Measure the concentration by A_{260} of the 50-fold diluted tRNA solution.

12. Add H_2O so that the concentration is adjusted to 300 µM $tRNA^{Lys}_{GAU}$, 300 µM $tRNA^{Ile}_{CUU}$, 300 µM $tRNA^{Glu}_{CUC}$, 50 µM $tRNA^{Asp}_{GUC}$, 50 µM $tRNA^{Tyr}_{GUA}$, and 10 µM each of the other tRNAs.

13. Dissolve 2500 pmol each of $tRNA^{Ini}_{CAU}$ precharged with $ClAc^DW$, $tRNA^{AsnE2}_{GAC}$ precharged with ^{Me}F, $tRNA^{AsnE2}_{GCG}$ precharged with ^{Me}S, and $tRNA^{AsnE2}_{GCC}$ precharged with ^{Me}G in 2.5 µL of 1 mM NaOAc (pH 5.2) to make a 1000 µM each solution.

14. Mix 2.0 µL of H_2O, 2.0 µL of 10× tRNA transcripts mixture, 2.0 µL of 10× TS (FD-) solution, 2.0 µL of 10× TCEP (*see* **Note 12**), 2.0 µL of 10× 18 amino acids mixture, 2.0 µL of 500 µM [^{14}C]-Asp or [^{12}C]-Asp (*see* **Note 7**), 2.0 µL of 2.5 µM DNA template coding for peptide, 2.0 µL of 10× AARS solution, 2.0 µL of 10× RP solution, and 2.0 µL of four acyl-tRNAs prepared in **step 13**.

15. Incubate the solution at 37 °C for 10 min, and then put it on ice.

16. Add 20 µL of 2× TBS.

17. Set the tube to mixing rotator and gently mix it by gentle inversion at room temperature for 1 h, while the macrocyclization proceeds [2].

18. The peptide product can be analyzed by either tricine-SDS-PAGE (Subheading 3.3.2) or MALDI-TOF-MS without FLAG purification procedure (**steps 21–40** in Subheading 3.3.3).

4   Notes

1. Amino acids containing aromatic side chains are activated by a cyanomethyl ester (CME) and charged to tRNAs by eFx, while amino acids containing non-aromatic side chains are activated by 3,5-dinitrobenzyl ester (DBE) and charged to tRNAs by dFx.

2. FD+ indicates that it contains the formyl donor (FD; 10-formyl-5,6,7,8-tetrahydrofolic acid). In the presence of FD, the α-amino group of Met-tRNA$^{Ini}_{CAU}$ is formylated by methionyl-tRNA formyltransferase (MTF), and the resulting fMet-tRNA$^{Ini}_{CAU}$ is selectively used for the initiation complex formation and starting the translation with fMet. Thus, the 10× TS (FD+) solution should be used when the AUG initiation codon is not reprogrammed (*see also* **Note 4**).

3. The time of freeze-thaw should be restricted no more than three times. Adjust the volume of the aliquots based on the experiment plan.

4. FD- indicates that it does not contain FD. When the AUG initiation is reprogrammed N-acylated Naa's such as ClAc^{D}W, FD is not required for translation initiation. In such cases, we use 10× TS (FD-) solution to avoid possible side reactions.

5. Orthogonal tRNAs, which are not aminoacylated by endogenous 20 AARSs, must be used for Naa incorporation. As such tRNAs, we usually utilize tRNA$^{AsnE2}_{NNN}$ (NNN denotes anticodon) [6].

6. If you want to charge unreported Naa's by flexizymes, we recommend quantifying the acylation efficiency using a microhelix RNA and acid PAGE analysis as described previously [1, 7]. Over 10% yield is usually sufficient to utilize the Naa in translation reaction. Optimization of reaction time (1–72 h), pH (7.5–9.5), amino acid concentration (5–40 mM), and DMSO concentration (20–40%) could improve the acylation reaction efficiency.

7. Use [^{14}C]-Asp and [^{12}C]-Asp for RI quantification and MALDI-TOF-MS analysis, respectively.

8. Since the ester bond of aminoacyl-tRNA is not stable in an aqueous solution, dissolve the precipitate of aminoacyl-tRNA just before preparing the translation reaction mixture. Add the aminoacyl-tRNA into the translation mixture in the last step, and start the incubation just after the mixing.

9. In this method the GGC codon is decoded by precharged ^{Iodo}F-tRNA$^{AsnE2}_{GCC}$, while the GGG codon is decoded by tRNA$^{Gly}_{CCC}$, which is in situ charged with glycine by GlyRS and continuously recycled after the charged glycine is transferred to the nascent polypeptide. In order to keep the decoding accuracy, the translation reaction should be quenched before the precharged Naa-tRNA is used up (or hydrolyzed), otherwise the misincorporation of glycine in response to the GGC codon could be observed. Quenching the reaction after 10 min incubation is usually short enough.

10. When you mix the tRNA transcripts, exclude the tRNA that corresponds to the codons designated to Naa assignment. For example, when you reprogram the AUG initiation codon and the GUC, CGC, GGC elongation codons, you should exclude tRNA$^{Ini}_{CAU}$, tRNA$^{Val}_{GAC}$, tRNA$^{Arg}_{GCG}$, and tRNA$^{Gly}_{GCC}$ and use a mixture of the other 28 tRNA transcripts. Note that tRNA$^{Ini}_{CAU}$ and tRNA$^{Met}_{CAU}$ decode the AUG initiation codon and the AUG elongation codon, respectively.

11. The optimal concentrations of three tRNA transcripts (tRNA$^{Lys}_{GAU}$, tRNA$^{Ile}_{CUU}$, and tRNA$^{Glu}_{CUC}$) are excess to the other tRNA transcripts since lack of the posttranscriptional nucleotide modifications of these tRNAs decreases their k_{cat}/K_M values for aminoacylation by the cognate AARSs [21, 30–32], resulting in less aminoacyl-tRNA formation.

12. When peptides containing cysteine residues are synthesized, TCEP should be added in order to prevent intramolecular and intermolecular disulfide bond formation.

Acknowledgment

This research was supported by the Japan Science and Technology Agency (JST) Core Research for Evolutional Science and Technology (CREST) of Molecular Technologies to H.S., Japan Society for the Promotion of Science (JSPS) Grant-in-Aid for Young Scientists (B) to Y.G. (22750145), and Grants-in-Aid for JSPS Fellows to Y.I. (26-9576).

References

1. Murakami H, Ohta A, Ashigai H, Suga H (2006) A highly flexible tRNA acylation method for non-natural polypeptide synthesis. Nat Methods 3(5):357–359

2. Goto Y, Ohta A, Sako Y, Yamagishi Y, Murakami H, Suga H (2008) Reprogramming the translation initiation for the synthesis of physiologically stable cyclic peptides. ACS Chem Biol 3(2):120–129

3. Goto Y, Murakami H, Suga H (2008) Initiating translation with *D*-amino acids. RNA 14(7):1390–1398

4. Kawakami T, Murakami H, Suga H (2008) Messenger RNA-programmed incorporation of multiple N-methyl-amino acids into linear and cyclic peptides. Chem Biol 15(1):32–42

5. Xiao H, Murakami H, Suga H, Ferre-D'Amare AR (2008) Structural basis of specific tRNA aminoacylation by a small *in vitro* selected ribozyme. Nature 454(7202):358–361

6. Ohta A, Murakami H, Higashimura E, Suga H (2007) Synthesis of polyester by means of genetic code reprogramming. Chem Biol 14(12):1315–1322

7. Goto Y, Katoh T, Suga H (2011) Flexizymes for genetic code reprogramming. Nat Protoc 6(6):779–790

8. Forster AC, Tan Z, Nalam MN, Lin H, Qu H, Cornish VW, Blacklow SC (2003) Programming peptidomimetic syntheses by translating genetic codes designed *de novo*. Proc Natl Acad Sci U S A 100(11):6353–6357

9. Josephson K, Hartman MC, Szostak JW (2005) Ribosomal synthesis of unnatural peptides. J Am Chem Soc 127(33):11727–11735

10. Yamagishi Y, Shoji I, Miyagawa S, Kawakami T, Katoh T, Goto Y, Suga H (2011) Natural product-like macrocyclic *N*-methyl-peptide inhibitors against a ubiquitin ligase uncovered from a ribosome-expressed de novo library. Chem Biol 18(12):1562–1570

11. Nemoto N, Miyamoto-Sato E, Husimi Y, Yanagawa H (1997) In vitro virus: bonding of mRNA bearing puromycin at the 3′-terminal end to the C-terminal end of its encoded protein on the ribosome in vitro. FEBS Lett 414(2):405–408

12. Roberts RW, Szostak JW (1997) RNA-peptide fusions for the *in vitro* selection of peptides and proteins. Proc Natl Acad Sci U S A 94(23):12297–12302

13. Terasaka N, Suga H (2014) Flexizymes-facilitated genetic code reprogramming leading to the discovery of drug-like peptides. Chem Lett 43(1):11–19

14. Tanaka Y, Hipolito CJ, Maturana AD, Ito K, Kuroda T, Higuchi T, Katoh T, Kato HE, Hattori M, Kumazaki K, Tsukazaki T, Ishitani R, Suga H, Nureki O (2013) Structural basis for the drug extrusion mechanism by a MATE multidrug transporter. Nature 496(7444):247–251

15. Hayashi Y, Morimoto J, Suga H (2012) *In vitro* selection of anti-Akt2 thioether-macrocyclic peptides leading to isoform-selective inhibitors. ACS Chem Biol 7(3):607–613

16. Morimoto J, Hayashi Y, Suga H (2012) Discovery of macrocyclic peptides armed with a mechanism-based warhead: isoform-selective inhibition of human deacetylase SIRT2. Angew Chem Int Ed Engl 51(14):3423–3427

17. Yamagata K, Goto Y, Nishimasu H, Morimoto J, Ishitani R, Dohmae N, Takeda N, Nagai R, Komuro I, Suga H, Nureki O (2014) Structural basis for potent inhibition of SIRT2 deacetylase by a macrocyclic peptide inducing dynamic structural change. Structure 22(2):345–352

18. Shimizu Y, Inoue A, Tomari Y, Suzuki T, Yokogawa T, Nishikawa K, Ueda T (2001) Cell-free translation reconstituted with purified components. Nat Biotechnol 19(8):751–755

19. Komine Y, Adachi T, Inokuchi H, Ozeki H (1990) Genomic organization and physical mapping of the transfer RNA genes in *Escherichia coli* K12. J Mol Biol 212(4):579–598

20. FVt M, Ramakrishnan V (2004) Structure of a purine-purine wobble base pair in the decoding center of the ribosome. Nat Struct Mol Biol 11(12):1251–1252

21. Gustilo EM, Vendeix FA, Agris PF (2008) tRNA's modifications bring order to gene expression. Curr Opin Microbiol 11(2):134–140. https://doi.org/10.1016/j.mib.2008.02.003

22. Crick FH (1966) Codon—anticodon pairing: the wobble hypothesis. J Mol Biol 19(2):548–555

23. Iwane Y, Hitomi A, Murakami H, Katoh T, Goto Y, Suga H (2016) Expanding the amino acid repertoire of ribosomal polypeptide synthesis via the artificial division of codon boxes. Nat Chem 8(4):317–325

24. Milligan JF, Groebe DR, Witherell GW, Uhlenbeck OC (1987) Oligoribonucleotide synthesis using T7 RNA polymerase and synthetic DNA templates. Nucleic Acids Res 15(21):8783–8798

25. Schagger H, von Jagow G (1987) Tricine-sodium dodecyl sulfate-polyacrylamide gel electrophoresis for the separation of proteins

in the range from 1 to 100 kDa. Anal Biochem 166(2):368–379

26. Schagger H (2006) Tricine-SDS-PAGE. Nat Protoc 1(1):16–22

27. Christian EL, McPheeters DS, Harris ME (1998) Identification of individual nucleotides in the bacterial ribonuclease P ribozyme adjacent to the pre-tRNA cleavage site by short-range photo-cross-linking. Biochemistry 37(50):17618–17628

28. Guo X, Campbell FE, Sun L, Christian EL, Anderson VE, Harris ME (2006) RNA-dependent folding and stabilization of C5 protein during assembly of the E. coli RNase P holoenzyme. J Mol Biol 360(1):190–203

29. Rappsilber J, Ishihama Y, Mann M (2003) Stop and go extraction tips for matrix-assisted laser desorption/ionization, nanoelectrospray, and LC/MS sample pretreatment in proteomics. Anal Chem 75(3):663–670

30. Tamura K, Himeno H, Asahara H, Hasegawa T, Shimizu M (1992) In vitro study of E.coli tRNA^{Arg} and tRNA^{Lys} identity elements. Nucleic Acids Res 20(9):2335–2339

31. Sylvers LA, Rogers KC, Shimizu M, Ohtsuka E, Soll D (1993) A 2-thiouridine derivative in tRNA^{Glu} is a positive determinant for aminoacylation by Escherichia coli glutamyl-tRNA synthetase. Biochemistry 32(15):3836–3841

32. Nureki O, Niimi T, Muramatsu T, Kanno H, Kohno T, Florentz C, Giege R, Yokoyama S (1994) Molecular recognition of the identity-determinant set of isoleucine transfer RNA from Escherichia coli. J Mol Biol 236(3):710–724

Cell-Free Protein Synthesis for Multiple Site-Specific Incorporation of Noncanonical Amino Acids Using Cell Extracts from RF-1 Deletion *E. coli* Strains

Eiko Seki, Tatsuo Yanagisawa, and Shigeyuki Yokoyama

Abstract

Cell-free protein synthesis (CFPS) is an effective method for the site-specific incorporations of noncanonical amino acids (ncAAs) into proteins. The nature of in vitro synthesis enables the use of experimental conditions that are toxic or reduce cellular uptake during in vivo site-specific incorporations of ncAAs. Using the *Escherichia coli* cell extract (S30) from the highly reproductive RF-1 deletion strains, B-60.ΔA::Z and B-95.ΔA, with orthogonal tRNA and aminoacyl-tRNA synthetase (aaRS) pairs from *Methanosarcina mazei*, we have developed CFPS methods for the highly productive and efficient multiple incorporation of ncAAs. In this chapter, we describe our methods for the preparation of the S30 and the orthogonal tRNA-^{Pyl} and PylRS pair, and two CFPS protocols for ncAA incorporation.

Key words Cell-free protein synthesis, Noncanonical amino acids, Release factor 1 (RF-1), Cell extract, S30, tRNA, Aminoacyl-tRNA synthetase

1 Introduction

Recently, more than 100 noncanonical amino acids (ncAAs) have been site specifically incorporated into proteins for various purposes [1–3]. With the help of orthogonal tRNA and aminoacyl-tRNA synthetase (aaRS) pairs from bacteria and archaea, the amber (UAG) stop codon has been reassigned as a target codon to incorporate ncAAs during protein synthesis.

Cell-free protein synthesis (CFPS) [4] has become a powerful tool for protein production, as it has numerous advantages over in vivo protein expression. The open nature of the in vitro system facilitates modifications and optimizations of reactions, and many proteins have been successfully synthesized for structural biology [5]. The relief from maintaining the cell viability enables CFPS to produce proteins that are toxic or require toxic substrates or conditions. Using specific inhibitors of intracellular enzymes, CFPS can

Edward A. Lemke (ed.), *Noncanonical Amino Acids: Methods and Protocols*, Methods in Molecular Biology, vol. 1728,
https://doi.org/10.1007/978-1-4939-7574-7_3, © Springer Science+Business Media, LLC 2018

successfully synthesize proteins that are inevitably altered by in vivo expression, such as posttranscriptionally acetylated proteins [6]. Using lipid bilayers, CFPS can generate large amounts of membrane proteins, which are difficult to produce by in vivo expression [7, 8].

Likewise, the site-specific incorporations of ncAAs into proteins are quite efficient in CFPS. The nature of a non-living, in vitro system makes it possible to use ncAAs, which are sometimes poor in cellular uptakes or toxic, and thus are fundamentally difficult to use for in vivo protein expression. The use of excess amounts of the orthogonal tRNA and aaRS pair for ncAA incorporation generally improves the incorporation efficiencies, but the in vivo overexpressions of tRNA and aaRS are limited, due to their cellular toxicity. In contrast, there are no limitations of tRNAs or aaRSs used for CFPS, and thus the ncAA incorporation efficiency can be increased. The substantially smaller amount of ncAAs used for CFPS, as compared to cell culture, is also a remarkable property, especially when ncAAs are expensive.

Release factor 1 (RF-1) is a cellular component that recognizes the UAG codon as a translation stop, and thus competes with the orthogonal tRNA used for ncAA incorporation. Several RF-1 deletion *Escherichia coli* strains have been developed to improve the incorporation efficiencies of ncAAs [9–12]. However, the mutations accompanying the RF-1 elimination severely hinder the growth of *E. coli*, resulting in low protein productivity in *E. coli* protein expression. To solve these problems, Mukai and his colleagues developed the highly reproductive, RF-1 deletion *E. coli* strains B-60.ΔA::Z and B-95.ΔA [13], in which 60 and 95 genomic UAG codons were respectively replaced synonymously prior to the disruption of the essential (*prfA*) gene, encoding RF-1.

The protein productivity of CFPS largely depends on the activity of the cell extract (S30 extract) used in the reaction mixture, and the activity of the S30 extract is highly influenced by the growth rate of *E. coli*. Thus, the RF-1 deletion *E. coli* strains B-60.ΔA::Z and B-95.ΔA are promising strains for S30 extract preparation, as they retain their rapid growth rates.

In this chapter, we describe our CFPS methods for the productive and efficient multiple incorporation of ncAAs, using S30 extracts from the *E. coli* strains B-60.ΔA::Z and B-95.ΔA. We describe two CFPS methods, a standard method in which all of the required components are supplemented as additives, and a convenient method using a specific S30 extract already containing the orthogonal tRNA and aaRS pair. We demonstrate the highly efficient incorporations of ncAAs with *Methanosarcina mazei* tRNA^{Pyl} and two types of orthogonal *M. mazei* PylRS mutants, which possess broader specificity to ncAAs than TyrRS and tRNA^{Tyr} pairs and are capable of incorporating various ncAAs [2, 3].

2   Materials

2.1   *E. coli* S30 Extract Preparation

2.1.1   *E. coli* Cell Preparation

1. *E. coli* strains: B-60.ΔA::Z and B-95.ΔA [13].
2. Plasmids: pMINOR [14] (*see* **Note 1**) and pCDF-Pyl-AFx2 [15] (*see* **Note 2**).
3. LB medium: 10 g/L tryptone, 5 g/L yeast extract, 10 g/L NaCl.
4. 2YT medium: 16 g/L tryptone, 10 g/L yeast extract, 5 g/L NaCl.
5. Antibiotics: 34 mg/mL chloramphenicol (Cm) and 20 mg/mL kanamycin (Km).
6. Induction: 100 mM IPTG.
7. S30 buffer: 10 mM Tris–acetate (pH 8.2), 14 mM magnesium acetate, 60 mM potassium acetate, 1 mM DTT.
8. Preincubation buffer: 300 mM Tris–acetate (pH 8.2), 9.3 mM magnesium acetate, 13 mM ATP (pH 7.0), 84 mM phosphoenolpyruvate (pH 7.0), 4 mM DTT, 40 μM each of 20 amino acids, and 6.7 U/mL pyruvate kinase.
9. 2 L baffled culture flasks.
10. Polytron cell homogenizer (KINEMATICA AG).

2.1.2   Cell Extract Preparation

1. Polytron cell homogenizer (KINEMATICA AG).
2. Pressure cell homogenizer (Stansted Fluid Power).
3. Dialysis tube: Spectra/Por7, MWCO = 15 kDa (SPECTRUM LABORATORIES).
4. Ultrafiltration unit: Amicon Ultra 15, MWCO = 10 kDa (Millipore).

2.2   tRNA^{Pyl} Preparation

1. Plasmid: pCR-T7P-tRNA^{Pyl} (*see* **Note 3**).
2. Primers:

   (a) Forward primer: 5′-GATAATACGACTCACTATAGGAA ACCTGATCATGTAGATCG.

   (b) Reverse primer: 5′-*t*gGCGGAAACCCCGGGAATC (*t, *g are 2′-O-methyl riboses).
3. Paq5000 DNA Polymerase Kit.
4. PCR solution: 0.2 ng/μg pCR-T7P-tRNA^{Pyl}, 0.2 μM each primers, 0.25 mM dNTP mixture, 1× reaction buffer, 0.05 U/μL Paq5000 DNA Polymerase.
5. tRNA transcription solution: 40 mM Tris–HCl (pH 8.0), 100 mM NaCl, 20 mM DTT, 2 mM spermidine, 5 mM ATP, 5 mM UTP, 5 mM GTP, 5 mM CTP, 20 mM GMP, 0.2 mg/mL bovine serum albumin, 20 mM MgCl$_2$, 0.02 mg/mL T7

  polymerase, 200 μL/mL PCR Solution (without purification).

6. Ion-exchange chromatography: ResourceQ (GE Healthcare).

7. Purification buffers:

  (2a) 20 mM Tris–HCl (pH 7.6).

  (2b) 20 mM Tris–HCl (pH 7.6), 280 mM NaCl.

  (2c) 20 mM Tris–HCl (pH 7.6), 1 M NaCl.

  (2d) 20 mM Tris–HCl (pH 7.6), 2 mM $MgCl_2$.

8. Reagents: Phenol/chloroform/isoamyl alcohol (25:24:1), pH 5.2, chloroform, ethanol, isopropanol, ammonium acetate.

2.3 PylRS Preparation

1. Plasmids: pET24_H6SUMO-MmPylRS(R61K/Y306A/Y384F), pET24_H6SUMO-MmPylRS(R61K/Y384F).

2. *E. coli* strain: BL21-Gold(DE3).

3. LB medium: 10 g/L tryptone, 5 g/L yeast extract, 10 g/L NaCl.

4. Antibiotics: 20 mg/mL Km.

5. Induction: 100 mM IPTG.

6. Lysis buffer: 20 mM Tris–HCl (pH 8.2), 1.5 M NaCl, 20 mM imidazole, 1% Tween 20, 0.1 mM DTT, 0.5 mM PMSF.

7. His-tag affinity column: HisTrap HP (GE Healthcare), 5 mL column.

8. Ion-exchange chromatography: ResourceS (GE Healthcare).

9. Purification buffers:

  (3a) 20 mM Tris–HCl (pH 8.0), 300 mM NaCl, 1 mM DTT, 20 mM imidazole.

  (3b) 20 mM Tris–HCl (pH 8.0), 300 mM NaCl, 1 mM DTT, 500 mM imidazole.

  (3c) 20 mM Tris–HCl (pH 8.0), 300 mM NaCl, 1 mM DTT.

  (3d) 20 mM Tris–HCl (pH 8.0), 1000 mM NaCl, 1 mM DTT.

10. SUMO protease.

11. 5 L baffled culture flasks.

12. Ultrafiltration unit: Amicon Ultra 15, MWCO = 50 kDa (Millipore).

2.4 Cell-Free Protein Synthesis

1. LMCPY mixture-PEG: 160 mM HEPES-KOH buffer (pH 7.5), 4.14 mM L-tyrosine, 534 mM potassium L-glutamate, 5 mM DTT, 3.47 mM ATP, 2.40 mM GTP, 2.40 mM CTP, 2.40 mM UTP, 0.203 mM folic acid, 1.78 mM cAMP, 74 mM ammonium acetate, and 214 mM creatine phosphate.

2. 17.5 mg/mL *E. coli* MRE600-derived tRNA mixture (Roche Diagnostics).

3. 5% (w/v) sodium azide.

4. 1.6 M magnesium acetate.

5. Amino acid mixture: 20 mM each of 19 amino acids, without L-tyrosine.

6. 3.75 mg/mL creatine kinase (Roche Diagnostics).

7. 10 mg/mL T7 RNA polymerase in 20 mM Tris–HCl (pH 8.0), 100 mM NaCl, 1 mM DTT, 1 mM EDTA, 50% glycerol.

8. S30 buffer: 10 mM Tris–acetate (pH 8.2), 14 mM magnesium acetate, 60 mM potassium acetate, 1 mM DTT.

9. *E. coli* S30 extract: prepared as described below.

10. Template DNA plasmids: pN11GFPS1sh [16] and pN11GFPS1sh-A17Amb (*see* **Note 4**).

11. Noncanonical amino acids:

    (a) 50 mM TCO*Lys in 0.2 M NaOH, 15% DMSO (*see* **Note 5**).

    (b) 50 mM BCNLys in 0.5 M HCl (*see* **Note 5**).

    (c) 50 mM ZLys in 0.5 M HCl (*see* **Note 5**).

    (d) 50 mM BocLys in water (*see* **Note 5**).

    (e) 50 mM PocLys in water (*see* **Note 5**).

12. tRNA^{Pyl}: prepared as described below.

13. *M. mazei* PylRS mutants: PylRS(R61K/Y384F) and PylRS(R61K/Y306A/Y384F), prepared as described below.

14. Dialysis tube: Spectra/Por7, MWCO = 15 kDa (SPECTRUM LABORATORIES).

15. KOH and HCl.

16. Dilution buffer: 20 mM Tris–HCl (pH 7.5), 150 mM NaCl.

17. 96-well black flat bottom polystyrene microplate (Corning).

18. Plate fluorescence reader: ARVO Victor2 V Multilabel Counter (PerkinElmer).

3 Methods

3.1 *E. coli* S30 Extract Preparation

Two types of S30 extract preparations are described in this section: (a) the standard ncAA incorporation CFPS method with all of the required components supplemented as additives, and (b) the convenient method using an S30 extract already containing the orthogonal tRNA^{Pyl} and PylRS pair (*see* **Notes 6–8**).

3.1.1 E. coli Cell
Preparation

1. Transform *E. coli* B-60.ΔA::Z or B-95.ΔA with pMINOR alone for (a), or co-transform with pMINOR and pCDF-Pyl-AFx2 for (b) (*see* **Note 9**).

2. Inoculate a single colony into 5 mL LB medium containing antibiotics (*see* **Note 10**), and incubate for several hours at 37 °C with continuous shaking.

3. Inoculate 0.1 mL cultured cells into 100 mL 2YT medium containing antibiotics (*see* **Note 10**), and incubate overnight at 30 °C with continuous shaking.

4. Inoculate 5 mL overnight cultured cells into each of eight 2 L baffled flasks containing 500 mL pre-warmed 2YT medium with antibiotics (*see* **Note 10**). Incubate at 30 °C with continuous shaking.

5. Monitor the OD_{600} frequently (*see* **Note 11**), and stop the culture when the OD_{600} reaches around 3 or when the growth rate starts decrease. Chill the flasks in ice-cold water (*see* **Note 12**).

6. Harvest the cells by centrifugation at 7,500 × *g* for 6 min at 4 °C, and suspend the cells in 1.5 L S30 buffer (*see* **Note 13**).

7. Centrifuge the suspended cells at 7,500 × *g* for 6 min at 4 °C, and resuspend the cells in 800 mL S30 buffer.

8. Centrifuge the suspended cells at 7,500 × *g* for 6 min at 4 °C, resuspend the cells in 200 mL S30 buffer, and collect the cells in one centrifuge tube.

9. Centrifuge the suspended cells at 7,500 × *g* for 23 min at 4 °C, discard the supernatant, measure the wet weight of the cells, and freeze the cells in liquid nitrogen. Store the cells in a −80 °C freezer.

3.1.2 Cell Extract
Preparation

1. Thaw the frozen cells and suspend them in the corresponding volume per weight amount of S30 buffer on ice (1 mL per 1 mg cells) (*see* **Note 13**).

2. Disrupt the cells using an ice-cold pressure cell homogenizer at a pressure around 150 to 260 MPa (*see* **Note 14**).

3. Centrifuge the cell lysate twice at 30,000 × *g* for 30 min at 4 °C (*see* **Note 15**). The supernatant will be a crude S30 extract.

4. Add 30% volume of preincubation buffer, and incubate for 70 min at 37 °C.

5. Transfer the crude extract into a dialysis membrane, and dialyze three times against a 30-fold volume of S30 buffer for 1 h at 4 °C.

6. Centrifuge the extract at 12,000 × *g* for 10 min at 4 °C, to obtain the supernatant as the S30 extract.

7. Measure the UV absorbance of the S30 extract at OD_{260}, and concentrate the extract with an ultrafiltration unit up to around $OD_{260} = 500$ (*see* **Note 16**).

8. Freeze aliquots of the S30 extract in liquid nitrogen, and store below −80 °C.

3.2 tRNA^{Pyl} Preparation

The orthogonal tRNA^{Pyl} is prepared by in vitro transcription.

1. Prepare the PCR reaction mixture, including pCR-T7P-tRNA^{Pyl} as the template, and the primers, in which the reverse primer has two 2′-O-methyl riboses at the 5′ end. Perform the PCR reaction (*see* **Note 17**).

2. Prepare the tRNA^{Pyl} transcription mixture (*see* Table 1) with the crude PCR product.

3. Perform the tRNA^{Pyl} transcription at 37 °C for 3 h.

4. Purify the transcript by standard phenol/chloroform extraction, followed by precipitation with 2 volumes of isopropanol and 10% of 3 M ammonium acetate. Wash the pellet with 70% ethanol.

Table 1
Components of reaction solution and external solution

Stock solution	Final	Reaction solution	External solution
LMCPY mixture-PEG		373.3 µL	3733 µL
17.5 mg/mL tRNA	0.175 mg/mL	10 µL	–
5% sodium azide	0.05%	10 µL	100 µL
1.6 M magnesium acetate	10 mM	6.25 µL	62.5 µL
20 mM amino acid mixture	1.5 mM	75 µL	750 µL
3.75 mg/mL creatine kinase	0.1 mg/mL	26.7 µL	–
10 mg/mL T7 RNA polymerase	0.067 mg/mL	6.7 µL	–
S30 extract	30%	300 µL	–
S30 buffer	30%	–	3000 µL
100 µg/mL template DNA	2 µg/mL	20 µL	–
1.25 mM tRNA^{Pyl}	10 µM	8 µL	–
75 µM PylRS mutant	10 µM	133.33 µL	–
KOH or HCl for pH adjustment	1/2 volume of ncAA	10 µL	150 µL
50 mM ncAA	1 mM	20 µL	300 µL
Milli-Q water		0.72 µL	1904.5 µL
Total		1 mL	10 mL

5. Resuspend in 1/3 transcription volume of (2a) buffer at 75 °C, and gradually cool down.

6. Apply the crude tRNA^{Pyl} to a ResourceQ ion-exchange column equilibrated with (2b) buffer at 3 mL/min. Elute the tRNA^{Pyl} with (2c) buffer in a 20 column volume linear gradient mode.

7. Collect the eluted fractions, precipitate with 2 volumes of isopropanol and 10% of 3 M ammonium acetate, and wash with 70% ethanol.

8. Resuspend with (2d) buffer, and measure the UV absorbance of the extract at OD$_{260}$ (*see* **Note 18**).

3.3 PylRS Preparation

We prepare the orthogonal PylRS by either *E. coli* in vivo expression or CFPS. Here, we describe the *E. coli* in vivo expression method.

1. Transform *E. coli* BL21-Gold(DE3) with the plasmid harboring the MmPylRS mutant.

2. Inoculate a single colony into 40 mL LB medium containing 20 mg/L Km, and incubate overnight at 37 °C with continuous shaking.

3. Inoculate 15 mL overnight cultured cells into each of two 5 L baffled flasks containing 1 L pre-warmed LB medium with 20 mg/L Km, and incubate at 30 °C with continuous shaking.

4. Monitor the OD$_{600}$, and chill the flasks in ice-cold water when the OD$_{600}$ reaches around 0.8 (approximately 3 h).

5. Add IPTG to a final concentration of 0.5 mM for induction, and incubate the cultures overnight at 20 °C with continuous shaking.

6. Harvest the cells by centrifugation at 5,000 × g for 10 min at 4 °C, and suspend the cells in 100 mL ice-cold lysis buffer.

7. Disrupt the cells by sonication, and centrifuge the cell lysate twice at 12,000 × g for 10 min at 4 °C.

8. Transfer the supernatant to a HisTrap HP 5 mL column, wash with (3a) buffer, and elute with (3b) buffer.

9. Concentrate the eluted fractions with an ultrafiltration unit, add 0.1 mg SUMO protease, and dialyze overnight against 1 L (3c) buffer at 4 °C.

10. Centrifuge the digested PylRS at 9,000 × g for 10 min at 4 °C. Apply the supernatant to a ResourceS ion-exchange column equilibrated with (3c) buffer, and elute with (3d) buffer in a 20 column volume linear gradient mode.

11. Collect the eluted fractions, concentrate with an ultrafiltration unit, and measure the UV absorbance of the extract at OD_{280} (*see* **Note 18**).

12. Freeze aliquots of the PylRS in liquid nitrogen, and store below −80 °C.

3.4 Site-Specific Incorporation of ncAA Using Cell-Free Protein Synthesis

The dialysis-mode CFPS is used for the incorporation of ncAAs. Any scale or dialysis-mode synthesis device [17] can be adopted, and here we describe a 1 mL scale CFPS as an example (*see* **Note 19**).

1. Prepare the reaction solution and the external solution on ice, according to Table 1 (*see* **Notes 20–23**).

2. Place the reaction solution into a dialysis tube and seal both ends with closures. Submerge the dialysis tube in the external solution, and incubate overnight at 25 °C with continuous gentle shaking (*see* **Notes 24** and **25**).

3. Collect the reaction solution from the dialysis membrane. Combine 1 μL reaction solution with 200 μL dilution buffer (20 mM Tris–HCl (pH 7.5), 150 mM NaCl) in a 96-well black flat bottom microplate. Measure the fluorescence using a plate fluorescence reader, with excitation at 485 nm and emission at 535 nm (*see* **Note 26**). Typical ncAA incorporation results are shown in Figs. 1 and 2 (*see* **Notes 27** and **28**).

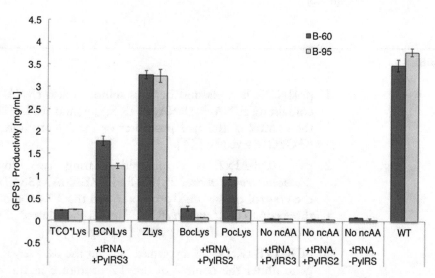

Fig. 1 Incorporations of ncAAs with two aaRS mutants, using two types of RF-1 deletion S30 extracts. The CFPS was conducted with the standard incorporation method, supplemented with tRNA^{pyl} and PylRS mutants as additives. PylRS(R61K/Y306A/Y384F), shown as PylRSm3, was used for TCO*Lys, BCNLys and ZLys incorporations. MmPylRS(R61K/Y384F), shown as PylRSm2, was used for BocLys and PocLys incorporations. The ncAAs were incorporated at the A17 site of N11GFPS1sh by the dialysis-mode CFPS at 25 °C overnight, with S30 extracts from B-60.ΔA::Z (B-60) or B-95.ΔA (B-95). The synthesis control, without an amber stop codon in the DNA template, is shown as WT. The reaction solutions were diluted and measured with a plate fluorescence reader, with excitation at 485 nm and emission at 535 nm. Bars indicate standard deviations (n = 3)

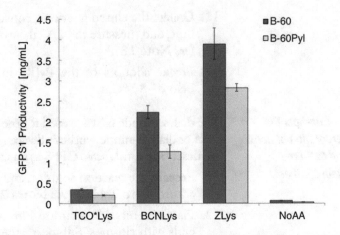

Fig. 2 Comparison of the two CFPS methods for ncAA incorporation. The standard CFPS method was conducted by supplementing the S30 from B-60.ΔA::Z harboring pMINOR (B-60) with tRNA^{pyl} and PylRS(R61K/Y306A/Y384F) as additives. The convenient CFPS method was conducted without supplementing tRNA^{pyl} and PylRS, with the S30 prepared from B-60.ΔA::Z harboring pMINOR and pCDF-Pyl-AFx2, in which PylRS(R61K/Y306A/Y384F) and tRNA^{Pyl} were induced during the S30 preparation (B-60Pyl). ZLys was incorporated at the A17 site of N11GFPS1sh, by the dialysis-mode CFPS at 25 °C overnight. The reaction solutions were diluted and measured with a plate fluorescence reader, with excitation at 485 nm and emission at 507 nm. Bars indicate standard deviations (n = 3)

---

4   Notes

1. pMINOR is a plasmid bearing minor codon tRNA clusters containing tRNA$_2^{Ile}$, tRNA$_3^{Arg}$, tRNA$_2^{Pro}$, and tRNA$_4^{Arg}$, under the control of the *tyrT* promoter or the *lpp* promoter in the pACYC184 vector [14].

2. pCDF-Pyl-AFx2 is a plasmid bearing two copies of *Methanosarcina mazei* PylRS(R61K/Y306A/Y384F) under the control of the *glnS'* promoter and the T7 promoter, and three copies of *M. mazei* tRNA^{Pyl} under the control of *lpp* promoters in the pCDF-1b vector (Novagen) [15].

3. pCR-T7P-tRNA^{Pyl} is a plasmid bearing the *M. mazei* tRNA^{Pyl} gene under the control of the T7 promoter in the pCR2.1 vector (Thermo Fisher Scientific).

4. pN11GFPS1sh is a plasmid bearing a poly-histidine N11-tag and a linker DNA, followed by a superfolder type GFP mutant (GFPS1) gene [16], under the control of the T7 promoter in the pCR2.1 vector. pN11GFPS1sh-A17Amb is a plasmid that has a TAG codon within the linker DNA, at the amino acid number Ala17 of pN11GFPS1sh. The amino acid sequence of

N11GFPS1sh-A17Amb is MKDHLIHNHHKHEHAH*
ENLYFQSKGEELFTGVVPILVELDGDVNGHKFSVS
GEGEGDATYGKLTLKFICTTGKLPVPWPTLVT
TLTYGVQCFSRYPDHMKRHDFFKSAMPEGYVQE
RTISFKDDGNYKTRAEVKFEGDTLVNRIELKGIDFKED
GNILGHKLEYNYNSHNVYITADKQKNGIK
ANFKIRHNIEDGSVQLADHYQQNTPIGDGP
VLLPDNHCLSTQSALSKDPNEKRDHMVLLE
FVTAAGITHGMDELYK (* is the TAG amber codon in the
DNA sequence).

5. Noncanonical amino acids are as follows.

   (a) TCO*Lys: N^{ε}-((((*E*)-cyclooct-2-en-1-yl)oxy)carbonyl)-
       L-lysine [18, 19], purchased from SiChem.

   (b) BCNLys: N^{ε}-(((((1*R*,8*S*)-bicyclo[6.1.0]non-4-yn-9-yl)
       methoxy)carbonyl)-L-lysine [20, 21], purchased from
       SYNAFFIX.

   (c) ZLys: N^{ε}-benzyloxycarbonyl-L-lysine [22, 23], purchased
       from Bachem.

   (d) BocLys: N^{ε}-(*tert*-butyloxycarbonyl)-L-lysine [15, 23],
       purchased from Watanabe Chemical.

   (e) PocLys: N^{ε}-propargyloxycarbonyl-L-lysine [3, 24, 25],
       purchased from SiChem.

6. We generally use different S30 extract preparation methods
   according to the purposes of the cell-free protein synthesis. In
   the case of incorporating ncAAs into proteins, two approaches
   are taken to improve the incorporation efficiency. First, RF-1
   deletion *E. coli* strains, B-60.ΔA::Z or B-95.ΔA [13], are used
   for the extract preparation (Fig. 3). Second, a pressure cell
   homogenizer, rather than a glass bead homogenizer, is used
   for cell disruption.

7. Since the endogeneous RF-1 is a competitor to the orthogonal tRNA used for ncAA incorporation, the productivity of
   ncAA-incorporated proteins increases with S30 extracts from
   RF-1 deletion *E. coli* strains, as compared to S30 extracts from
   standard *E. coli* BL21 strains (Fig. 3). As the number of amber
   codons to incorporate ncAAs increases, the protein productivities decrease as compared to the control synthesis without
   an amber codon (WT). The productivities of the S30 extracts
   from the RF-1 deletion B-60 and B-95 strains were about
   three times greater than that from the BL21 strain with one
   amber codon, and 6 times and 18 times greater with two and
   three amber codons, respectively. These results indicate that it
   is more efficient to use RF-1 deletion *E. coli* strains for S30
   extract preparation.

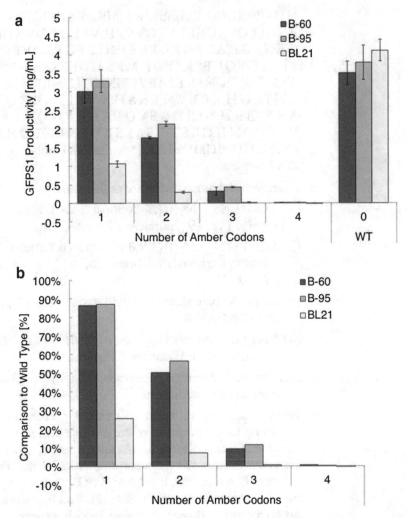

Fig. 3 Comparison of the S30 extracts for multiple ncAA incorporation. For the standard strain, BL21/pMINOR2 was used (BL21). For the RF-1 deletion strain, B-60.ΔA::Z (B-60) and B-95.ΔA (B-95) were used. ZLys was incorporated into 1–4 amber stop codon sites of N11-GFPS1, at the positions of A17, S24, E154, and E235. The dialysis-mode CFPS was conducted with tRNA^{Pyl} and PylRS(R61K/Y306A/Y384F) at 25 °C overnight. The synthesis control, without an amber stop codon in the DNA template, is shown as WT. The reaction solutions were diluted and measured with a plate fluorescence reader, with excitation at 485 nm and emission at 535 nm. (**a**) The total amount of synthesized N11-GFPS1, with bars indicating standard deviations (n = 3). (**b**) Comparison of the ncAA-incorporated synthesis to the wild-type synthesis without an amber codon

8. Although the disruption method, using a pressure cell homogenizer, is coarse, it allows the preparation of active extracts from various types of *E. coli* mutants with different cell wall rigidities, without precise optimization.

9. The pMINOR plasmid is transformed into all of the *E. coli* strains, because it increases the amounts of *E. coli* minor codon tRNAs in the S30 extract. The pCDF-Pyl-AFx2 plasmid is transformed to express *Methanosarcina mazei* tRNA^{Pyl} and

PylRS(R61K/Y306A/Y384F) during the cell culture, for the S30 extract preparation.

10. Antibiotics:

    (a) *E. coli* with pMINOR: 34 mg/L Cm.

    (b) *E. coli* with pMINOR and pCDF-Pyl-AFx2: 4 mg/L Cm and 20 mg/L Km.

11. For (b) with pCDF-Pyl-AFx2, when the OD_{600} reaches around 0.8, add IPTG to a final concentration of 0.25 mM for induction.

12. It is important for the cells to be kept ice-cold during the centrifugation and suspension steps.

13. To facilitate the cell suspension steps, we use a Polytron cell homogenizer.

14. To avoid the excess disruption of cells, which will lead to lower cell extract activity in CFPS, we use a pressure cell homogenizer from Stansted Fluid Power LTD. Using this homogenizer, the disruption pressure can be controlled by adjusting the size of the orifice, and the cells and the lysate can be maintained under ice-cold conditions. These features lead to highly reproducible disruption of the cells.

15. There will be two layers of cellular debris, packed hard and soft. Take care not to contaminate the supernatant with the debris, or the activity of the S30 extract will decrease.

16. The S30 extracts from the B-60.ΔA::Z and B-95.ΔA strains exhibit an approximate concentration dependence of the protein productivity at OD_{260} (Fig. 4). The concentration of the

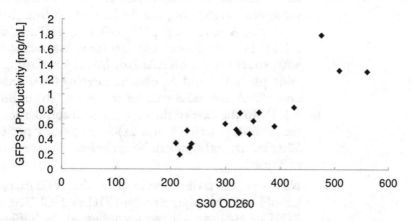

Fig. 4 The protein productivity dependency on the concentration of the B-60.ΔA::Z and B-95.ΔA S30 extracts. N11GFPS1sh was synthesized by the dialysis-mode CFPS at 25 °C overnight, using S30 extracts from B-60.ΔA::Z and B-95.ΔA. The reaction solutions were diluted and measured with a plate fluorescence reader, with excitation at 485 nm and emission at 535 nm. The OD_{260} of each S30 extract was measured and its N11GFPS1sh productivity is plotted

B-95.ΔA extract obtained after the cell disruption tends to be low, which may be due to changes in the cell wall rigidity induced by multiple genomic mutations. The average concentration rate for a highly productive B-60.ΔA::Z S30 extract is twofold, and that for the B-95.ΔA S30 extract is threefold.

17. Any PCR reaction kit may be used. The 5′ end of the reverse primer has two 2′-O-methyl riboses. This will prevent undesired nucleotide addition during the transcription with T7 RNA polymerase.

18. It is better to decrease the liquid volume added to the CFPS reaction mixture. Thus, the concentration of the resuspended tRNA^{pyl} should be greater than 1 mM. The preferable concentration of PylRS is around 80 μM, as higher concentrations will lead to precipitation.

19. The standard ncAA incorporation CFPS method (a) is supplemented with all of the required components as additives, using the widely applicable S30 extract from an RF-1 deletion *E. coli* strain. The convenient method (b) uses a specific S30 extract containing an orthogonal tRNA and PylRS pair, expressed within the cultured cells during the S30 extract preparation. For the investigation of the incorporation of an unfamiliar ncAA, the standard method is recommended for its optimization convenience, and once the best synthesis conditions are defined, the convenient method is recommended to reduce the number of steps for tRNA and aaRS preparation.

20. All of the components should be kept ice-cold. LMCPY-PEG should be mixed vigorously before use. Other components should be mixed gently. For the standard ncAA incorporation method, use the S30 extract from B-60.ΔA::Z or B-95.ΔA harboring pMINOR only ((a) in Subheading 3.1.1). For the convenient incorporation method, use the S30 extract from B-60.ΔA::Z or B95.ΔA harboring two plasmids, pMINOR and the plasmid carrying the specific orthogonal tRNA and aaRS pair for the ncAA ((b) in Subheading 3.1.1). In the case of the convenient incorporation method, the orthogonal tRNA and aaRS are already present in the S30 extract, and they can be excluded from the components in Table 1.

21. NcAAs are often dissolved in many kinds of buffers, and thus the pH must be adjusted with KOH or HCl. The volume of KOH or HCl may change according to the buffers used for ncAAs.

22. The optimal amounts of tRNA^{Pyl} and PylRS may vary with different ncAAs. The standard amounts to start the validation are 10 μM each.

23. PEG8000, which is used in our standard CFPS method, is excluded from LMCPY for the ncAA incorporation method [26]. The misincorporation rate of ncAA significantly decreases by excluding PEG8000.

24. The temperature used for ncAA incorporation is 25 °C, which is lower than that in our standard CFPS method [26]. The ncAA misincorporation rate significantly decreases at 25 °C. To compensate for the lower temperature, we incubate the reaction overnight. Depending on the nature of the target protein, the incubation time may be shortened.

25. Pictures and precise descriptions of the devices that may be used for the dialysis-mode CFPS are provided in the reference [17].

26. To determine the productivity of the CFPS, 1 mg/mL purified N11GFPS1sh was used as a control during the fluorescence measurements.

27. Protein productivity varies with different ncAAs. Some ncAAs, such as ZLys, are incorporated as efficiently as other natural amino acids, and the protein productivity can be as high as the wild-type protein synthesis. The S30 extract from B-60.ΔA::Z tends to be more productive than that from B-95.ΔA, especially with ncAAs using MmPylRS(R61K/Y384F) for incorporation. Considering the fact that the S30 extract preparation yield is higher with the B-60.ΔA::Z strain (from **Note 16**), it may be more convenient to use the B-60.ΔA::Z strain as the first choice.

28. The convenient ncAA incorporation CFPS method tends to be less productive than the standard method, in compensation for the alleviation of the tRNA and aaRS preparation.

Acknowledgments

We thank Dr. Kensaku Sakamoto and Dr. Takahito Mukai for providing the *E. coli* strains B-60.ΔA::Z and B-95.ΔA. We thank Dr. Kazushige Katsura and Dr. Chie Takemoto for assisting with the S30 extract preparation. This work was supported by the Platform Project for Supporting in Drug Discovery and Life Science Research (Platform for Drug Discovery, Informatics, and Structural Life Science) from the Ministry of Education, Culture, Sports, Science and Technology (MEXT) and the Japan Agency for Medical Research and Development (AMED).

References

1. Liu CC, Schultz PG (2010) Adding new chemistries to the genetic code. Annu Rev Biochem 79:413–444. https://doi.org/10.1146/annurev.biochem.052308.105824

2. Wan W, Tharp JM, Liu WR (2014) Pyrrolysyl-tRNA synthetase: an ordinary enzyme but an outstanding genetic code expansion tool. Biochim Biophys Acta 1844:1059–1070. https://doi.org/10.1016/j.bbapap.2014.03.002

3. Yanagisawa T, Umehara T, Sakamoto K, Yokoyama S (2014) Expanded genetic code technologies for incorporating modified lysine at multiple sites. Chembiochem 15:2181–2187. https://doi.org/10.1002/cbic.201402266

4. Kigawa T, Yabuki T, Matsuda N et al (2004) Preparation of Escherichia coli cell extract for highly productive cell-free protein expression. J Struct Funct Genom 5:63–68. https://doi.org/10.1023/B:JSFG.0000029204.57846.7d

5. Terada T, Murata T, Shirouzu M, Yokoyama S (2014) Cell-free expression of protein complexes for structural biology. Methods Mol Biol:151–159

6. Mukai T, Yanagisawa T, Ohtake K et al (2011) Genetic-code evolution for protein synthesis with non-natural amino acids. Biochem Biophys Res Commun 411:757–761. https://doi.org/10.1016/j.bbrc.2011.07.020

7. Schwarz D, Junge F, Durst F et al (2007) Preparative scale expression of membrane proteins in Escherichia coli-based continuous exchange cell-free systems. Nat Protoc 2:2945–2957. https://doi.org/10.1038/nprot.2007.426

8. Shinoda T, Shinya N, Ito K et al (2016) Cell-free methods to produce structurally intact mammalian membrane proteins. Sci Rep 6:30442. https://doi.org/10.1038/srep30442

9. Mukai T, Hayashi A, Iraha F et al (2010) Codon reassignment in the Escherichia coli genetic code. Nucleic Acids Res 38:8188–8195. https://doi.org/10.1093/nar/gkq707

10. Johnson DBF, Xu J, Shen Z et al (2011) RF1 knockout allows ribosomal incorporation of unnatural amino acids at multiple sites. Nat Chem Biol 7:779–786. https://doi.org/10.1038/nchembio.657

11. Heinemann IU, Rovner AJ, Aerni HR et al (2012) Enhanced phosphoserine insertion during Escherichia coli protein synthesis via partial UAG codon reassignment and release factor 1 deletion. FEBS Lett 586:3716–3722. https://doi.org/10.1016/j.febslet.2012.08.031

12. Lajoie MJ, Rovner AJ, Goodman DB et al (2013) Genomically recoded organisms expand biological functions. Science 342:357–360. https://doi.org/10.1126/science.1241459

13. Mukai T, Hoshi H, Ohtake K et al (2015) Highly reproductive Escherichia coli cells with no specific assignment to the UAG codon. Sci Rep 5:9699. https://doi.org/10.1038/srep09699

14. Chumpolkulwong N, Sakamoto K, Hayashi A et al (2006) Translation of "rare" codons in a cell-free protein synthesis system from Escherichia coli. J Struct Funct Genom 7:31–36. https://doi.org/10.1007/s10969-006-9007-y

15. Yanagisawa T, Takahashi M, Mukai T et al (2014) Multiple site-specific installations of Nε-monomethyl-L-lysine into histone proteins by cell-based and cell-free protein synthesis. Chembiochem 15:1830–1838. https://doi.org/10.1002/cbic.201402291

16. Seki E, Matsuda N, Yokoyama S, Kigawa T (2008) Cell-free protein synthesis system from Escherichia coli cells cultured at decreased temperatures improves productivity by decreasing DNA template degradation. Anal Biochem 377:156–161. https://doi.org/10.1016/j.ab.2008.03.001

17. Kigawa T, Matsuda T, Yabuki T, Yokoyama S (2008) Bacterial cell-free system for highly efficient protein synthesis. In: Cell-free protein synth. Methods Protoc, pp 83–97

18. Plass T, Milles S, Koehler C et al (2012) Amino acids for diels-alder reactions in living cells. Angew Chem Int Ed Engl 51:4166–4170. https://doi.org/10.1002/anie.201108231

19. Nikić I, Plass T, Schraidt O et al (2014) Minimal tags for rapid dual-color live-cell labeling and super-resolution microscopy. Angew Chem Int Ed Engl 53:2245–2249. https://doi.org/10.1002/anie.201309847

20. Borrmann A, Milles S, Plass T et al (2012) Genetic encoding of a bicyclo[6.1.0]nonyne-charged amino acid enables fast cellular protein imaging by metal-free ligation. Chembiochem 13:2094–2099. https://doi.org/10.1002/cbic.201200407

21. Lang K, Davis L, Wallace S et al (2012) Genetic encoding of bicyclononynes and trans-cyclooctenes for site-specific protein labeling in vitro and in live mammalian cells via rapid fluorogenic diels-alder reactions. J Am

Chem Soc 134:10317–10320. https://doi.org/10.1021/ja302832g

22. Yanagisawa T, Ishii R, Fukunaga R et al (2008) Multistep engineering of pyrrolysyl-tRNA synthetase to genetically encode N(epsilon)-(o-azidobenzyloxycarbonyl) lysine for site-specific protein modification. Chem Biol 15:1187–1197. https://doi.org/10.1016/j.chembiol.2008.10.004

23. Mukai T, Kobayashi T, Hino N et al (2008) Adding l-lysine derivatives to the genetic code of mammalian cells with engineered pyrrolysyl-tRNA synthetases. Biochem Biophys Res Commun 371:818–822. https://doi.org/10.1016/j.bbrc.2008.04.164

24. Flügel V, Vrabel M, Schneider S (2014) Structural basis for the site-specific incorporation of lysine derivatives into proteins. PLoS One 9:1–7. https://doi.org/10.1371/journal.pone.0096198

25. Nguyen DP, Lusic H, Neumann H et al (2009) Genetic encoding and labeling of aliphatic Azides and alkynes in recombinant proteins via a Pyrrolysyl-tRNA Synthetase/tRNA CUA pair and click chemistry. J Am Chem Soc 131:8720–8721. https://doi.org/10.1021/ja900553w

26. Kigawa T, Yabuki T, Yoshida Y et al (1999) Cell-free production and stable-isotope labeling of milligram quantities of proteins. FEBS Lett 442:15–19

Chapter 4

Tub-Tag Labeling; Chemoenzymatic Incorporation of Unnatural Amino Acids

Jonas Helma, Heinrich Leonhardt, Christian P.R. Hackenberger, and Dominik Schumacher

Abstract

Tub-tag labeling is a chemoenzymatic method that enables the site-specific labeling of proteins. Here, the natural enzyme tubulin tyrosine ligase incorporates noncanonical tyrosine derivatives to the terminal carboxylic acid of proteins containing a 14-amino acid recognition sequence called Tub-tag. The tyrosine derivative carries a unique chemical reporter allowing for a subsequent bioorthogonal modification of proteins with a great variety of probes. Here, we describe the Tub-tag protein modification protocol in detail and explain its utilization to generate labeled proteins for advanced applications in cell biology, imaging, and diagnostics.

Key words Chemoenzymatic labeling, Tubulin tyrosine ligase, Site-specific protein functionalization, Tyrosine derivatives, Bioorthogonal reaction

1 Introduction

Within recent years, the functionalization of proteins with organic fluorophores, drugs, and other probes has enabled powerful tools for cellular biology, molecular and medical imaging, as well as the development of new pharmaceuticals for the treatment of severe illnesses [1–3]. While early strategies utilized activated carboxylic acids that randomly react with any solvent exposed lysine residue of a given protein, it became apparent that the uncontrolled attachment of probes to proteins can result in loss of function [4]. Therefore, methods that allow the controlled and site-specific modification of proteins are on the rise [5, 6]. Among them are chemical methods in which a probe chemoselectively reacts with a terminal amino acid of a protein [7], expressed protein ligation [8, 9], as well as auxotrophic expression and amber suppression [10, 11]. Moreover, several naturally occurring enzymes have been engineered to site-specifically modify peptides and proteins [12,

Edward A. Lemke (ed.), *Noncanonical Amino Acids: Methods and Protocols*, Methods in Molecular Biology, vol. 1728, https://doi.org/10.1007/978-1-4939-7574-7_4, © Springer Science+Business Media, LLC 2018

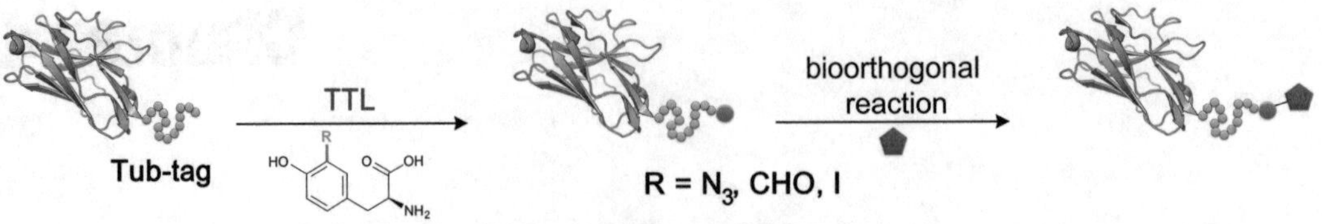

Fig. 1 Protein functionalization by tubulin tyrosine ligase (TTL). Illustration of the two-step Tub-tag labeling of an antigen binding nanobody (PDB Code 3K1K) [28]. The enzyme ligates a tyrosine derivative functionalized with a bioorthogonal handle to the terminal carboxylic acid of Tub-tagged protein. A subsequent bioorthogonal reaction mediates the covalent attachment of a fluorophore, biotin, or probe

13]. We have recently expanded the tool-box of chemoenzymatic technologies and introduced Tub-tag labeling for the site-specific C-terminal functionalization of proteins utilizing the ligation of noncanonical amino acids. In this, the enzyme tubulin tyrosine ligase (TTL) binds to a short tubulin-derived peptide tag fused to any protein of interest (VDSVEGEGEEEGEE, the so-called Tub-tag) and catalyzes the ligation of a tyrosine derivative carrying a unique chemical entity [14]. One of the many well-established bioorthogonal reactions [15] is subsequently used for the site-specific attachment of a fluorophore or probe to the ligated tyrosine (*see* Fig. 1). Here, we describe the experimental steps needed for Tub-tag labeling of proteins. We start with the genetic addition of the Tub-tag peptide to a protein of interest (POI, exemplarily described for an antigen binding nanobody, VHH [16]), explain the expression and purification of the protein and the TTL, and show the ligation of azide-, aldehyde-, and halide-containing tyrosine derivatives. We demonstrate the use of Strain-promoted Azide Alkyne Cycloaddition (SPAAC) [17], Staudinger phosphite reaction [18–25], hydrazone and oxime forming reaction [26], as well as Sonogashira cross-coupling [27] as prominent examples for bioorthogonal reactions. Finally, we offer analytical methods to determine the success of protein functionalization by Tub-tag labeling.

2   Materials

Prepare all solutions using ultrapure water (18.2 MΩ•cm at 25 °C), analytical grade reagents at ambient temperature (unless indicated otherwise). 3-azido-L-tyrosine can be obtained from ALFA CHEMISTRY (Holtsville USA), 3-iodo-L-tyrosine can be obtained from Sigma Aldrich (St. Louis, USA).

2.1   Cloning

1. Microwave.

2. 50× rotiphorese buffer: 2 M Tris–HCl, 1 M acetic acid, 50 mM EDTA, pH 8.5 (*see* **Note 1**).

3. Sterile PCR tubes (autoclaved for at least 20 min at 120 °C and 2 bar).

4. 1.5 mL sterile microcentrifuge tubes (autoclaved for at least 20 min at 120 °C and 2 bar).

5. Centrifuge.

6. Heating block.

7. DNA binding column.

8. Incubator shaker.

9. Incubator.

10. dNTPs (10 mM).

11. Forward primer: VHH N-His-Tag NcoI.
    (5′-GGGGCCATGGCCCATCATCACCATCACCATG
    ATGTGCAGCTGCAGGA GTCTGGGGGGAG-3′).

12. Reverse Primer: VHH Tub-Tag 14mer EcoRI
    3′-GGTCACCGTCTCCTCAGTGGATAGCGTG
    GAAGGCGAAGGCGAAGAAGA
    AGGCGAAGAATAAGAATTCGGG -5′.

13. VHH template DNA.

14. Phusion DNA Polymerase.

15. Phusion buffer (5×).

16. Agarose.

17. Midori Green Direct or similar EtBr supplement.

18. PCR-amplificant.

19. Smartladder (Marker).

20. ddH2O (preheated at 50 °C).

21. Target vector DNA (pHEN6c; 1 μg/μL).

22. Restriction Enzymes: NcoI and EcoRI.

23. 10× digestion buffer.

24. 10× T4 DNA Ligase Buffer.

25. T4 DNA Ligase.

26. Ampicillin stock solution: Dissolve100 mg of D-α-aminobenzylpenicillin sodium salt in 1 mL water and sterile-filter with a 0.2 μm syringe filter.

27. LB-agar plates supplemented with ampicillin: Add about 400 mL water to a 1 L graduated cylinder. Weigh 10 g peptone, 5 g NaCl and 5 g yeast extract, 16 g agar and transfer to the cylinder. Add water to a volume of 1 L and mix thoroughly. Fill the solution in a 1 L Schott glass bottle and sterilize it for at least 20 min at 120 °C and 2 bar in an autoclave (see **Note 2**). Wait until the Schott bottle cooled down to ~40 °C and add 1 mL of ampicillin stock solution. Mix and pour agar plates directly. Wait until the agar is solid and store at 4 °C.

2.2 Protein Expression and Purification

1. Baffled Erlenmeyer flask, 1 L.

2. *E. coli* strain JM109, BL21(DE3) or similar, chemically competent (*see* **Note 3**).

3. Ampicillin stock solution: Dissolve 100 mg of D-*α*-aminobenzylpenicillin sodium salt in 1 mL water and sterile-filter with a 0.2 μm syringe filter.

4. LB medium: Add about 400 mL water to a 1 L graduated cylinder. Weigh 10 g peptone, 5 g NaCl, and 5 g yeast extract and transfer to the cylinder. Add water to a volume of 1 L and mix thoroughly. Fill the medium in a 1 L Schott glass bottle and sterilize it for at least 20 min at 120 °C and 2 bar in an autoclave (*see* **Note 2**). Store at 4 °C.

5. Spectrophotometer to measure optical density (600 nm).

6. IPTG stock solution: 1 M isopropyl-*β*-D-thiogalactopyranosid (IPTG). Dissolve 238 mg of IPTG in 1 mL water and sterile-filter with a 0.2 μm syringe filter. Store at 4 °C.

7. Incubator shaker.

8. High-speed centrifuge with cooling unit (~$50.000 \times g$).

9. Centrifugation tubes.

10. Lysozyme.

11. DNAse.

12. PMSF stock solution: 0.2 M phenylmethanesulfonyl fluoride (PMSF). Dissolve 34.8 mg PMSF in 1 mL isopropanol. Store at 4 °C.

13. Ultrasonic cell disruptor.

14. TTL IMAC buffer: 50 mM Tris–HCl, pH 7.4, 0.5 M NaCl, 10 mM DTT. Add about 400 mL water to a 1 L graduated cylinder. Weigh 6.06 g Tris, 29.22 g NaCl and 1.5 g dithiothreitol (DTT) and transfer to the cylinder. Add water to a volume of 900 mL and mix thoroughly. Adjust the pH to 7.4 with HCl (*see* **Note 4**). Adjust to 1 L with water. Store at 4 °C.

15. Fast protein liquid chromatography system (FPLC system).

16. Ni-NTA column (*see* **Note 5**).

17. Protein desalting column (*see* **Note 6**).

18. TTL IMAC elution buffer: 50 mM Tris–HCl, pH 7.4, 0.5 M NaCl, 500 mM imidazole, 10 mM DTT. Add about 400 mL water to a 1 L graduated cylinder. Weigh 6.06 g Tris, 29.22 g NaCl, 34 g imidazole and 1.5 g dithiothreitol (DTT) and transfer to the cylinder. Add water to a volume of 900 mL and mix thoroughly. Adjust the pH to 7.4 with HCl (*see* **Note 4**). Adjust to 1 L with water. Store at 4 °C.

19. Protein IMAC buffer: 50 mM Tris–HCl, pH 7.4, 0.5 M NaCl. Add about 400 mL water to a 1 L graduated cylinder. Weigh

6.06 g Tris and 29.22 g NaCl and transfer to the cylinder. Add water to a volume of 900 mL and mix thoroughly. Adjust the pH to 7.4 with HCl (*see* **Note 4**). Adjust to 1 L with water. Store at 4 °C.

20. Protein IMAC elution buffer: 50 mM Tris–HCl, pH 7.4, 0.5 M NaCl, 500 mM imidazole. Add about 400 mL water to a 1 L graduated cylinder. Weigh 6.06 g Tris, 29.22 g NaCl and 34 g imidazole and transfer to the cylinder. Add water to a volume of 900 mL and mix thoroughly. Adjust the pH to 7.4 with HCl (*see* **Note 4**). Adjust to 1 L with water. Store at 4 °C.

21. Protein storage buffer: 20 mM MES-KOH, 100 mM KCl, 10 mM $MgCl_2$, pH 7.0. Add about 400 mL water to a 1 L graduated cylinder. Weigh 3.90 g MES, 7.45 g KCl and 0.95 g $MgCl_2$ and transfer to the cylinder. Add water to a volume of 900 mL and mix thoroughly. Adjust the pH to 7.0 with 1 M KOH. Adjust to 1 L with water. Store at 4 °C.

22. LB-agar plates supplemented with ampicillin: Add about 400 mL water to a 1 L graduated cylinder. Weigh 10 g peptone, 5 g NaCl and 5 g yeast extract, 16 g agar and transfer to the cylinder. Add water to a volume of 1 L and mix thoroughly. Fill the solution in a 1 L Schott glass bottle and sterilize it for at least 20 min at 120 °C and 2 bar in an autoclave (*see* **Note 2**). Wait until the Schott bottle cooled down to ~40 °C and add 1 mL of ampicillin stock solution. Mix and pour agar plates directly. Wait until the agar is solid and store at 4 °C.

2.3 Synthesis of 3-Formyl-L-Tyrosine

1. 250 mL round-bottom flask with conical ground joint (NS) 29/32.

2. 100 mL round-bottom flask with conical ground joint (NS) 14/23.

3. 25 mL round-bottom flask with conical ground joint (NS) 14/23.

4. Reflux condenser for (NS) 14/23.

5. Rotary evaporator.

6. High vacuum pump.

7. Chromatography column with fused-in frit (porosity 0) and PTFE stopcock.

8. Silica gel (60 Å, ~0.035–0.070 mm).

9. Preparative HPLC system.

10. Preparative C18 column like Macherey-Nagel Nucleodur C18 HTec Spum column.

11. Round-bottom flask freeze-dryer for lyophilizing.

12. L-tyrosine.

13. Dioxane.

14. Triethylamine.

15. Ice/water bath.

16. Di-*tert*-butyl dicarbonate.

17. Ethyl acetate.

18. 1 N HCl.

19. Saturated NaCl solution.

20. Saturated $NaHCO_3$.

21. Dry $MgSO_4$.

22. $CHCl_3$.

23. Powdered NaOH.

24. MeOH.

25. Acetic acid.

26. CH_2Cl_2.

27. Trifluoroacetic acid.

28. Acetonitrile.

2.4 TTL Reaction

1. Dry block incubator.

2. Dialysis tube with a 1 kDa cutoff.

3. 1.5 mL microcentrifuge tube.

4. Ligation buffer: 20 mM MES-KOH, 100 mM KCl, 10 mM $MgCl_2$, pH 7.0. Add about 400 mL water to a 1 L graduated cylinder. Weigh 3.90 g MES, 7.45 g KCl and 0.95 g $MgCl_2$ and transfer to the cylinder. Add water to a volume of 900 mL and mix thoroughly. Adjust the pH to 7.0 with 1 M KOH. Adjust to 1 L with water. Store at 4 °C.

5. TTL stock solution at 2 mg/mL (*see* Subheading 3.2).

6. Protein-Tub-tag stock solution at 4 mg/mL (*see* Subheading 3.3).

7. ATP stock solution: 100 mM adenosine triphosphate (ATP). Weigh 50.72 mg ATP and dissolve in 1 mL water. Store at −20 °C (*see* **Note 7**).

8. Tyrosine derivative stock solution: 11 mM tyrosine derivative. Dissolve the corresponding amount of the respective tyrosine derivative in ligation buffer. Store at 4 °C (*see* **Note 8**).

9. DTT stock solution: 2 M dithiotreitol (DTT). Weigh 308.5 mg DTT and dissolve in 1 mL water. Store at −20 °C (*see* **Note 9**).

10. Glutathione stock solution: 100 mM reduced glutathione. Weigh 30.7 mg reduced glutathione and dissolve in 1 mL water. Store at −20 °C (*see* **Note 10**).

11. 1 N KOH.

12. 0.1 N KOH.

2.5 Sonogashira Cross-Coupling

1. Dry block incubator.

2. 1.5 mL microcentrifuge tube.

3. 10 mL round-bottom flask with conical ground joint (NS) 14/23.

4. 2-amino-4,6-dihydroxypyridine.

5. 0.1 M NaOH.

6. Pd(OAc)$_2$.

7. Sodium ascorbate.

8. Cross-coupling reaction buffer: 20 mM KH$_2$PO$_4$-HCl, pH 7.4. Add about 400 mL water to a 1 L graduated cylinder. Weigh 3.48 g KH$_2$PO$_4$ and transfer to the cylinder. Add water to a volume of 900 mL and mix thoroughly. Adjust the pH to 7.4 with HCl (*see* **Note 4**). Adjust to 1 L with water. Store at 4 °C.

9. Alkyne containing fluorophore, biotin, drug, or other probe [14].

2.6 Strain-Promoted Azide Alkyne Click Reaction

1. Dry block incubator.

2. 1.5 mL microcentrifuge tube.

3. SPAAC reaction buffer: 1.8 mM KH$_2$PO$_4$, 10 mM Na$_2$HPO$_4$, 2.7 mM KCl and 137 mM NaCl, pH 7.4. Add about 400 mL water to a 1 L graduated cylinder. Weigh 244 mg KH$_2$PO$_4$, 1.78 g Na$_2$HPO$_4$ • 2 H$_2$O, 201 mg KCl and 8 g NaCl and transfer to the cylinder. Add water to a volume of 900 mL and mix thoroughly. Adjust the pH to 7.4 with HCl (*see* **Note 4**). Adjust to 1 L with water. Store at 4 °C.

4. Dibenzocyclooctyne containing fluorophore, biotin, drug, or other probe [14].

2.7 Staudinger-Phosphite Reaction

1. Dry block incubator.

2. 1.5 mL microcentrifuge tube.

3. Staudinger reaction buffer: 50 mM Tris–HCl, 100 mM KCl, pH 8.5. Add about 400 mL water to a 1 L graduated cylinder. Weigh 6.06 g Tris, 7.4 mg KCl and transfer to the cylinder. Add water to a volume of 900 mL and mix thoroughly. Adjust the pH to 8.5 with HCl (*see* **Note 4**). Adjust to 1 L with water. Store at 4 °C.

4. Phosphite containing fluorophore, biotin, drug, or other probe [19].

2.8 Hydrazone and Oxime Forming Reactions

1. Dry block incubator.

2. 1.5 mL microcentrifuge tube.

3. Condensation reaction buffer: 20 mM MES-KOH, 100 mM KCl, pH 6.0. Add about 400 mL water to a 1 L graduated cylinder. Weigh 3.90 g MES, 7.45 g KCl and 0.95 g MgCl$_2$

and transfer to the cylinder. Add water to a volume of 900 mL and mix thoroughly. Adjust the pH to 6.0 with 1 M KOH. Adjust to 1 L with water. Store at 4 °C.

4. Hydrazide (for hydrazone forming reaction) or hydroxylamine (for oxime forming reaction) containing fluorophore, biotin, drug, or other probe [14].

2.9 SDS Polyacrylamide Gel and in-Gel Fluorescence Analysis

1. SDS-PAGE chamber.

2. Bench-top centrifuge.

3. Dry block incubator.

4. 1.5 mL microcentrifuge tube.

5. Imager with possibility to perform fluorescence analysis.

6. Microwave (*see* **Note 11**).

7. Prestained molecular weight standard.

8. 30% acrylamide solution with 0.8% bisacrylamide in water (ratio of 37.5:1).

9. Ammonium persulfate: 10% solution in water. Leave one 1 mL aliquot at 4 °C. Store remaining aliquots at −20 °C.

10. Sodium dodecyl sulfate (SDS): 10% solution in water (*see* **Note 12**).

11. N,N,N,N'-Tetramethyl-ethylenediamine (TEMED): Store at 4 °C.

12. Resolving gel buffer: 1.5 M Tris–HCl, pH 8.8. Add about 400 mL water to a 1 L graduated cylinder. Weigh 181.7 g Tris and transfer to the cylinder. Add water to a volume of 900 mL and mix thoroughly. Adjust the pH to 8.8 with HCl (*see* **Note 4**). Adjust to 1 L with water. Store at 4 °C.

13. Stacking gel buffer: 0.5 M Tris–HCl, pH 6.8. Add about 400 mL water to a 1 L graduated cylinder. Weigh 60.6 g Tris and transfer to the cylinder. Add water to a volume of 900 mL and mix thoroughly. Adjust the pH to 6.8 with HCl (*see* **Note 4**). Adjust to 1 L with water. Store at 4 °C.

14. SDS running buffer (10×): 0.25 M Tris, 1.92 M Glycin and 1% SDS (*see* **Note 13**).

15. Sample buffer (5×): 0.3 M Tris–HCl, pH 6.8, 10% SDS, 25% ß-mercaptoethanol, 0.1% bromophenol blue, 45% glycerol. Leave one 1 mL aliquot at 4 °C. Store remaining aliquots at −20 °C.

16. Staining solution: Add about 400 mL water to a 1 L graduated cylinder and dissolve 2.5 g Coomassie Brilliant Blue, 300 mL MeOH, and 100 mL acetic acid. Mix thoroughly and adjust to 1 L with water.

17. Destaining solution: Add about 400 mL water to a 1 L graduated cylinder and combine 300 mL MeOH and 100 mL acetic acid. Mix thoroughly and adjust to 1 L with water.

2.10 Tryptic Digest and MSMS Analysis

1. Bench-top centrifuge.

2. Vacuum centrifuge.

3. Acetonitrile.

4. Ultrasonic bath.

5. 1.5 mL microcentrifuge tube.

6. Equilibration and digestion buffer: 50 mM $(NH_4)HCO_3$, ~ pH 7.8. Dissolve 1.975 g $(NH_4)HCO_3$ in 500 mL water. There is no need to further adjust the pH (*see* **Note 14**).

7. Wash and dehydration buffer: 50 mM ammonium bicarbonate/acetonitrile 1:1. Add 100 mL acetonitrile to 100 mL of equilibration and digestion buffer.

8. Trypsin stock solution: Dissolve 20 µg trypsin in 100 µL 1 mM HCl and store aliquots of 0.4 µg trypsin at −20 °C (*see* **Note 15**).

9. Stop solution: 0.5% trifluoroacetic acid in acetonitrile. Add 500 µL of trifluoroacetic acid to 90 mL acetonitrile. Mix and add acetonitrile to a volume of 100 mL.

10. Reducing buffer: 10 mM dithiotreitol in equilibration and digestion buffer. Dissolve 15.4 mg DTT in 9 mL of equilibration and digestion buffer. Adjust pH to 7.8 with 1 N HCl. Add water to a volume of 10 mL. Store aliquots of 20 µL at −20 °C (*see* **Note 15**).

11. Alkylation buffer: 55 mM iodoacetamide in equilibration and digestion buffer. Dissolve 102 mg iodoacetamide in 9 mL of equilibration and digestion buffer. Adjust pH to 7.8 with 1 N HCl. Add water to a volume of 10 mL. Store aliquots of 20 µL at −20 °C (*see* **Note 15**).

12. MSMS solution: Water, 5% acetonitrile and 0.1% trifluoroacetic acid. Add 5 mL acetonitrile and 100 µL trifluoroacetic acid to 90 mL water. Mix and add water to reach a volume of 100 mL.

2.11 Immunoblotting of Biotinylated Proteins

1. Western blot chamber.

2. Chemiluminescence imaging system.

3. Chemiluminescence substrate for horseradish peroxidase and western blot development.

4. Western blot buffer: Add about 400 mL water to a 1 L graduated cylinder. Combine 100 mL SDS running buffer (10×) and 200 mL MeOH. Mix thoroughly and adjust to 1 L with water (*see* **Note 16**).

5. Nitrocellulose membrane.

6. Blotting paper.

7. Washing buffer: 1.8 mM KH_2PO_4, 10 mM Na_2HPO_4, 2.7 mM KCl and 137 mM NaCl, pH 7.4. Add about 400 mL water to

a 1 L graduated cylinder. Weigh 244 mg KH_2PO_4, 1.78 g $Na_2HPO_4 \cdot 2 H_2O$, 201 mg KCl and 8 g NaCl and transfer to the cylinder. Add water to a volume of 900 mL and mix thoroughly. Adjust the pH to 7.4 with HCl (*see* **Note 4**). Adjust to 1 L with water.

8. Washing buffer containing 0.05% Tween-20.

9. Blocking solution: 5% milk in washing buffer. Store at 4 °C.

10. Streptavidin-HRP (horseradish peroxidase) conjugate at 0.5 μg/μL.

3 Methods

Carry out all the procedures at room temperature unless otherwise specified.

3.1 Cloning

1. The cloning procedure is exemplified for an antigen binding nanobody (VHH). Experimental steps can be transferred to any protein of interest. Amplify and equip the VHH sequence with a N-terminal H_6-Tag and C-terminal Tub-tag using the forward and reverse primer via a standard PCR protocol.

2. Cast a 1% agarose-gel by adding 1 g of agarose in 100 mL 1× rotiphorese buffer in a 250 mL Erlenmeyer flask and heat until boiling using microwave. Let the solution cool down for 5 min and transfer to the casting apparatus of your DNA electrophoresis system.

3. Load the PCR reaction mixture on the agarose-gel next to 5 μL of smartladder and perform electrophoresis at a constant voltage of 100 V for 30–40 min.

4. Excise designated PCR product under 365 nm UV light from the gel and transfer into a 1.5 mL microcentrifuge tube (*see* **Note 17**).

5. Purify the PCR product using standard DNA purification columns and resolve it in ddH_2O.

6. Digest both PCR product and vector backbone (pHEN6c) using *Nco*I and *Eco*RI according to the manufacturer's protocol to provide sticky ends for plasmid ligation (*see* **Note 18**).

7. Purify the digested DNA fragments using standard DNA purification columns and resolve them in ddH_2O at 10–20 ng/μL (PCR fragment) and 50–100 ng/μL (vector backbone).

8. Ligate the VHH fragment into the pHEN6c bacterial expression vector at a molar ratio of 3:1 using T4 DNA Ligase for 18 h at 4 °C, then inactivate T4 DNA Ligase at 65 °C.

9. Thaw a 50 μL aliquot of chemically competent *E. coli* JM109 on ice.

10. Add 1 μL of ligation product and incubate for 30 min on ice. Heat shock the cells at 42 °C for 45 s, add 500 μL of pre-warmed LB medium, and incubate for 1 h at 37 °C with shaking (180 rpm).

11. Streak 100 μL of the transformed cells on a LB agar plate supplemented with ampicillin (*see* **Notes 19** and **20**).

12. Incubate the agar plate overnight at 37 °C in an incubator.

13. Pick 20 colonies and inoculate 5 mL of LB medium supplemented with 5 μL of ampicillin stock solution with a single colony picked from the agar plate and incubate at 37 °C, 180 rpm overnight in an incubator shaker.

14. Isolate plasmid DNA using DNA Mini-Prep Kit

15. Verify the right insert via analytical restriction digest and DNA sequencing.

3.2 TTL Expression and Purification

1. Thaw a 50 μL aliquot of chemically competent *E. coli* BL21(DE3) on ice.

2. Add 1 μL of plasmid encoding H_6-Sumo-TTL and incubate 30 min on ice. Heat shock the cells at 42 °C for 45 s, add 500 μL of pre-warmed LB medium, and incubate for 1 h at 37 °C with shaking (180 rpm).

3. Streak 100 μL of the transformed cells on a LB agar plate supplemented with ampicillin (*see* **Note 19**).

4. Incubate the agar plate overnight at 37 °C in an incubator shaker.

5. Inoculate 5 mL of LB medium supplemented with 5 μL of ampicillin stock solution with a single colony picked from the agar plate and incubate at 37 °C, 180 rpm overnight (pre-culture) in an incubator shaker.

6. Inoculate 250 mL of LB medium supplemented with 250 μL of ampicillin stock solution in a 1 L baffled Erlenmeyer flask with 2 mL of pre-culture and incubate at 37 °C, 180 rpm.

7. Use a spectrophotometer (600 nm) to measure the optical density of the cell suspension. As soon as an optical density of ~ 0.7 is reached, induce the expression of the TTL by the addition of 125 μL IPTG stock solution.

8. Incubate the cells for 18 h at 18 °C, 180 rpm in an incubator shaker.

9. Harvest the cells by centrifugation ($18,000 \times g$, 4 °C, 30 min) and suspend in 10 mL TTL IMAC buffer supplemented with 100 μg/mL lysozyme, 25 μg/mL DNAse, and 100 μL of PMSF stock solution.

10. Perform ultrasonic disruption of the cells via three consecutive 90 s cycles of short ultrasonic pulses using an ultrasonic cell

disruptor. Keep the cells on ice during the whole process and make sure to allow the solution to cool down between single cycles (*see* **Note 21**).

11. Remove cell debris by centrifugation (49,000 × g, 4 °C, 30 min) and directly apply the lysate to a Ni-NTA column using a FPLC system (*see* **Note 22**).

12. Wash the column with 10 column volumes of TTL IMAC buffer and elute the immobilized H_6-Sumo-TTL by slowly increasing the amount of TTL IMAC elution buffer to 100% within 2 column volumes.

13. Analyze the purity of eluting protein fractions by SDS-PAGE (*see* Subheading 3.10), combine pure fractions, and desalt the purified protein with a protein desalting column and protein storage buffer (*see* **Note 23**).

14. Shock-freeze 20 μL TTL aliquots at ~ 2 mg/mL using liquid nitrogen and store at −80 °C until further use.

3.3 Tub-Tagged Protein Expression and Purification

1. Thaw a 50 μL aliquot of chemically competent *E. coli* JM109 on ice.

2. Add 1 μL of plasmid encoding H_6-Nanobody-Tub-tag and incubate 30 min on ice. Heat shock the cells at 42 °C for 45 s, add 500 μL of pre-warmed LB medium, and incubate for 1 h at 37 °C with shaking (180 rpm).

3. Streak 100 μL of the transformed cells on a LB agar plate supplemented with ampicillin (*see* **Notes 19** and **20**).

4. Incubate the agar plate overnight at 37 °C in an incubator.

5. Inoculate 5 mL of LB medium supplemented with 5 μL of ampicillin stock solution with a single colony picked from the agar plate and incubate at 37 °C, 180 rpm overnight (pre-culture) in an incubator shaker.

6. Inoculate 250 mL of LB medium supplemented with 250 μL of ampicillin stock solution in a 1 L baffled Erlenmeyer flask with 2 mL of pre-culture and incubate at 37 °C, 180 rpm.

7. Use a spectrophotometer to measure the optical density of the cell suspension. As soon as an optical density of ~0.7 is reached, induce the expression of the TTL by the addition of 125 μL IPTG stock solution.

8. Incubate the cells for 18 h at 18 °C, 180 rpm in an incubator shaker (*see* **Note 24**).

9. Harvest the cells by centrifugation (18,000 × g, 4 °C, 30 min) and suspend in 10 mL Protein IMAC buffer supplemented with 100 μg/mL lysozyme, 25 μg/mL DNAse, and 100 μL of PMSF stock solution.

10. Perform ultrasonic disruption of the cells via three consecutive 90 s cycles of short ultrasonic pulses using an ultrasonic cell disruptor. Keep the cells on ice during the whole process and make sure to allow the solution to cool down between single cycles (*see* **Note 21**).

11. Remove cell debris by centrifugation (49,000 × *g*, 4 °C, 30 min) and directly apply the lysate to a Ni-NTA column using a FPLC system (*see* **Note 22**).

12. Wash the column with 10 column volumes of protein IMAC buffer and elute the immobilized H_6-POI-Tub-tag by slowly increasing the amount of protein IMAC elution buffer to 100% within 2 column volumes.

13. Analyze the purity of eluting protein fractions by SDS-PAGE (*see* Subheading 3.10), combine pure fractions, and desalt the purified protein with a protein desalting column and protein storage buffer.

14. In case of unsatisfactory purity after His-tag purification, perform a following size exclusion chromatography with protein storage buffer. Analyze the purity of eluting protein fractions by SDS-PAGE (*see* Subheading 3.10) and combine pure fractions.

15. Shock-freeze 20 μL protein aliquots at ~ 4 mg/mL using liquid nitrogen and store at −80 °C until further use.

3.4 Synthesis of 3-Formyl-ʟ-Tyrosine

1. Weigh 2 g of ʟ-tyrosine and transfer to a 250 mL round-bottom flask with conical ground joint (NS) 29/32 equipped with a magnetic stirring bar. Start the synthesis of 3-formyl-ʟ-tyrosine (Scheme 1) by weighing 2 g of ʟ-tyrosine and transferring to a 250 mL round-bottom flask with conical ground joint (NS) 29/32 equipped with a magnetic stirring bar. Place an ice/water bath on a magnetic stirrer and mount the round-bottom flask half covered with ice using a stand clamp.

2. Completely dissolve ʟ-tyrosine in 100 mL of a 1:1 mixture of dioxane and water.

3. While stirring, slowly add 2.32 mL of the base trimethylamine to deprotonate the amine of ʟ-tyrosine.

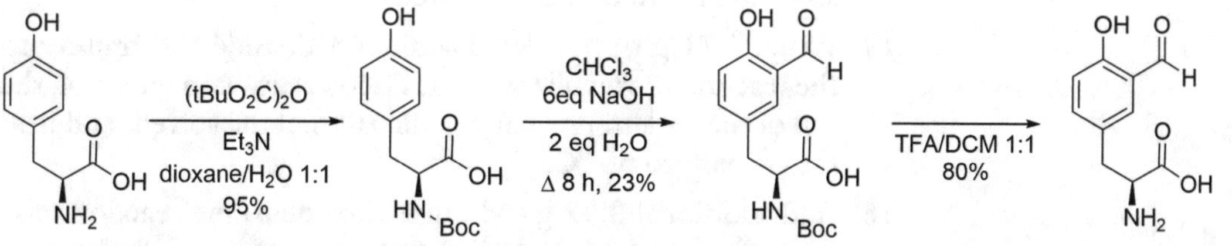

Scheme 1 Synthesis of 3-formyl-ʟ-tyrosine according to a published protocol [14, 29, 30]

4. Weigh 2.64 g of *di-tert*-butyl dicarbonate and slowly add it to the L-tyrosine mixture within two steps (*see* **Note 25**). Close the reaction flask with a septum.

5. Keep stirring for 1 h and further cool the reaction flask with an ice/water bath (*see* **Note 26**).

6. Keep stirring for further 24 h and allow the ice to thaw at ambient temperature. Slowly increasing the temperature ensures high amine selectivity of the reaction.

7. Remove dioxane under reduced pressure with a rotary evaporator set to 40 °C and 90 mbar (*see* **Note 27**).

8. Add 50 mL of saturated $NaHCO_3$ to the remaining aqueous solution and wash the mixture with 50 mL of ethyl acetate to remove side-products using a separation funnel.

9. After the removal of the ethyl acetate layer (*see* **Note 28**) acidify the aqueous solution to pH 1 using 1 N HCl. You can do this in the separation funnel. At acidic pH the Boc protected L-tyrosine is insoluble in aqueous solutions and a white precipitate is forming.

10. Add 25 mL of ethyl acetate to the separation funnel. The white participate is solving in the ethyl acetate layer. Discard the aqueous phase (*see* **Note 28**).

11. Wash the ethyl acetate phase with 50 mL saturated NaCl solution to further remove water soluble components and impurities. Add anhydrous $MgSO_4$ to the organic layer to remove residual water. Filter off the $MgSO_4$ and remove ethyl acetate under reduced pressure with a rotary evaporator set to 40 °C and 200 mbar. Further decrease the pressure once no additional solvent is evaporating (*see* **Note 29**). A white foam is forming (Boc protected L-tyrosine, 95%) which can be used in the following Reimer-Tiemann reaction to synthesize Boc protected 3-formyl-L-tyrosine.

12. Weigh 2 g of Boc protected L-tyrosine and transfer to a 100 mL round-bottom flask with conical ground joint (NS) 14/23 equipped with a magnetic stirring bar. Place an oil bath for heating on a magnetic stirrer and mount the round-bottom flask half covered with oil using a stand clamp.

13. Completely dissolve Boc protected L-tyrosine in 30 mL $CHCl_3$ suspended with 0.256 mL water.

14. Weigh 1.71 g of powdered sodium hydroxide and transfer to the reaction flask while stirring. Place a reflux condenser on the top of the flask to prevent volatilization of the solvent and heat the oil bath to 68 °C.

15. Add additional 0.42 g sodium hydroxide to the reaction mixture after 1 and 2 h to further promote product formation.

16. After 8 h of reflux, cool the mixture to ambient temperature and add 15 mL of water and 15 mL of ethyl acetate. Transfer the mixture to a separation funnel.

17. Discard the organic layer (*see* **Note 28**) and acidify the aqueous layer to pH 1 using 1 N HCl. You can do this in the separation funnel. At acidic pH the Boc protected 3-formyl-L-tyrosine is insoluble in aqueous solutions and a yellow precipitate is forming.

18. Add 15 mL of ethyl acetate to the separation funnel. The yellow precipitate is solving in the ethyl acetate layer. Discard the aqueous phase (*see* **Note 28**).

19. Wash the ethyl acetate phase with 20 mL saturated NaCl solution to further remove water soluble components and impurities. Add anhydrous $MgSO_4$ to the organic layer to remove residual water. Filter off the $MgSO_4$ and remove the solvent under reduced pressure with a rotary evaporator set to 40 °C and 200 mbar. Further decrease the pressure once no additional solvent is evaporating (*see* **Note 29**). A yellow-brown oil is forming (impure Boc protected 3-formyl-L-tyrosine).

20. Purify Boc protected 3-formyl-L-tyrosine by flash column chromatography. For this, dissolve silica gel in a 12:1 mixture of $CHCl_3$ and MeOH suspended with 1% acetic acid and transfer to a chromatography column with fused-in frit (porosity 0) and PTFE stopcock. Carefully place the yellow-brown oil on the top of the silica gel and cover with a small layer of sand. Perform chromatographic purification using a 12:1 mixture of $CHCl_3$ and MeOH suspended with 1% acetic acid and collect the eluting fractions with test tubes and check for the elution of the product by TLC ($R_f = 0.46$).

21. Combine fractions containing product and remove solvent under reduced pressure with a rotary evaporator set to 40 °C and 430 mbar. Further decrease the pressure once no additional solvent is evaporating (*see* **Note 24**). A yellow oil is forming (pure Boc protected 3-formyl-L-tyrosine, 23%) (*see* **Note 30**). Store the oil at −20 °C until final deprotection with TFA.

22. For Boc deprotection weigh 0.5 g of Boc protected 3-formyl-L-tyrosine and transfer to a 25 mL round-bottom flask with conical ground joint (NS) 14/23 equipped with a magnetic stirring bar. Place an ice/water bath on a magnetic stirrer and mount the round-bottom flask half covered with ice using a stand clamp.

23. Dissolve the compound in 4 mL CH_2Cl_2 and slowly add 4 mL TFA while stirring.

24. Slowly increase the temperature to ambient temperature within 2 h.

25. Carefully remove the solvent and TFA using a high vacuum pump. A yellow oil is forming (3-formyl-L-tyrosine).

26. Purify 3-formyl-L-tyrosine by preparative HPLC with a C18 column (like Macherey-Nagel Nucleodur C18 HTec Spum column) using the following gradient: A = H$_2$O + 0.1% TFA, B = MeCN +0.1% TFA) flow rate 32 mL/min, 10% B 0,5 min, 10,100% B 5,35 min, 100% B 35,40 min. Collect fractions and analyze by MS analysis. Combine pure fractions and remove solvent using a round-bottom flask freeze-dryer for lyophyliza-tion. Store 3-formyl-L-tyrosine (yellow powder) at −20 °C until further use.

3.5   TTL Reaction

1. Add 100 μL of ligation buffer to a 1.5 mL microcentrifuge tube.

2. Add 5 μL of ATP stock solution (final concentration 2.5 mM) and 13.64 μL of tyrosine derivate stock solution (final concentration 1 mM).

3. In case of 3-azido-L-tyrosine add 7.50 μL glutathione stock solution, in case of all other tyrosine derivatives add 0.38 μL DTT stock solution (final concentration for both 5 mM, *see* **Note 31**).

4. Adjust pH to 7.0 using 1 N and 0.1 N KOH (*see* **Note 32**).

5. Add ligation buffer to 134.72 μL.

6. Add 11.68 μL of protein-Tub-tag stock solution to a final concentration of 10 μM and 3.6 μL TTL stock solution to a final concentration of 1 μM.

7. Incubate the protein mixture at 37 °C for 3 h in a dry block incubator (*see* **Note 33**).

8. Remove excess of reducing agent and tyrosine derivative using dialysis at 4 °C for 12 h (cutoff 1 kDa) against SPAAC reaction buffer, Staudinger reaction buffer, cross-coupling reaction buffer, or condensation reaction buffer (depending on the fol-lowing bioorthogonal reaction). The success of ligation can be analyzed by tryptic digest and MSMS analysis (*see* Subheading 3.11). The tyrosinated protein can be short-term stored at 4 °C (*see* **Note 34**).

3.6   Sonogashira Cross-Coupling

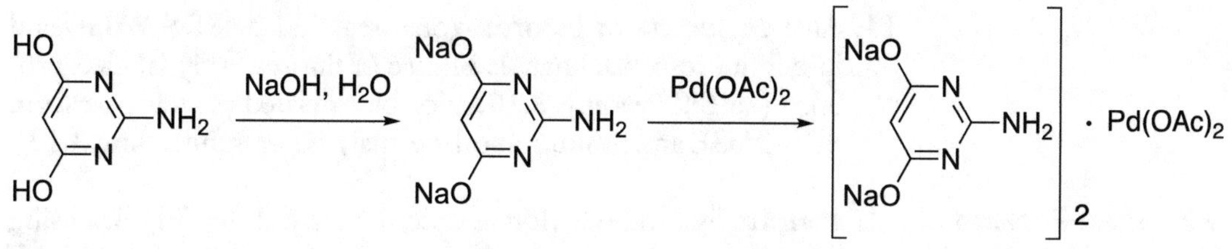

Scheme 2 Synthesis of the water soluble Pd catalyst $Pd(OAc)_2 \cdot [DMADHP]_2$ according to a published protocol [31]

1. Start by synthesizing the water soluble Pd(II) catalyst $Pd(OAc)_2 \cdot [DMADHP]_2$ in accordance with a published protocol (Scheme 2) [31].

2. For this, weigh 13 mg of 2-amino-4,6-dihydroxypyridine and transfer to a 10 mL round-bottom flask with conical ground joint (NS) 14/23 equipped with a magnetic stirring bar. Place an oil bath for heating on a magnetic stirrer and mount the round-bottom flask half covered with oil using a stand clamp.

3. Completely dissolve 2-amino-4,6-dihydroxypyridine in 2 mL of 0.1 M NaOH.

4. Heat the reaction mixture to 65 °C and add 11 mg of $Pd(OAc)_2$ while stirring vigorously. After 30 min at 65 °C the solution turns orange.

5. Cool the mixture to ambient temperature and dilute with 5 mL water to give a final concentration of 0.01 M $Pd(OAc)_2 \cdot [DMADHP]_2$.

6. The catalyst can be stored at ambient temperature for weeks and directly used for cross-coupling reactions.

7. Perform ligation reaction as described in Subheading 3.5 using 3-iodo-L-tyrosine and dialyze the ligation mixture to cross-coupling reaction buffer at 4 °C for 12 h (cutoff 1 kDa).

8. Transfer protein mixture to a 1.5 mL microcentrifuge tube and add 600 mol eq. of alkyne containing fluorophore, biotin, drug, or other probe.

9. Activate the catalyst by adding 4.4 mol eq. (in relation to the catalyst) of sodium-ascorbate in a separate 1.5 mL microcentrifuge tube (*see* **Note 35**).

10. Transfer 4 mol eq. of activated catalyst (in relation to the Tub-tagged protein) to the protein mixture.

11. Incubate at 37 °C for 20 min (*see* **Note 36**).

12. Add a second batch of 4 mol eq. activated catalyst and 2 mol eq. of alkyne containing fluorophore, biotin, drug, or other probe to the mixture.

13. Incubate at 37 °C for additional 20 min (*see* **Note 36**).

14. Analyze success of bioorthogonal reaction by SDS-PAGE and in-gel fluorescence analysis in case of fluorescently labeled proteins (*see* Subheading 3.10). For biotinylated proteins perform SDS-PAGE and immunoblotting analysis (*see* Subheading 3.12).

3.7 Strain-Promoted Azide Alkyne Click Reaction

1. Perform ligation reaction as described in Subheading 3.5 using 3-azido-L-tyrosine and dialyze the ligation mixture to SPAAC reaction buffer at 4 °C for 12 h (cutoff 1 kDa).

2. Transfer protein mixture to a 1.5 mL microcentrifuge tube and add 30 mol eq. of dibenzocyclooctyne containing fluorophore, biotin, drug, or other probe.

3. Incubate at 30 °C for 4 h (*see* **Note 36**).

4. Analyze success of bioorthogonal reaction by SDS-PAGE and in-gel fluorescence analysis in case of fluorescently labeled proteins (*see* Subheading 3.10). For biotinylated proteins perform SDS-PAGE and immunoblotting analysis (*see* Subheading 3.12).

3.8 Staudinger-Phosphite Reaction

1. Perform ligation reaction as described in Subheading 3.5 using 3-azido-L-tyrosine and dialyze the ligation mixture to Staudinger reaction buffer at 4 °C for 12 h (cutoff 1 kDa).

2. Transfer protein mixture to a 1.5 mL Eppendorf tube and add 40 mol eq. of phosphite containing fluorophore, biotin, drug, or other probe.

3. Incubate at 37 °C for 24 h (*see* **Note 36**).

4. Analyze success of bioorthogonal reaction by SDS-PAGE and in-gel fluorescence analysis in case of fluorescently labeled proteins (*see* Subheading 3.10). For biotinylated proteins perform SDS-PAGE and immunoblotting analysis (*see* Subheading 3.12).

3.9 Hydrazone and Oxime Forming Reactions

1. Perform ligation reaction as described in Subheading 3.5 using 3-formyl-L-tyrosine and dialyze the ligation mixture to condensation reaction buffer at 4 °C for 12 h (cutoff 1 kDa).

5. Transfer protein mixture to a 1.5 mL microcentrifuge tube and add 30 mol eq. of hydrazide (for hydrazone forming reaction) or hydroxylamine (for oxime forming reaction) containing fluorophore, biotin, drug, or other probe.

2. Incubate at 37 °C for 4 h (*see* **Note 36**).

3. Analyze success of bioorthogonal reaction by SDS-PAGE and in-gel fluorescence analysis in case of fluorescently labeled proteins (*see* Subheading 3.10). For biotinylated proteins perform SDS-PAGE and immunoblotting analysis (*see* Subheading 3.12).

3.10 SDS Polyacrylamide Gel Electrophoresis and in-Gel Fluorescence Analysis

1. Mix 1.7 mL of resolving gel buffer, 5 mL of acrylamide mixture, and 3.1 mL water in a 15 mL falcon tube. Add 100 μL of SDS solution, 100 μL of ammonium persulfate solution, and 10 μL of TEMED and cast gel within a ~7.25 cm × 10 cm × 1.5 mm gel cassette. Allow space for stacking gel and gently

overlay with isopropanol (*see* **Note 37**). Polymerize the gel at ambient temperature for 20 min.

2. Carefully remove isopropanol after polymerization.

3. Prepare the stacking gel by mixing 1.25 mL of stacking gel buffer, 0.65 mL of acrylamide mixture, and 3 mL water in a 15 mL falcon tube. Add 50 μL of SDS solution, 50 μL of ammonium persulfate solution, and 5 μL of TEMED. Transfer to the resolving gel and immediately insert a 10-well gel comb (*see* **Note 38**).

4. Add 1.5 μL labeled protein to 2 μL sample buffer and add water to 10 μL. Heat the sample at 95 °C for 10 min and centrifuge the samples at ~3000 × g for 1 min.

5. Load the sample on a gel in a pocket next to one loaded with 5 μL prestained molecular weight standard (*see* **Note 39**). Fill the chamber with SDS running buffer (1×) and perform electrophoresis at a constant voltage of 250 V until the bromophenol blue dye reaches the end of the gel. Separate the glass plates and remove the stacking gel.

6. In case of fluorescently labeled proteins, analyze successful incorporation of the fluorophore by a fluorescence imager.

7. Incubate the SDS-gel in staining solution (completely covered). Heat the staining solution gel mixture in a microwave until it starts boiling (*see* **Note 40**). Further incubate the gel for approx. 10 min.

8. Remove the staining solution and incubate the SDS-gel in destaining solution (completely covered). Heat the destaining solution gel mixture in a microwave until it starts boiling (*see* **Note 40**). Incubate the gel further until the destaining solution is fully saturated with Coomassie Brilliant Blue. Repeat destaining until background staining vanished and protein bands are visible.

9. Analyze SDS-gel using an imager (*see* **Note 41**).

3.11 Tryptic Digest and MSMS Analysis

1. Mix 1.7 mL of resolving gel buffer, 5 mL of acrylamide mixture and 3.1 mL water in a 15 mL falcon tube. Add 100 μL of SDS solution, 100 μL of ammonium persulfate solution and 10 μL of TEMED and cast gel within a ~7.25 cm × 10 cm × 1.5 mm gel cassette. Allow space for stacking gel and gently overlay with isopropanol (*see* **Note 37**). Polymerize the gel at ambient temperature for 20 min.

2. Carefully remove isopropanol after polymerization.

3. Prepare the stacking gel by mixing 1.25 mL of stacking gel buffer, 0.65 mL of acrylamide mixture, and 3 mL water in a 15 mL falcon tube. Add 50 μL of SDS solution, 50 μL of ammonium persulfate solution, and 5 μL of TEMED. Transfer

to the resolving gel and immediately insert a 10-well gel comb (*see* **Note 38**).

4. Add 1.5 µL labeled protein to 2 µL sample buffer and add water to 10 µL. Boil the sample at 95 °C for 10 min and centrifuge the samples at ~3000 × g for 1 min.

5. Load the sample on a gel in a pocket next to one loaded with 5 µL prestained molecular weight standard (*see* **Note 39**). Fill the chamber with SDS running buffer (1×) and perform electrophoresis at a constant voltage of 250 V until the bromophenol blue dye reaches the end of the gel. Separate the glass plates and remove the stacking gel.

6. Incubate the SDS-gel in staining solution (completely covered). Incubate the gel for approx. 30 min.

7. Remove the staining solution and incubate the SDS-gel in destaining solution (completely covered). Incubate the gel further until the destaining solution is fully saturated with Coomassie Brilliant Blue. Repeat destaining until background staining vanished and protein bands are visible.

8. Wash the gel with water 3–4 times. Excise the bands of interest with a clean scalpel (*see* **Note 42**).

9. Cut excised bands in small cubes (~1 × 1 cm), transfer them into a microcentrifuge tube, and spin them down at 3000 × g for 30 s.

10. Add 100 µL of wash and dehydration buffer and incubate the pieces for ~30 min at 37 °C with occasional vortexing to destain gel pieces (*see* **Note 43**).

11. Add 500 µL acetonitrile and incubate at ambient temperature. Gel pieces will become white and shrink as acetonitrile withdraws remaining water from the pieces.

12. Add 50 µL of reducing buffer (pieces completely covered) and incubate at 56 °C for 30 min in a dry block incubator to reduce all disulfides in the protein. Chill the tube to ambient temperature and add 500 µL of acetonitrile. Incubate for 10 min and remove the liquid.

13. Add 50 µL of alkylation buffer (pieces completely covered) and incubate at ambient temperature for 30 min to alkylate all free thiols. Add 500 µL of acetonitrile. Incubate for 10 min and remove the liquid.

14. Add 500 µL acetonitrile and incubate at ambient temperature. Gel pieces will become white and shrink as acetonitrile withdraws remaining water from the pieces. Remove the acetonitrile using a vacuum centrifuge.

15. Equilibrate gel pieces with 100 µL of equilibration and digestion buffer for 10 min. Remove all liquid and cover gel pieces

with 40 μL of equilibration and digestion buffer. Add 5 μL of trypsin solution followed by additional 15 μL of equilibration and digestion buffer. Digest the protein at 37 °C in a dry block incubator overnight.

16. Spin down the mixtures at 3000 × g for 30 s. Add 20 μL of stop solution.

17. Leave the samples for 10 min in an ultrasonic bath and transfer the solution in a new microcentrifuge tube.

18. Add 50 μL of acetonitrile to the gel pieces and wait until they shrink completely. Transfer the acetonitrile to the other solution in the new microcentrifuge tube.

19. Remove all solvent using a vacuum centrifuge.

20. Dissolve the digested protein in 6 mL MSMS solution and perform MSMS analysis.

3.12 Immunoblotting of Biotinylated Proteins

1. Mix 1.7 mL of resolving gel buffer, 5 mL of acrylamide mixture and 3.1 mL water in a 15 mL falcon tube. Add 100 μL of SDS solution, 100 μL of ammonium persulfate solution, and 10 μL of TEMED and cast gel within a ~7.25 cm × 10 cm × 1.5 mm gel cassette. Allow space for stacking gel and gently overlay with isopropanol (*see* **Note 37**). Polymerize the gel at ambient temperature for 20 min.

2. Carefully remove isopropanol after polymerization.

3. Prepare the stacking gel by mixing 1.25 mL of stacking gel buffer, 0.65 mL of acrylamide mixture, and 3 mL water in a 15 mL falcon tube. Add 50 μL of SDS solution, 50 μL of ammonium persulfate solution, and 5 μL of TEMED. Transfer to the resolving gel and immediately insert a 10-well gel comb (*see* **Note 38**).

4. Add 1.5 μL labeled protein to 2 μL sample buffer and add water to 10 μL. Boil the sample at 95 °C for 10 min and centrifuge the samples at ~3000 × g for 1 min.

5. Load the sample on a gel in a pocket next to one loaded with 5 μL prestained molecular weight standard (*see* **Note 39**). Fill the chamber with SDS running buffer (1×) and perform electrophoresis at a constant voltage of 250 V until the bromophenol blue dye reaches the end of the gel. Separate the glass plates and remove the stacking gel.

6. Wash the gel twice with western blot buffer.

7. Incubate blotting paper and nitrocellulose membrane in western blot buffer for 10 min.

8. Gently lay the nitrocellulose membrane on the top of the gel and enclose with blotting paper. Fill the blotting chamber with western blot buffer and add the gel-membrane sandwich to it.

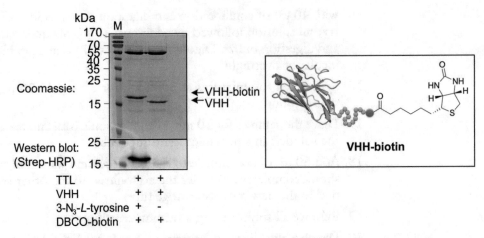

Fig. 2 Biotin labeling of VHH-3-azido-L-tyrosine. Shown is the biotin labeling of an antigen binding nanobody (VHH) with DBCO-biotin. 3-azido-L-tyrosine was enzymatically incorporated to the C-terminus of the VHH using TTL. A following incubation with 30 mol eq. of DBCO-biotin shows selective labeling of the azide containing VHH by strain promoted azide-alkyne cycloaddition (SPAAC)

9. Perform the western blot at a constant current of 250 mA for 1 h while cooling and stirring with a magnetic stirrer.

10. Isolate the nitrocellulose membrane and wash it with water and washing buffer.

11. Incubate the membrane for 1 h at ambient temperature with blocking solution (completely covered).

12. Wash the membrane three times with washing solution containing tween.

13. Add 2 μL of streptavidin-HRP conjugate to 10 mL of washing buffer with tween (1:5000 dilution) and incubate the membrane at 4 °C for 90 min.

14. Wash as in **step 12**. Wash three times with washing solution.

15. Add chemiluminescence substrate for visualization of biotinylated proteins as described in the manufacturer's protocol.

16. Analyze biotinylated proteins using a chemiluminescence imaging system (*see* Fig. 2).

4   Notes

1. 50× rotiphorese buffer is commercially available (Carl Roth GmbH & Co. Kg, Germany).

2. Make sure to not completely close the cap of the Schott bottle to allow for pressure equalization. After the termination of the sterilization process, immediately close the cap using heat-resistant gloves.

3. It is advisable to use an *E. coli* strain with the T7 polymerase promoter system to achieve high regulation of expression. Chemically competent cells can be obtained commercially or prepared using published protocols [32].

4. Start by using concentrated HCl (12 N) until getting close to the required pH. From then on use diluted HCl solutions for final adjustment of the pH.

5. We use a His-Trap column (GE Healthcare, USA).

6. We use a PD10 column (GE Healthcare).

7. Prepare 20 μL aliquots of ATP stock solution and store at −20 °C. Do not re-use an aliquot after thawing once.

8. To fully dissolve 3-azido-L-tyrosine at 11 mM in water, thoroughly mix and sonicate the solution. Make sure to readjust the pH of any tyrosine stock solution to pH 7.0.

9. Prepare 10 μL aliquots of DTT stock solution and store at −20 °C. Do not re-use an aliquot after thawing once.

10. Prepare 20 μL aliquots of glutathione stock solution and store at −20 °C. Do not re-use an aliquot after thawing once.

11. This is an optional equipment. Heating during staining and destaining of SDS-gels decreases time needed for staining of proteins.

12. Make sure to store 10% SDS solution at ambient temperature only.

13. Commercially available.

14. Prepare $(NH_4)HCO_3$ containing buffers freshly and discard after use.

15. Do not re-use an aliquot after thawing once.

16. We find that it is best to prepare western blot buffer fresh.

17. Try to keep the exposure of your DNA to UV-light short. UV light is damaging the DNA.

18. Do not exceed 4 h of digestion time to prevent star activity of the enzymes.

19. Make sure to streak out the cell suspension evenly to allow for single colony formation.

20. Depending on the protein you are expressing and the vector you are using, you might need to change the antibiotic.

21. The cell suspension will become grayish as soon as the cells have been disrupted successfully.

22. It is not mandatory to use an automated FPLC system and the purification of the protein can be performed manually as described in the literature [33]. However, if you perform the Ni-NTA purification of the H_6 tagged TTL manually, make

sure that you elute the immobilized protein by slowly increasing the TTL IMAC elution buffer and carefully analyze the purity of the eluting fractions by SDS-PAGE.

23. To prevent protein aggregation, avoid TTL concentrations higher than 4 mg/mL.

24. Depending on the POI you may need to change IPTG concentration, temperature, and expression time. If your protein seems to be toxic for *E. coli* cells, you may decrease the IPTG concentration to 50 μM.

25. *Di-tert*-butyl dicarbonate melts at ~22 °C. We recommend to store this compound at 4 °C and keep it cold until weighing and usage.

26. Replace the ice/water bath in case that the ice melted completely.

27. Slowly decrease the pressure until you reach 90 mbar. This way you prevent boiling retardation. Keep the pressure at 90 mbar until no additional solvent is evaporating.

28. The ethyl acetate phase is on the top of the aqueous solution.

29. Slowly decrease the pressure until you reach 200 mbar. This way you prevent boiling retardation. Keep the pressure at 200 mbar until no additional solvent is evaporating. Further decrease the pressure to 10 mbar (very slowly). The compound forms a white foam and boiling retardation is very likely if you decrease the pressure too fast. Keep the pressure at 10 mbar for at least 20 min.

30. The yield of this reaction is very low (usually 20–30%). Try to collect unreacted starting material ($R_f = 0.29$) and repeat reaction several times.

31. DTT reduces 3-azido-L-tyrosine to 3-amino-L-tyrosine.

32. Start by using 1 N KOH until getting close to pH 7.0. From then on use 0.1 N KOH for final adjustment of the pH.

33. Depending on the POI you can increase the incubation time to achieve full conversion. However, make sure not to incubate mixtures including 3-azido-L-tyrosine for more than 10 h to prevent reduction to the amine. Do not shake while incubating to prevent aggregation of protein.

34. It is advisable to directly perform a subsequent bioorthogonal reaction after ligation of the tyrosine derivative.

35. In our hands, pre-activation of the catalyst in a separate microcentrifuge tube gives higher cross-coupling yields.

36. Do not shake while incubating to prevent aggregation of protein. For more examples of protein functionalization utilizing the Staudinger phosphite reaction, *see* references [18–22].

37. We use a Mini-PROTEAN Tetra Cell Casting stand (BioRad, USA). Isobutanol prevents contact with polymerization inhibiting oxygen and levels the resolving gel.

38. Make sure to not introduce air bubbles when adding the gel comb. In our hands, storing the gel in a wet tissue at 4 °C for one night results in better protein separation.

39. Adding 5 μL sample buffer to any empty pocket of the gel ensures a uniform running front and prevents the "smiling effect."

40. If you want to perform tryptic digest and MSMS analysis, do not boil the gel during staining and destaining to increase sensitivity of the MS analysis.

41. The incorporation of single amino acids cannot be resolved in an SDS-gel. However, we have successfully separated proteins (up to 30 kDa) that are functionalized with a fluorophore or biotin (usually 800–1000 Da) from the non-functionalized protein.

42. It is important to wear gloves to reduce keratin background.

43. Although most of the Coomassie should be removed, it is not necessary to destain the pieces completely.

Acknowledgments

This work was supported by grants from the Deutsche Forschungsgemeinschaft (SPP1623) to C.P.R.H. (HA 4468/9-1), and H.L. (LE 721/13-1), the Nano- systems Initiative Munich (NIM) to H.L., the Einstein Foundation Berlin (Leibniz-Humboldt Professorship) and the Boehringer-Ingelheim Foundation (Plus 3 award) to C.P.R.H. and the Fonds der Chemischen Industrie (FCI) to C.P.R.H. and to D.S. (Kekulé-scholarship).

References

1. Hackenberger CP, Schwarzer D (2008) Chemoselective ligation and modification strategies for peptides and proteins. Angew Chem Int Ed 47(52):10030–10074. https://doi.org/10.1002/anie.200801313

2. Massa S, Xavier C, De Vos J, Caveliers V, Lahoutte T, Muyldermans S, Devoogdt N (2014) Site-specific labeling of cysteine-tagged camelid single-domain antibody-fragments for use in molecular imaging. Bioconjug Chem 25(5):979–988. https://doi.org/10.1021/bc500111t

3. Schumacher D, Hackenberger CP (2014) More than add-on: chemoselective reactions for the synthesis of functional peptides and pro-teins. Curr Opin Chem Biol 22:62–69. https://doi.org/10.1016/j.cbpa.2014.09.018

4. Pleiner T, Bates M, Trakhanov S, Lee C-T, Erik Schliep JE, Chug H, Böhning M, Stark H, Urlaub H, Görlich D (2015) Nanobodies: site-specific labeling for super-resolution imaging, rapid epitope-mapping and native protein complex isolation. elife 4:e11349. https://doi.org/10.7554/eLife.11349

5. Schumacher D, Hackenberger CP, Leonhardt H, Helma J (2016) Current status: site-specific antibody drug conjugates. J Clin Immunol 36:100. https://doi.org/10.1007/s10875-016-0265-6

6. Massa S, Xavier C, Muyldermans S, Devoogdt N (2016) Emerging site-specific bioconjugation strategies for radioimmunotracer development. Expert Opin Drug Deliv 13(8):1149–1163. https://doi.org/10.1080/17425247.2016.1178235

7. Agrawal D, Hackenberger CPR (2013) Site-specific chemical modifications of proteins. Indian J Chem A 52(8–9):973–991

8. Dawson PE, Muir TW, Clark-Lewis I, Kent SB (1994) Synthesis of proteins by native chemical ligation. Science 266(5186):776–779

9. Muir TW (2003) Semisynthesis of proteins by expressed protein ligation. Annu Rev Biochem 72:249–289. https://doi.org/10.1146/annurev.biochem.72.121801.161900

10. Budisa N (2004) Prolegomena to future experimental efforts on genetic code engineering by expanding its amino acid repertoire. Angew Chem Int Ed Engl 43(47):6426–6463. https://doi.org/10.1002/anie.200300646

11. Liu CC, Schultz PG (2010) Adding new chemistries to the genetic code. Annu Rev Biochem 79:413–444. https://doi.org/10.1146/annurev.biochem.052308.105824

12. Lotze J, Reinhardt U, Seitz O, Beck-Sickinger AG (2016) Peptide-tags for site-specific protein labelling in vitro and in vivo. Mol BioSyst 12:1731. https://doi.org/10.1039/c6mb00023a

13. Rashidian M, Dozier JK, Distefano MD (2013) Enzymatic labeling of proteins: techniques and approaches. Bioconjugate Chem 24(8):1277–1294. https://doi.org/10.1021/Bc400102w

14. Schumacher D, Helma J, Mann FA, Pichler G, Natale F, Krause E, Cardoso MC, Hackenberger CP, Leonhardt H (2015) Versatile and efficient site-specific protein functionalization by tubulin tyrosine ligase. Angew Chem Int Ed Engl 54(46):13787–13791. https://doi.org/10.1002/anie.201505456

15. Patterson DM, Nazarova LA, Prescher JA (2014) Finding the right (bioorthogonal) chemistry. ACS Chem Biol 9(3):592–605. https://doi.org/10.1021/Cb400828a

16. Muyldermans S (2013) Nanobodies: natural single-domain antibodies. Annu Rev Biochem 82:775–797. https://doi.org/10.1146/annurev-biochem-063011-092449

17. Baskin JM, Prescher JA, Laughlin ST, Agard NJ, Chang PV, Miller IA, Lo A, Codelli JA, Bertozzi CR (2007) Copper-free click chemistry for dynamic in vivo imaging. Proc Natl Acad Sci U S A 104(43):16793–16797. https://doi.org/10.1073/pnas.0707090104

18. Serwa R, Wilkening I, Del Signore G, Muhlberg M, Claussnitzer I, Weise C, Gerrits M, Hackenberger CP (2009) Chemoselective staudinger-phosphite reaction of azides for the phosphorylation of proteins. Angew Chem Int Ed Engl 48(44):8234–8239. https://doi.org/10.1002/anie.200902118

19. Bohrsch V, Serwa R, Majkut P, Krause E, Hackenberger CP (2010) Site-specific functionalisation of proteins by a Staudinger-type reaction using unsymmetrical phosphites. Chem Commun (Camb) 46(18):3176–3178. https://doi.org/10.1039/b926818a

20. Hoffmann E, Streichert K, Nischan N, Seitz C, Brunner T, Schwagerus S, Hackenberger CP, Rubini M (2016) Stabilization of bacterially expressed erythropoietin by single site-specific introduction of short branched PEG chains at naturally occurring glycosylation sites. Mol BioSyst 12(6):1750–1755. https://doi.org/10.1039/c5mb00857c

21. Nischan N, Chakrabarti A, Serwa RA, Bovee-Geurts PH, Brock R, Hackenberger CP (2013) Stabilization of peptides for intracellular applications by phosphoramidate-linked polyethylene glycol chains. Angew Chem Int Ed Engl 52(45):11920–11924. https://doi.org/10.1002/anie.201303467

22. Nischan N, Kasper MA, Mathew T, Hackenberger CP (2016) Bis(arylmethyl)-substituted unsymmetrical phosphites for the synthesis of lipidated peptides via Staudinger-phosphite reactions. Org Biomol Chem 14(31):7500–7508. https://doi.org/10.1039/c6ob00843g

23. Vallee MR, Artner LM, Dernedde J, Hackenberger CP (2013) Alkyne phosphonites for sequential azide-azide couplings. Angew Chem Int Ed Engl 52(36):9504–9508. https://doi.org/10.1002/anie.201302462

24. Vallee MR, Majkut P, Krause D, Gerrits M, Hackenberger CP (2015) Chemoselective bioconjugation of triazole phosphonites in aqueous media. Chemistry 21(3):970–974. https://doi.org/10.1002/chem.201404690

25. Vallee MR, Majkut P, Wilkening I, Weise C, Muller G, Hackenberger CP (2011) Staudinger-phosphonite reactions for the chemoselective transformation of azido-containing peptides and proteins. Org Lett 13(20):5440–5443. https://doi.org/10.1021/ol2020175

26. Dirksen A, Dawson PE (2008) Rapid oxime and hydrazone ligations with aromatic aldehydes for biomolecular labeling. Bioconjug Chem 19(12):2543–2548. https://doi.org/10.1021/bc800310p

27. Li N, Lim RK, Edwardraja S, Lin Q (2011) Copper-free sonogashira cross-coupling for functionalization of alkyne-encoded proteins in

aqueous medium and in bacterial cells. J Am Chem Soc 133(39):15316–15319. https://doi.org/10.1021/ja2066913

28. Kirchhofer A, Helma J, Schmidthals K, Frauer C, Cui S, Karcher A, Pellis M, Muyldermans S, Casas-Delucchi CS, Cardoso MC, Leonhardt H, Hopfner KP, Rothbauer U (2010) Modulation of protein properties in living cells using nanobodies. Nat Struct Mol Biol 17(1):133–138. https://doi.org/10.1038/nsmb.1727

29. Jung ME, Lazarova TI (1997) Efficient synthesis of selectively protected L-Dopa derivatives from L-tyrosine via Reimer-Tiemann and Dakin reactions. J Org Chem 62(5):1553–1555. https://doi.org/10.1021/Jo962099r

30. Banerjee A, Panosian TD, Mukherjee K, Ravindra R, Gal S, Sackett DL, Bane S (2010) Site-specific orthogonal labeling of the carboxy terminus of alpha-tubulin. ACS Chem Biol 5(8):777–785. https://doi.org/10.1021/cb100060v

31. Chalker JM, Wood CS, Davis BG (2009) A convenient catalyst for aqueous and protein Suzuki-Miyaura cross-coupling. J Am Chem Soc 131(45):16346–16347. https://doi.org/10.1021/ja907150m

32. Inoue H, Nojima H, Okayama H (1990) High efficiency transformation of Escherichia coli with plasmids. Gene 96(1):23–28

33. Bornhorst JA, Falke JJ (2000) Purification of proteins using polyhistidine affinity tags. Methods Enzymol 326:245–254

Part II

Engineering in Prokaryotes

Chapter 5

Directed Evolution of Orthogonal Pyrrolysyl-tRNA Synthetases in *Escherichia coli* for the Genetic Encoding of Noncanonical Amino Acids

Moritz J. Schmidt and Daniel Summerer

Abstract

The directed evolution of orthogonal aminoacyl-tRNA synthetases (aaRS) for the genetic encoding of noncanonical amino acids (ncAA) has paved the way for the site-specific incorporation of >170 functionally diverse ncAAs into proteins in a large number of organisms [1, 2]. Here, we describe the directed evolution of orthogonal pyrrolysyl-tRNA synthetase (PylRS) mutants with new amino acid selectivities from libraries using a two-step selection protocol based on chloramphenicol and barnase reporter systems. Although this protocol focuses on the evolution of PylRS variants, this procedure can be universally employed to evolve orthogonal aaRS.

Key words Pyrrolysyl-tRNA synthetase, Noncanonical amino acids, Directed evolution, Genetic selection, Amber suppression, Expanded genetic code

1 Introduction

Studying protein function in life cells and organisms is a central goal of molecular biology. Using organisms with an expanded genetic code to site-specifically incorporate noncanonical amino acids (ncAA) into proteins has recently emerged as a promising approach to study modified proteins with minimal perturbation of their structure in living cells, where posttranslational modifications and molecular crowding significantly influence protein function [1, 2]. Key to successfully add new chemistries to the genetic code is the evolution of orthogonal aminoacyl-tRNA synthetases (aaRS)/tRNA pairs that can accept an ncAA as a substrate and direct its site-specific incorporation in response to a unique codon, e.g., the amber stop codon TAG. Several orthogonal aaRS/tRNA pairs have been successfully introduced that feature orthogonality with respect to different host organisms. Orthogonality can be achieved by heterologous expression of aaRS/tRNA pairs from an organism of another domain of life than the employed organism and has diverged through natural evolution.

Edward A. Lemke (ed.), *Noncanonical Amino Acids: Methods and Protocols*, Methods in Molecular Biology, vol. 1728,
https://doi.org/10.1007/978-1-4939-7574-7_5, © Springer Science+Business Media, LLC 2018

Alternatively, pyrrolysyl-tRNA synthetase pairs (tRNA/PylRS) [3] have been shown to be orthogonal in organisms from all domains of life [4]. Specificity of the aaRS for the ncAA of interest can subsequently be generated by evolutionary design. Typically, five to six amino acid residues within the amino acid binding pocket of the aaRS are identified on the basis of aaRS crystal structures and are randomized to all possible combinations (Fig. 1a) [5, 6]. This is typically achieved by iterative rounds of enzymatic inverse PCRs using degenerated primers with NNK codons followed by a final ligation step to insert the PylRS library into the corresponding selection plasmid. Major limitation in library construction is the transformation efficiency of *E. coli*, which limits the number of randomized positions

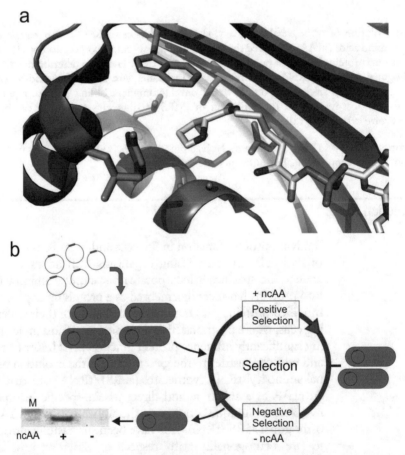

Fig. 1 (**a**) Focused PylRS libraries are constructed on the basis of crystal structures of the C-terminal catalytic fragment of PylRS [5, 17, 20, 21]. Active-site residues forming the binding pocket of the amino acid substrate are selected and randomized using site-saturation mutagenesis via enzymatic inverse PCR. (**b**) Pooled libraries are subjected to a two-step selection protocol, linking the ability to suppress the amber stop codon with the expression of an essential (positive selection) or toxic gene (negative selection) to enrich for functional and ncAA-specific PylRS variants [12]

to a maximum of six, when a saturating NNK randomization approach is used [7]. To circumvent bias, smart libraries based on unique, codon-optimized primers can be constructed [8, 9]. Alternatively, the randomized gene can be readily obtained commercially, generated by synthetic approaches. A preselection of promising target residues that constitute the binding pocket can be made on the basis of crystal structures and previous selection experiments [4]. The constructed and transformed libraries are subsequently subjected to an in vivo selection in *E. coli* to enrich for orthogonal aaRS/tRNA pairs in alternating rounds of positive and negative selection. In rounds of positive selection, the ability of an tRNA/aaRS mutant to direct suppression of the amber stop codon in the presence of the ncAA is linked to the survival of the cell by the use of an antibiotic resistance gene, e.g., ß-lactamase [10] or chloramphenicol acetyl transferase [11] bearing an amber stop codon at a permissive site, so that the truncated expression products are non-functional. To eliminate aaRS from the pool that recognize a canonical amino acid as a substrate, rounds of negative selection are subsequently performed. Here, the ability to suppress the amber stop codon is coupled to the expression of a toxic gene product (e.g., ribonuclease barnase [12] or DNA topoisomerase inhibitor ccdB [13]), and the cells are grown in the absence of the ncAA of interest (Fig. 1b). Here, we provide detailed guidelines for conducting a selection protocol that has been successfully used for the directed evolution of orthogonal PylRS variants for a variety of structurally diverse ncAA.

2 Materials

All buffers should be prepared using ultrapure water and biochemical grade chemicals. Buffers, antibiotic stocks, and chemicals should be sterile-filtered using a 0.22 μM syringe filter, or sterilized by autoclaving and stored individually according to the manufacturer. Plasmid maps are provided in Fig. 2 and important sequences are highlighted in Fig. 3.

2.1 Plasmids

1. *Positive Selection Plasmid (pREP_PylRS_CAT):* The expression of the *Methanosarcina mazei* PylRS variants is maintained by a constitutive glnS' promotor [14], the expression of the tRNA^{Pyl} is controlled by a proK promotor, and the chloramphenicol acetyltransferase (CAT) gene is expressed constitutively under the control of a trp promotor. The CAT gene bears an in-frame amber stop codon at position Q98. The plasmid features an additional tetracycline resistance for plasmid propagation (Fig. 2a and Fig. 3).

2. *Negative Selection Plasmid (pMinus_Barnase_Q2TAG_G65TAG):* The barnase gene is controlled by an arabinose-inducible (araBAD) promotor and bears two in-frame amber stop codons at

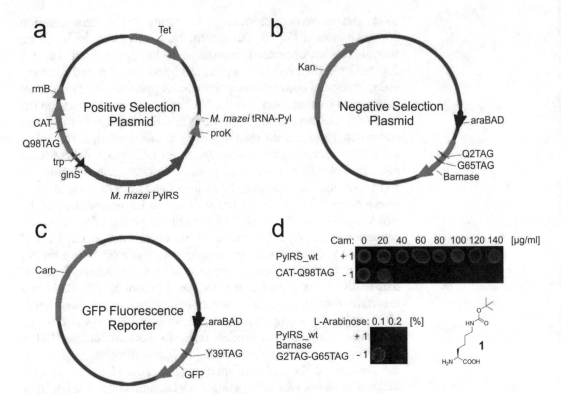

Fig. 2 (**a**) Plasmid for the positive selection (*pREP_PylRS_CAT*) encoding a *M. mazei* PylRS variant under the control of the constitutive glnS' promotor, a proK regulated transcription of tRNA^{Pyl}, a trp controlled expression of the reporter gene CAT (CAT: chloramphenicol-acetyltransferase, bearing an in-frame amber stop codon at position Q98) and a gene encoding tetracycline resistance (Tet) for plasmid propagation. (**b**) Plasmid for the negative selection (*pMinus_Barnase_Q2TAG_G65TAG*) encoding an L-arabinose inducible (araBAD) barnase gene, which is interrupted by two amber stop codons at position Q2 and G65. The plasmid encodes a kanamycin resistance for plasmid propagation. (**c**) Plasmid for fluorescence-based detection of amber suppression (*pBAD_Flag_GFP-Y39TAG_6His*), encoding an araBAD-controlled gene of GFP (bearing an in-frame amber stop codon at position Y39) and a carbenicillin resistance for plasmid propagation. The GFP gene features an N-terminal Flag-tag and a C-terminal his6-tag. (**d**) Growth assays of *E. coli* harboring positive selection plasmid pREP_PylRS_CAT, encoding for wtPylRS. Growth was detected in the presence or absence of a reference ncAA (H-Lys(Boc)-OH **1**, a known substrate for wtPylRS) and varying concentrations of chloramphenicol on GMML-Agar plates. *E. coli* harboring plasmid pMinus_Barnase_Q2TAG_G65TAG was tested for growth of an LB-agar plate, resembling conditions of rounds of negative selection in the presence and absence of H-Lys(Boc)-OH **1** and varying concentrations of L-arabinose. Chemical structure of H-Lys(Boc)-OH **1**

positions Q2 and G65. The plasmid features an additional kanamycin resistance for plasmid propagation (Fig. 2b).

3. *GFP Reporter (pBAD_Flag_GFP_Y39TAG_6His)* [15]: The gene encoding green fluorescent protein is controlled by an araBAD promotor and harbors an in-frame amber stop codon at position Y39, an N-terminal flag tag, and a C-Terminal His6-tag. The plasmid features an additional carbenicillin resistance for plasmid propagation (Fig. 2c).

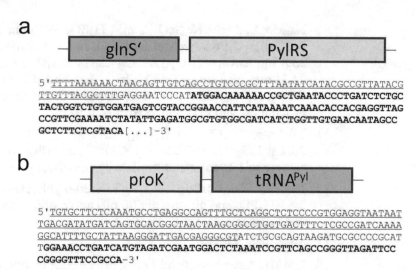

Fig. 3 (**a**) Schematic illustration and sequence information of glnS' promotor (underlined) and PylRS gene (bold) as applied in the selection plasmid. (**b**) proK (underlined) driven transcription of tRNA^{Pyl} (bold) shown as scheme with sequence information

2.2 Glycerole-Based Minimal Medium with Leucine Supplementation (GMML) Agar Plates

1. 5× M9 Salts: Weigh in 64 g $Na_2HPO_4 \cdot 7H_2O$, 15 g KH_2PO_4, 5 g NH_4Cl and 2.5 g NaCl, dissolve in 1 L water and autoclave for 15 min at 121 °C for sterilization. Store at room temperature.

2. 1 M $MgSO_4$ solution: Weigh in 246.48 g $MgSO_4 \cdot 7H_2O$, adjust the volume to 1 L with water and sterilize by autoclaving as in the previous step.

3. 1 M $MgCl_2$ solution: Weigh in 203.3 g $MgCl_2 \cdot 6H_2O$, adjust the volume to 1 L with water, and sterilize by autoclaving as in the previous step.

4. 1 M Glucose solution: Weigh in 180.16 g glucose, dissolve in a final volume of 1 L, and sterilize by filtration (*see* **Note 1**).

5. 50% (v/v) Glycerole solution: Dilute 500 mL glycerole with 500 mL water and sterilize by autoclaving.

6. 100 mM $CaCl_2$ solution: Weigh in 11.1 g $CaCl_2$, dissolve in a final volume of 1 L water, and sterilize by autoclaving.

7. 850 mM NaCl solution: Weigh in 49.7 g NaCl, dissolve in a final volume of 1 L water, and sterilize by autoclaving.

8. 30 mM L-Leucine solution: Weigh in 3.94 g L-Leucine dissolve in a final volume of 1 L water and sterilize by autoclaving.

9. 5 mM $FeSO_4$ solution: Weigh in 1.39 g of $FeSO_4 \cdot 7H_2O$ dissolve in a final volume of 1 L water and sterilize by filtration.

10. 1000× Trace Metal Solution:50 mM $FeCl_3$, 10 mM $MnCl_2$, 10 mM $ZnSO_4$, 2 mM $CoCl_2$, 2 mM $CuCl_2$, 2 mM $NiCl_2$, 2 mM

Na_2MoO_4, 2 mM Na_2SeO_3, 2 mM H_3BO_3. Weigh in 675.5 mg $FeCl_3 \bullet 6H_2O$, 98.9 mg $MnCl_2 \bullet 4H_2O$, 143.7 mg $ZnSO_4 \bullet 7H_2O$, 23.8 mg $CoCl_2 \bullet 6H_2O$, 17.1 mg $CuCl_2 \bullet 2H_2O$, 13 mg $NiCl_2$, 24.2 mg $Na_2MoO_4 \bullet 2H_2O$, 26.3 mg $Na_2SeO_3 \bullet 5H_2O$ and 6.2 mg H_3BO_3, dissolve in a final volume of 50 mL water, sterilize by filtration, and store aliquoted at -20 °C.

11. TB-Buffer: 0.17 M KH_2PO_4, 0.72 M K_2HPO_4. Weigh in 23.1 g KH_2PO_4 and 125.4 g K_2HPO_4 and dissolve in a final volume of 1 L water and sterilize by autoclaving.

12. 10× PBS: Weigh in 80 g NaCl, 2 g KCl, 25.6 g $Na_2HPO_4 \bullet 7H_2O$, 2 g KH_2PO_4 and dissolve in 800 mL water. Adjust the pH to 8 and bring to a final volume of 1 L.

13. Agar-Agar, Kobe 1.

14. GMML-Agar: Resuspend 13.5 g of agar-agar in 499 mL water. Add 0.5 mL PBS (4×) and 0.5 mL of TB buffer to adjust the pH to 7 and sterilize by autoclaving (*see* **Note 2**).

15. 1000× Chloramphenicol stock: 34 mg/mL. Dissolve 340 mg chloramphenicol in a final volume of 10 mL ethanol and sterilize by filtration.

16. 1000× Carbenicillin stock: 50 mg/mL. Dissolve 500 mg carbenicillin in a final volume of 10 mL ethanol/water (1:1, v/v) and sterilize by filtration.

17. 1000× Tetracycline stock: 12.5 mg/mL. Dissolve 125 mg tetracycline in a final volume of 10 mL ethanol and sterilize by filtration.

18. ncAA solution: dissolve ncAA in water to obtain stock concentration of 50–100 mM (*see* **Note 3**).

19. Sterile petri dish, Ø 14.5 cm and Ø 9 cm.

20. Magnetic stirrer.

2.3 Transformation and Growth Assays

1. LB-Medium: Weigh in 10 g tryptone, 5 g yeast extract and 5 g NaCl, dissolve in 900 mL water, adjust pH to 7 and bring to a final volume of 1 L, and sterilize by autoclaving.

2. LB-Agar: Weigh in 10 g tryptone, 5 g yeast extract, 5 g NaCl, and 15 g agar-agar, dissolve in 900 mL water, adjust pH to 7 and bring to a final volume of 1 L and sterilize by autoclaving.

3. SOC-Medium: Weigh in 20 g tryptone, 5 g yeast extract, 0.58 g NaCl, 0.19 g KCl and dissolve in 900 mL water and sterilize by autoclaving. Add 10 mL 1 M $MgSO_4$, 10 mL 1 M $MgCl_2$ and 20 mL 1 M glucose solution and adjust volume to 1 L.

4. Nε-Boc-(L)-lysine (H-Lys(Boc)-OH, Fig. 2d): commercially available, or synthesize according to published procedures [16].

5. 96-deep well plates, 2 mL.

6. Nunc sealing tape, breathable, sterile (Thermo Fisher).

7. Electroporation device (Eppendorf—Eporator).

8. Electroporation cuvettes, 1 mm (Carl Roth).

9. Glas beads Ø 5 mm (Carl Roth).

10. 20% (w/v) L-Arabinose solution. Sterilized by filtration.

11. Photometer (Eppendorf—Biopohotometer).

12. Fluorescence micro plate reader (Tecan).

13. Micro plate incubator (Heidolph—Titramax).

14. Centrifuge (Eppendorf—5810).

2.4 Protein Expression and Purification

1. B-Per Lysis Buffer: Bacterial Protein Extraction Reagent (Thermo Fisher).

2. 100 mM PMSF (phenylmethanesulfonyl fluoride) in ethanol.

3. Wash buffer:50 mM NaH_2PO_4, 300 mM NaCl, 20 mM Imidazole, pH = 8. Weigh in 6.9 g $NaH_2PO_4 \bullet H_2O$, 17.54 g NaCl and 1.36 g imidazole dissolve in 900 mL water and adjust the pH to 8. Bring to a final volume of 1 L.

4. Elution buffer: 50 mM NaH_2PO_4, 300 mM NaCl, 500 mM Imidazol, pH = 8. Weigh in 6.9 g $NaH_2PO_4 \bullet H_2O$, 17.54 g NaCl and 34 g imidazole dissolve in 900 mL water and adjust the pH to 8. Bring to a final volume of 1 L.

2.5 Plasmid Isolation and Preparation

1. Plasmid Isolation Kit.

2. PCR Purification Kit.

3. Restriction Enzyme: *MluI*.

4. NEB buffer 3.1 (10×): 1 M NaCl, 500 mM Tris–HCl, 100 mM $MgCl_2$, 1000 µg/mL BSA, pH = 7.9 @ 25 °C.

3 Methods

All the reaction steps should be performed at room temperature unless otherwise noted. When working with bacterial cultures, a sterile environment has to be maintained and all waste regularities have to be followed.

PylRS libraries for the following experiments were constructed using enzymatic inverse PCR with NNK-degenerated primers and traceless ligation using *BsaI* restriction sites, as described elsewhere [17, 18], and were cloned into the positive selection plasmid *pREP_PylRS_CAT* using *NdeI/PstI* restriction sites (Fig. 2). The randomized PylRS gene can alternatively be obtained commercially. The transformations were pooled, grown on agar plates, scraped, and stored as *E. coli* glycerol stocks at −80 °C. For the preparation of GMML agar plates a minimal volume of 20 mL is suggested for Ø 14.5 cm petri dishes, 10 mL for Ø 9 cm petri dishes respectively.

3.1 Selection

1. Prepare GMML agar plates (*see* **Note 4**) for the first positive selection step by melting the sterile agar suspension in a microwave at 200 W for 35 min until all particles are completely dissolved. Let the agar solution cool down for 15 min on a magnetic stirrer.

2. Mix 400 µL glycerol solution (50%, v/v), 4 mL M9 salts (5×), 200 µL NaCl solution (850 mM), 200 µL L-Leucine (30 mM), 20 µL $MgSO_4$ (1 M), 20 µL $CaCl_2$ (100 mM), 20 µL $FeSO_4$ (5 mM), and 20 µL trace metal solution (1000×) in a sterile 50 mL falcon tube. Add 20 µL tetracycline (12.5 mg/mL) and 47.1 µL chloramphenicol (34 mg/mL), 200 µL ncAA solution (100 mM) (*see* **Note 5**) and fill up to a final volume of 20 mL with the warm agar solution. Mix by vortexing and pour into Ø 14.5 cm sterile petri dish and let it cool to room temperature to obtain a GMML agar plate with final concentrations of: 1× M9 salt, 1% (v/v) glycerol, 1 mM $MgSO_4$, 0.1 mM $CaCl_2$, 17.1 mM NaCl, 0.3 mM L-leucine, 5 µM $FeSO_4$, 1× trace metals, 1 mM ncAA (*see* **Note 6**), 12.5 µg/mL tetracycline and 80 µg/mL chloramphenicol (*see* **Note 7**).

3. Plate libraries on GMML plates to cover the theoretical diversity at least by a factor of 3. Use as many Ø 14.5 cm petri dishes as possible to reduce direct competition of clones (ideally, colonies should be separated, however, for large libraries this is not realistic and generally, the amount of ncAA will be limiting). Spread the bacterial suspension using glass beads for homogenous distribution. Incubate the plates at 37 °C for 24–72 h.

4. Harvest the cells by scraping the colonies using a *Drigalski spatula* and 20 mL LB medium. Collect the cell suspension in a falcon tube and pellet the cells by centrifugation for 10 min. at $3320 \times g$ (*see* **Note 8**).

5. Isolate the plasmids using a commercial plasmid isolation kit and determine the concentration using a photometer (*see* **Note 9**).

6. Transform 100 ng of isolated plasmid DNA into 100 µL highly electro-competent *E. coli* GH371 cells (carrying plasmid pMinus_Barnase_Q2TAG_G65TAG; *see* **Note 10**) using electroporation and recover the cells with 1 mL pre-warmed (37 °C) SOC-medium.

7. Incubate the cells in a 1.5 mL Eppendorf tube at 37 °C and $93 \times g$ shaking in a thermomixer for 1 h.

8. Prepare Ø 14.5 cm LB-agar plates supplemented with 50 µg/mL kanamycin, 12.5 µg/mL tetracycline, and 0.1% (w/v) L-arabinose (*see* **Note 11**).

9. Streak cell suspension on LB-agar plates and incubate for 12–16 h at 37 °C (*see* **Note 12**).

10. Harvest the cells by scraping and isolate plasmid DNA (as mentioned in **steps 4** and **5**).

11. Digest the isolated plasmid mixture to linearize plasmid pMinus_Barnase_Q2TAG_G65TAG with restriction enzyme *MluI* in 1× NEB 3.1 buffer at 37 °C for 3 h (*see* **Note 13**).

12. Heat inactivate *MluI* at 80 °C for 20 min in a thermomixer.

13. Purify the restriction digest using a commercial PCR purification kit and determine the concentration of the DNA solution.

14. Transform 50 ng of plasmid solution into 100 µL electro-competent *E. coli* GH371 cells harboring pBAD_Flag_GFP_Y39TAG_6His (*see* **Note 14**).

15. Recover the cells in 1 mL pre-warmed (37 °C) SOC-medium and incubate in a 1.5 mL Eppendorf tube at 37 °C and $93 \times g$ shaking in a thermomixer.

16. Plate dilutions of the transformation suspension to obtain single clones on GMML-agar plates (supplemented with 50 µg/mL carbenicillin, 12.5 µg/mL tetracycline, 80 µg/mL chloramphenicol and ±1 mM ncAA).

17. Pick 96 individual clones with sterile pipette tips and individually inoculate 1 mL LB medium (supplemented with 50 µg/mL carbenicillin and 12.5 µg/mL tetracycline) in a 96-deep well plate. Incubate at 37 °C with $93 \times g$ shaking for 18 h (*see* **Notes 14** and **15**).

18. Prepare Ø 14.5 cm GMML-agar plates (supplemented with 50 µg/mL carbenicillin, 12.5 µg/mL tetracycline, 80 µg/mL chloramphenicol, 0.2% L-arabinose and ±1 mM ncAA or no ncAA supplementation as a control).

19. Dilute overnight cultures 500× by mixing 2 µL of the bacterial suspension with 1 mL PBS (1×) (*see* **Note 16**).

20. Print 5 µL of the diluted cell suspension on the GMML-agar plates and incubate at 37 °C for 24–72 h (*see* **Note 17**).

21. Analyze the plates for ncAA-dependent growth by qualitatively comparing growth intensities at different time points. Due to the supplementation with L-arabinose, the plates can also be analyzed according to their GFP-fluorescence under a standard hand-held UV-lamp (*see* Fig. 4a, b for representative results of a successful selection experiment).

To verify the hits and determine the incorporation efficiency and fidelity, a fine-tuned growth assay at varying chloramphenicol concentrations as well as cellular fluorescence measurements should be conducted (*see* below).

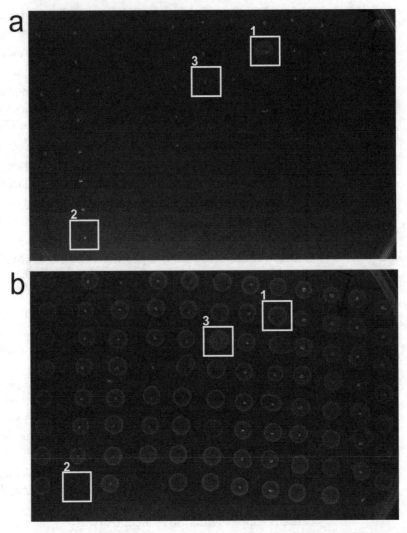

Fig. 4 Representative screening experiment after three rounds of alternating positive and negative selection. 96 clones are screened individually on chloramphenicol-containing agar plates in the absence (**a**) and presence (**b**) of ncAA. Reference 1 shows a false positive clone, reference 2 shows a negative clone, whereas reference clone 3 shows ncAA-dependent growth on chloramphenicol plates, indicating the selection of a specific and functional PylRS variant

3.2 Growth Assay

1. Pour LB-agar plates supplemented with 12.5 μg/mL tetracycline, 50 μg/mL carbenicillin, 60–140 μg/mL chloramphenicol, and ±1 mM ncAA.

2. Inoculate 5 mL LB medium (50 μg/mL carbenicillin, 12.5 μg tetracycline) with hit colonies and incubate at 37 °C and shake for 18 h.

3. Dilute the overnight cultures 1:500 into fresh LB-medium and print 5 μL on LB-agar plates. Incubate at 37 °C for 24–48 h and analyze growth differences by visual inspection (*see* Fig. 2d for representative result).

3.3 Cellular Fluorescence/ SDS-PAGE Analysis

1. Dilute the aforementioned overnight cultures 1:50 into 5 mL LB medium (50 μg/mL carbenicillin, 12.5 μg/mL tetracycline) and incubate at 37 °C and shake until an OD_{600} of 0.4 is reached. Add ncAA (*see* **Note 18**) and 0.2% L-arabinose to induce the expression of GFP and incubate for 16 h at 37 °C and shake (*see* **Note 19**).

2. Harvest the cells by centrifugation for 10 min. at $3320 \times g$ and resuspend the cell pellet with 5 mL PBS (1×). Repeat this washing step three times and resuspend the cell pellet finally in 5 mL PBS (1×) and subject to fluorescence measurements on a micro plate reader (*see* **Note 20**).

3. Pellet expression culture again and lyse the cells using 1 mL B-Per lysis buffer (supplemented with 1 mM PMSF). Incubate for 20 min at 4 °C while shaking.

4. Spin-down the suspension at $20,817 \times g$ for 5 min at 4 °C and transfer the supernatant to a 1.5 mL Eppendorf tube and add 100 μL Ni-NTA slurry.

5. Incubate the suspension for 20 min at 4 °C and $182 \times g$ shaking in a Thermomixer.

6. Transfer the suspension into a filter spin column and spin down at $900 \times g$ for 1 min.

7. Wash the Ni-beads five times by resuspending with 700 μL wash buffer and centrifugation at $900 \times g$ for 1 min.

8. Finally elute his-tagged protein by resuspending the Ni-beads with 100 μL elution buffer, incubate for 5 min, and centrifuge at $20,817 \times g$ for 1 min. Repeat this step once.

9. Analyze protein solution using SDS-PAGE (*see* **Note 21**).

4   Notes

1. To avoid degradation, glucose should be filtered for sterilization and not be subjected to autoclaving.

2. Agar suspension has to be buffered to prevent acid-induced degradation during autoclaving.

3. If problems with the solubility of the ncAA occur, the solvent can be changed to 100 mM HCl or 100 mM NaOH. For very unpolar ncAAs, 1 M solutions in DMSO can be prepared alternatively. In this case the solubility in solid and liquid media should be investigated.

4. Instead of GMML agar plates, LB-agar plates can be used, but it is recommended to perform rounds of positive selection using minimal medium to limit the amount of natural amino acids present in the medium to decrease false positives during selection.

5. Always pour a control plate without the ncAA supplemented in rounds of positive selection, or without L-arabinose in rounds of negative selection, as a reference for qualitative growth effects.

6. Increasing the ncAA concentration up to 5 mM can sometimes help to enrich for weaker PylRS variants. Check the toxicity of the ncAA before in growth assays using *E. coli*.

7. The stringency of the selection can be controlled by varying the chloramphenicol concentration. Decreasing the concentration down to 60 µg/mL can be helpful when weaker hits should be also considered. On the other hand, the concentration can be increased up to 140 µg/mL for more stringent selection conditions. This is considered the most critical parameter for successful PylRS enrichment in rounds of positive selection and should be varied in case of unsuccessful experiments. An identification of weak hits can additionally be facilitated by applying an increased ncAA concentration.

8. Cell pellets can be stored at −20 °C for several months.

9. Elute plasmid DNA using nuclease-free water. Elution using salted buffers, such as commercial elution buffers, will interfere with the subsequent electroporation transformation and should thus be avoided.

10. *pMinus_Barnase_Q2TAG_G65TAG* is featuring two in-frame amber stop codons for amber suppression-dependent elimination of PylRS variants that accept canonical amino acids as substrate. Similar barnase constructs featuring only one amber stop codon were not applicable in maintaining the survival of the cells due to leaky expression and excessive toxicity.

11. The amount of L-arabinose added during rounds of negative selection can be varied to control the stringency of the selection and can be increased up to 0.2% (w/v). Though control of this promoter is nonlinear [19], this parameter should be adapted if selection fails.

12. Prolonged incubation at this selection step can increase the number of false positives and is therefore not recommended.

13. *MluI* will selectively linearize *pMinus_Barnase_Q2TAG_G65TAG*, but not *pREP_PylRS_CAT*.

14. Depending on the activity distribution toward the target ncAA in the PylRS library, it can be helpful to transform *E. coli* GH371 cells and directly perform additional rounds of positive and subsequent negative selections before the individual screening. Control plates lacking ncAA supplementation can indicate successful selection by qualitatively assessing the growth differences. Typically, 3–5 rounds of selection are per-

formed, depending on the distribution of aminoacylation activities present in the mutant library

15. By supplementing the LB medium with 0.2% (w/v) L-arabinose, the overnight cultures can additionally be screened for ncAA-dependent GFP fluorescence. This allows rapidly identifying false positive clones, when ncAA is not added to the LB-medium (this can substantially decrease the number of individual screenings, saving time, and material). If ncAA is added to the LB-medium at this stage, potential hits can be identified by preparing a control deep-well plate without ncAA. To be able to accurately compare the fluorescence values, the cells should be washed with PBS (1×) three times by centrifugation and resuspension. Additionally, the cellular fluorescence should be normalized according to cell density obtained from OD_{600} measurements. Moreover, a positive control experiment should be integrated, e.g., by including clones encoding wild-type PylRS grown in the presence of the positive control ncAA Nε-Boc-(L)-lysine (H-Lys(Boc)-OH). This will help in setting a fluorescent threshold to define background noise arising from autofluorescence of cells and allows a relative quantitative judgment of the selected PylRS mutant's activities.

16. It is necessary to dilute the bacterial cultures before printing for growth assays to avoid growth artifacts.

17. It is recommended to use a multichannel pipette for printing (8 or 12 channels) and to use a printed template to accurately print the colonies on the plates.

18. ncAA can also be added upon inoculation, which might increase the cellular concentration when the expression is induced. It is always necessary to set up a control culture in parallel without supplementation of ncAA.

19. The expression time can be decreased to 4–8 h, especially when the hits show some background incorporation of natural amino acids.

20. Cellular fluorescence measurements should be normalized using the cell number/density obtained from OD_{600} measurements. To detect GFP fluorescence, use an excitation wavelength of 475 nm and an emission wavelength of 510 nm. Furthermore, a positive control experiment should be integrated (*see* **Note 14**).

21. To obtain high protein expression yields, it is recommended to clone the identified PylRS variants into a pEVOL-based expression plasmid. Here, an additional copy of the PylRS gene is under the control of an arabinose-inducible promotor, which significantly increases the amber suppression and overall protein yield.

Acknowledgments

We acknowledge support by the TU Dortmund, the University of Konstanz and the Konstanz Research School Chemical Biology. This work was supported by grants from the Deutsche Forschungsgemeinschaft (SU-726/2-2 in SPP1623 and SU-726/4-2 in SPP1601).

References

1. Liu CC, Schultz PG (2010) Adding new chemistries to the genetic code. Annu Rev Biochem 79:413. https://doi.org/10.1146/Annurev.Biochem.052308.105824

2. Chin JW (2014) Expanding and reprogramming the genetic code of cells and animals. Annu Rev Biochem 83:379–408. https://doi.org/10.1146/annurev-biochem-060713-035737

3. Blight SK, Larue RC, Mahapatra A, Longstaff DG, Chang E, Zhao G, Kang PT, Church-Church KB, Chan MK, Krzycki JA (2004) Direct charging of tRNA(CUA) with pyrrolysine in vitro and in vivo. Nature 431(7006):333. https://doi.org/10.1038/Nature02895

4. Wan W, Tharp JM, Liu WR (2014) Pyrrolysyl-tRNA synthetase: an ordinary enzyme but an outstanding genetic code expansion tool. Biochim Biophys Acta 1844(6):1059–1070. https://doi.org/10.1016/j.bbapap.2014.03.002

5. Kavran JM, Gundllapalli S, O'Donoghue P, Englert M, Soll D, Steitz TA (2007) Structure of pyrrolysyl-tRNA synthetase, an archaeal enzyme for genetic code innovation. Proc Natl Acad Sci U S A 104(27):11268–11273. https://doi.org/10.1073/pnas.0704769104

6. Schmidt MJ, Weber A, Pott M, Welte W, Summerer D (2014) Structural basis of furanamino acid recognition by a polyspecific aminoacyl-tRNA-synthetase and its genetic encoding in human cells. Chembiochem 15(12):1755–1760. https://doi.org/10.1002/cbic.201402006

7. Cropp TA, Anderson JC, Chin JW (2007) Reprogramming the amino-acid substrate specificity of orthogonal aminoacyl-tRNA synthetases to expand the genetic code of eukaryotic cells. Nat Protoc 2(10):2590–2600. https://doi.org/10.1038/nprot.2007.378

8. Lacey VK, Louie GV, Noel JP, Wang L (2013) Expanding the library and substrate diversity of the Pyrrolysyl-tRNA Synthetase to incorporate unnatural amino acids containing conjugated rings. Chembiochem 14(16):2100–2105. https://doi.org/10.1002/cbic.201300400

9. Tang LX, Gao H, Zhu XC, Wang X, Zhou M, Jiang RX (2012) Construction of "small-intelligent" focused mutagenesis libraries using well-designed combinatorial degenerate primers. Biotechniques 52(3):149–158. https://doi.org/10.2144/000113820

10. Liu DR, Schultz PG (1999) Progress toward the evolution of an organism with an expanded genetic code. Proc Natl Acad Sci U S A 96(9):4780–4785

11. Pastrnak M, Magliery TJ, Schultz PG (2000) A new orthogonal suppressor tRNA/aminoacyl-tRNA synthetase pair for evolving an organism with an expanded genetic code. Helv Chim Acta 83(9):2277–2286. https://doi.org/10.1002/1522-2675(20000906)83:9<2277::Aid-Hlca2277>3.0.Co;2-L

12. Wang L, Brock A, Herberich B, Schultz PG (2001) Expanding the genetic code of Escherichia Coli. Science 292(5516):498–500

13. Umehara T, Kim J, Lee S, Guo LT, Soll D, Park HS (2012) N-acetyl lysyl-tRNA synthetases evolved by a CcdB-based selection possess N-acetyl lysine specificity in vitro and in vivo. FEBS Lett 586(6):729–733. https://doi.org/10.1016/J.Febslet.2012.01.029

14. Plumbridge J, Soll D (1987) The effect of dam methylation on the expression of Glns in Escherichia-Coli. Biochimie 69(5):539–541. https://doi.org/10.1016/0300-9084(87)90091-5

15. Plass T, Milles S, Koehler C, Schultz C, Lemke EA (2011) Genetically encoded copper-free click chemistry. Angew Chem Int Ed Engl 50(17):3878. https://doi.org/10.1002/anie.201008178

16. Dose A, Liokatis S, Theillet FX, Selenko P, Schwarzer D (2011) NMR profiling of histone deacetylase and acetyl-transferase activities in real time. ACS Chem Biol 6(5):419–424. https://doi.org/10.1021/cb1003866

17. Schmidt MJ, Borbas J, Drescher M, Summerer D (2014) A genetically encoded spin label for electron paramagnetic resonance distance measurements. J Am Chem Soc 136(4):1238–1241. https://doi.org/10.1021/ja411535q

18. Schmidt MJ, Summerer D (2013) Red-light-controlled protein-RNA crosslinking with a genetically encoded furan. Angew Chem Int Ed Engl 52(17):4690–4693. https://doi.org/10.1002/anie.201300754

19. Siegele DA, Hu JC (1997) Gene expression from plasmids containing the araBAD promoter at subsaturating inducer concentrations represents mixed populations. Proc Natl Acad Sci U S A 94(15):8168–8172

20. Yanagisawa T, Ishii R, Fukunaga R, Kobayashi T, Sakamoto K, Yokoyama S (2008) Crystallographic studies on multiple conformational states of active-site loops in pyrrolysyl-tRNA synthetase. J Mol Biol 378(3):634–652

21. Schneider S, Gattner MJ, Vrabel M, Flugel V, Lopez-Carrillo V, Prill S, Carell T (2013) Structural insights into incorporation of Norbornene amino acids for click modification of proteins. Chembiochem 14:2114. https://doi.org/10.1002/cbic.201300435

Chapter 6

Genetic Code Expansion in Enteric Bacterial Pathogens

Huangtao Zheng, Shixian Lin, and Peng R. Chen

Abstract

The genetic code expansion strategy has become an elegant method for site-specific incorporation of non-canonical amino acids with diverse functionalities into proteins of interest in bacteria, yeast, mammalian cells, and even animals. This technique allows precise labeling as well as manipulation of a given protein to dissect its physiological and/or pathological roles under living conditions. Here, we demonstrate the extension of a recently emerged pyrrolysine-based genetic code expansion strategy for encoding noncanonical amino acids into enteric bacterial pathogens.

Key words Genetic code expansion, Noncanonical amino acids, Site-specific incorporation, Enteric bacterial pathogen, Pyrrolysine

1 Introduction

Enteric bacterial pathogens including virulent *Escherichia coli* (*E. coli*) species (e.g., *enteropathogenic E. coli* (EPEC), *enterohemorrhagic E. coli* (EHEC), etc.), *Shigella* species, and *Salmonella* species are continuously possessing as a formidable threat to human health [1, 2]. Unlike laboratory *E. coli* strains, these pathogens have evolved highly efficient and sophisticated defense systems to escape from host immune responses, as well as diverse bacterial toxins to establish infections within hosts [3–5]. However, the molecular mechanisms of these systems remain elusive, largely due to the lack of tools to directly monitor and manipulate essential protein components during bacterial pathogenesis.

The genetic code expansion strategy has become a powerful strategy for site-specific incorporation of noncanonical amino acids (ncAAs) carrying diverse functionalities into proteins of interest (POI) in both prokaryotic and eukaryotic cells [6–8]. In particular, the pyrrolysine (Pyl)-based genetic code expansion system has become a one-stop-shop for ncAA incorporation in bacteria, yeast, mammalian cells, and even animals [9–12]. However, although this system has been successfully applied to laboratory *E. coli*

Edward A. Lemke (ed.), *Noncanonical Amino Acids: Methods and Protocols*, Methods in Molecular Biology, vol. 1728,
https://doi.org/10.1007/978-1-4939-7574-7_6, © Springer Science+Business Media, LLC 2018

strains, there are considerable genetic differences among these enteric bacterial pathogens that render the Pyl system incompatible [13–15]. Here, we describe a general strategy for successful expression of the pyrrolysinyl-tRNA synthetase (PylRS) and Pyl-tRNA ($tRNA_{CUA}^{Pyl}$) pair in these enteric pathogens for site-specifically ncAA incorporation. In this protocol, we provide a step-by-step, straightforward procedure to expand the genetic code of EPEC through an engineered expressing system for PylRS-$tRNA_{CUA}^{Pyl}$ pair. Two representative Pyl analogs DiZPK (3-(3-methyl-3H-diazirine-3-yl)-propamino-carbonyl-N^ε-L-Lysine) and ACPK ((((1R, 2R)-2-azidocyclopentyloxy)-carbonyl)-N^ε-L-lysine) are used as the model ncAAs [16]. We also briefly discuss the key points for genetic code expansion in *Shigella* and *Salmonella species*. More generally, our protocol serves as a template for genetic code expansion in diverse bacterial species.

2 Materials

2.1 Instruments

1. 1300 Series Class II, Type A2 Biological Safety Cabinet Package (Thermo Scientific™).

2. C1000™ Thermal Cycler (Bio-Rad).

3. NanoDrop™ 2000 Spectrophotometer (Thermo Scientific™).

4. Gene Pulser Xcell™ Electroporation System (Bio-Rad).

5. ChemiDoc™ XRS⁺ System (Bio-Rad).

6. AKTA™ Purifier UPC 100 (FPLC) (GE Healthcare).

7. QSTAR XL mass spectrometer (AB Sciex, Foster City, CA, USA).

2.2 Bacterial Strains

1. EPEC 2348/69 (serotype O127:H6 belonging to *E. coli* phylogroup B2) abbreviated here as EPEC.

2. *Shigella flexneri 2a pcp301*, abbreviated here as *Shigella*.

3. *Salmonella typhimurium*, abbreviated here as *Salmonella*.

2.3 General Materials

All the reagents are dissolved in Milli-Q water unless otherwise specified.

1. ncAAs: DiZPK and ACPK are synthesized as described previously [17, 18], and dissolved as 100 mM stock solution.

2. 10% V/V glycerol solution is autoclaved, and stored at 4 °C.

3. 1000× antibiotic stock solutions:

    (a) 100 mg/mL Ampicillin (Amp).

    (b) 50 mg/mL Streptomycin.

    (c) 34 mg/mL Chloramphenicol (Cm) in 100% ethanol solution.

4. 20% m/V L-arabinose stock.

5. Phosphate-Buffered Saline (PBS): 150 mM NaCl, 10 mM phosphate, pH 7.4.

6. Anti-His antibody (Santa Cruz Biotechnology).

7. Ni-nitrilotriacetic acid (NTA) agarose column (GE Healthcare).

8. HiTrap™ Desalting column (GE Healthcare).

9. Amicon Ultra-15 centrigugal filter units (Millipore).

2.4 Growth Medium Preparation

1. LB medium: add 2 g Bacto Tryptone, 1 g Bacto Yeast Extract, 2 g NaCl to a 250 mL bluebottle, mix and make up to 200 mL with water.

2. 2×YT medium: add 5 g Bacto Tryptone, 2.5 g Bacto Yeast Extract, 20 mL glycerol to a 250 mL bluebottle, mix and make up to 200 mL with water.

3. CYT medium: add 0.5 g Bacto Tryptone, 0.25 g Bacto Yeast Extract, 20 mL glycerol to a 250 mL bluebottle, mix and make up to 200 mL with water.

4. The medium is autoclaved (*see* **Note 1**) and stored at 4 °C.

2.5 Agar Plate Preparation

1. Add 2 g Bacto Tryptone, 1 g Bacto Yeast Extract, 2 g NaCl, and 3 g agar powder to a 200 mL conical flask, mix and make up to 200 mL with water.

2. Autoclave agar medium.

3. After sterilization, cool down agar media to 55 °C at water bath, add the proper antibiotics, and gently mix agar media.

4. Dispense 20 mL agar media into each 90 mm plate and wait until solidification (*see* **Note 2**).

5. Store the LB agar plates at 4 °C and protect from light.

2.6 PCR-based Site-Directed Mutagenesis

1. Phusion® High-Fidelity DNA polymerase (Thermo Scientific).

2. dNTP (10 mM each) (Thermo Scientific).

3. Restriction endonuclease *Dpn*I (Thermo Scientific).

4. PCR primers are synthesized by Invitrogen.

3 Methods

Caution: all enteric bacterial pathogen-related procedures should be conducted in biosafety level 2 labs. Laboratory personnel should complete biosafety training before handling any opportunistic bacterial pathogen (*see* **Note 3**).

3.1 Choose Appropriate Components (Promoters and Replication Origin)

1. Choose plasmids for genetic code expansion of enteric bacterial pathogen. Consider the compatibility of chosen plasmid backbone for target bacterial strain, such as replication origin, antibiotics resistance marker, species-specific expression condition, and so on (*see* **Note 4**). The pSupAR plasmid developed by Schultz lab is widely used to express the PylRS- tRNA$^{Pyl}_{CUA}$ pair in laboratory *E. coli* strains [17–19]. The overexpressed wild-type PylRS- tRNA$^{Pyl}_{CUA}$ pair were shown to be directly used by the translation machinery of laboratory *E. coli*. Therefore, the pSupAR plasmid is used as the starting template for further plasmid construction for all the three enteric bacterial pathogens: EPEC, *Shigella*, and *Salmonella* (*see* **Note 5**).

2. Choose appropriate plasmid replication origin for enteric bacterial pathogen. The p15A origin is chosen because it is suitable for plasmid replication in all the three enteric bacterial pathogens and it shows good compatibility to plasmids with pBR322, ColE1 origin, which are commonly used for expression of POI (*see* **Note 6**).

3. Choose appropriate PylRS gene transcription promoter. In this protocol, PylRS gene is under the control of the L-arabinose inducible promoter (Fig. 1a), which has been routinely used to express proteins in all the three enteric bacterial pathogens [20, 21] (*see* **Note 7**).

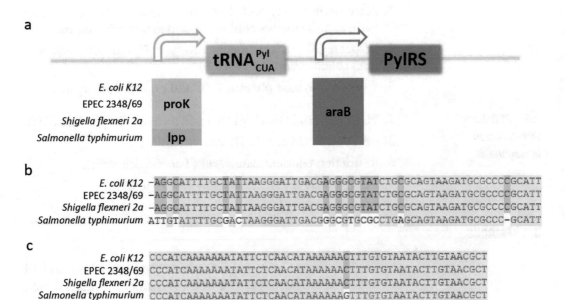

Fig. 1 (**a**) Suitable promoter pairs for PylRS- tRNA$^{Pyl}_{CUA}$ pair expression in *E.coli* K12, EPEC 2348/69, *Shigella flexneri* 2a, and *Salmonella typhimurium*. Alignment of proK tRNA promoters (**b**) and lpp tRNA promoters (**c**) from these bacterial strains. The alignments were performed on the software Vector NTI

4. Choose appropriate $tRNA_{CUA}^{Pyl}$ transcription promoter and terminator. The $tRNA_{CUA}^{Pyl}$ gene is under the control of tRNA transcription promoter and terminator. The tRNA promoter turns out to be the key component of genetic code expansion of enteric bacterial pathogens. Inappropriate tRNA promoter will end up with poor expression level of $tRNA_{CUA}^{Pyl}$, resulting in unsuccessful genetic code expansion. The proK tRNA promoter and lpp tRNA promoter are chosen in this protocol due to their high usage rates in laboratory *E. coli* and high transcription expression level of corresponding tRNA gene (Fig. 1a) (*see* **Note 8**). The tRNA terminator is also an essential feature controlling tRNA transcription termination. Ribosomal RNA transcription terminator rrnC is chosen for all enteric bacterial pathogens in this protocol (*see* **Note 9**).

5. To search for the sequences of prok tRNA promoter of all enteric bacterial pathogens, go to National Center for Biotechnology Information (NCBI) website and click the gene page (linker: http://www.ncbi.nlm.nih.gov/gene) and search proK tRNA of EPEC, *Shigella* and *Salmonella*, respectively. The promoter sequence is about 80 nucleotides upstream of its tRNA gene (*see* **Note 10**).

6. Copy all these promoter sequences into Vector NTI software (Thermal Scientific) or any software that is suitable for DNA sequence alignment. Perform promoter sequence alignment according to the software's instruction. The result is shown in Fig. 1b. The proK tRNA promoter is conserved in laboratory *E. coli*, EPEC, and *Shigella*. Laboratory *E. coli* prok-tRNA promoter is thus used for EPEC and *Shigella*. While a more appropriate tRNA promoter is needed for *salmonella*, as its prok-tRNA is not conserved with that of prok promoter in laboratory *E. coli* (*see* **Note 11**).

7. Repeat **steps 5** and **6** for lpp tRNA transcription promoter alignment. The result is illustrated in Fig. 1c. The *Salmonella* lpp tRNA promoter is highly conserved to that of laboratory *E. coli* K12. Therefore, it will be easier to construct pSupAR-*Salmonella* vector though mutation of pSupAR-*E. coli*-lpp vector (*see* **Note 12**).

3.2 Plasmid Construction

1. After consideration of all the key components (promoter, terminator, and replication origin), construct plasmid carrying all of these components through the library of pSupAR series of plasmids. For *EPEC and Shigella*, directly pick out the plasmid from pSupAR library, label the pSupAR plasmids carrying the appropriate components as pSupAR-EPEC and pSupAR-*Shigella*, and use these two plasmids to expand the genetic code of EPEC and *Shigella*, respectively.

2. For *Salmonella*, generate pSupAR-*Salmonella* plasmid through pSupAR-*E. coli*-lpp plasmid by mutating *E. coli*-lpp promoter to *Salmonella* lpp promoter using PCR-based site-directed mutagenesis. Then apply this plasmid to expand the genetic code of *Salmonella* (*see* **Note 13**).

3. Design a pair of primers for site-directed mutagenesis according to the standard primer design protocol (*see* **Note 14**).

4. Dissolve mutagenesis primers with a final concentration of 10 μM as stocks, store the primers at −20 °C.

5. Pipet mutagenesis PCR reaction with the following condition (for 25 μL PCR system): 5 μL Phusion HF Buffer, 0.5 μL dNTPs (10 mM stock), 1 μL Forward primer, 1 μL Reverse primer, 25 ng Template DNA, 0.25 μL Phusion DNA Polymerase (2 U/μL stock) and make water up to 25 μL. Also pipet negative control reaction without adding DNA polymerase.

6. Perform mutagenesis PCR reaction with the following condition: Initial Denaturation 98 °C for 30 s, Main process 25 cycles: Denaturation 98 °C for 10 s, Annealing 55–72 °C for 30 s, Extension 72 °C for 4 min, Final extension 72 °C for 5 min and 4 °C forever (*see* **Note 15**).

7. Load 5 μL of the 25 μL PCR reaction product to a DNA agarose gel to check the quality of the mutagenesis PCR product.

8. Add 2 μL 10× Tango buffer and 0.5 μL *Dpn*I endonuclease (a restriction enzyme digests methylated template DNA only) to the remaining 20 μL PCR product. Digest overnight at 37 °C to remove DNA template. The digested product can be directly transformed or stored at −20 °C.

9. Transform to laboratory *E. coli* (DH10B) with the following protocol: Add 5 μL *Dpn*I digested product to 100 μL competent cell suspension, keep on ice for 30 min. Incubate the reaction in 42 °C water bath for 60–90 s, directly ice-chill the tube, and keep on ice for 1–2 min.

10. Add 900 μL LB medium, incubate at 37 °C for 45 min. Centrifuge the tube at $1500 \times g$ for 2 min, carefully pour out the supernatant, and gently resuspend the pellet in the remaining ~100 μL medium.

11. Transfer all the suspension to a Cm selection LB agar plate, allow the liquid to be absorbed at room temperature, then invert the plate and incubate at 37 °C.

12. Pick up single colonies after ~12 h and sequence the plasmid to verify positive colonies.

13. Label the plasmid as pSupAR-*Salmonella* and apply this plasmid to expand the genetic code of *Salmonella*.

3.3 EPEC Electro-competent Cell Preparation

1. Pick up single EPEC colony from streptomycin selection LB agar plate (*see* **Note 16**) and grow the cell at 4 mL LB medium with streptomycin overnight at 37 °C, 220 rpm/min.

2. Dilute 1 mL of overnight grown EPEC medium to 100 mL of fresh medium in a conical flask with streptomycin at 37 °C, 220 rpm/min until $OD_{600} = 0.5$ (*see* **Note 17**).

3. Ice-chill EPEC culture for 0.5 h, precool the centrifuge to 4 °C.

4. Harvest the EPEC cells at $5000 \times g$ for 5 min (*see* **Note 18**).

5. Discard the supernatant and gently resuspend bacterial pellet in 100 mL cold pure water (*see* **Note 19**).

6. Spin down the bacteria for 5 min at $5000 \times g$ in a centrifuge precooled at 4 °C.

7. Discard the supernatant and gently resuspend bacterial pellet in a 50 mL cold 10% glycerol solution.

8. Spin down the bacteria for 5 min at $5000 \times g$ in a centrifuge precooled at 4 °C.

9. Discard the supernatant and gently resuspend the bacterial pellet in a 2 mL cold 10% glycerol solution.

10. Spin down the bacteria for 5 min at $5000 \times g$ in a centrifuge precooled at 4 °C.

11. Discard the supernatant and try to remove all the remaining solution with a pipette. Gently resuspend the bacterial pellet in 200 μL precooled GYT medium.

12. Measure the OD600 of cell suspension by 1:100 dilution. Dilute the cell suspension to $2 \times 10^{10} \sim 3 \times 10^{10}$ cell/mL (1.0 OD600 = ~2.5×10^{10} cell/mL).

13. Aliquot 100 μL to each prechilled Eppendorf tube, fast freeze the competent cell with liquid nitrogen, and store at −80 °C. The electro-competent cell can be stored as long as 1 year (*see* **Note 20**).

14. Remember to autoclave all EPEC contaminated items and waste after experiments.

3.4 Co-transformation of Two Plasmids

1. Thaw 100 μL frozen competent EPEC cell, plasmid DNA pSupAR-EPEC (carrying chloramphenicol resistance gene) and pBAD-GFP-149TAG (carrying ampicillin resistance gene) (*see* **Note 21**) on ice. Place 15 mL conical tube containing 10 mL of 2xYT media without antibiotics and 0.2 cm cuvette on ice.

2. Pipet 100 ng pSupAR-EPEC vector and 100 ng pBAD-GFP-149TAG vector into a new Eppendorf tube, add 50 μL thawed electro-competent EPEC cells onto DNA mix, keep the DNA-EPEC mixture on ice for 5 min.

3. Transfer the DNA-EPEC mixture along the wall of 0.2 cm cuvette into the bottom, flick cuvette to settle all DNA-EPEC mixture into the bottom of cuvette (*see* **Note 22**).

4. Turn on electroporator power source and locate the cuvette holder. Set the conditions for transformation according to strain. For EPEC, use 25 mFD, 200 W, and 2.5 kV (time constant = 4.6–4.8 ms).

5. Dry off any moisture from cuvette outside and immediately place cuvette in a white plastic holder. Slide holder into position and zap cells.

6. Immediately transfer 1 mL 2×YT medium to cells after hearing a high constant tone. Pipet the cells from cuvette into a 1.5 mL prechilled Eppendorf tube and keep on ice for 5 min.

7. Recover transformed cells in Eppendorf tubes by incubating the tubes in 37 °C water bath for 1–1.5 h. Centrifuge the tube at $1500 \times g$ for 2 min, carefully pour out the supernatant, and gently resuspend the pellet in the remaining ~100 μL medium.

8. Transfer all the suspension to Amp/Cm selection LB agar plate, allow the liquid to be absorbed at room temperature, then invert the plate and incubate at 37 °C (*see* **Note 23**).

3.5 ncAA Incorporation into Proteins and Immunoblotting Analysis

1. Pick up a co-transformed single EPEC colony and culture overnight in a fresh LB medium containing Amp and Cm at 37 °C, 220 rpm/min.

2. In parallel, dilute the overnight EPEC culture in three tubes of fresh 4 mL LB medium at a ratio of 1:100, then grow the diluted EPEC cultures at 37 °C, 220 rpm/min for about 2.5 h until the OD_{600} reach around 0.5.

3. Add DiZPK stock (100 mM) to one tube to a final concentration of 1 mM, add ACPK stock (100 mM) to another tube to a final concentration of 1 mM, and leave the third tube untreated as a negative control. Label the tubes and incubate the cell cultures at 37 °C for another 30 min, 220 rpm/min (*see* **Note 24**).

4. Add L-arabinose stock (20% m/V) to a final concentration of 0.2% to induce the expression of both GFP-149TAG and PylRS.

5. Culture the induced EPEC cell at 30 °C for another 8 h.

6. Harvest 1 mL EPEC cell in a 1.5 mL Eppendorf tube by centrifugation at $13{,}000 \times g$ for 1 min. Wash the EPEC cells three times with PBS buffer by consecutive resuspension and centrifugation at $13{,}000 \times g$ for 1 min.

7. Resuspend the EPEC pellet with 100 μL PBS buffer, then add 25 μL 5× protein sample buffer. Lyse EPEC cell at 95 °C for at least 20 min with metal bath (*see* **Note 25**).

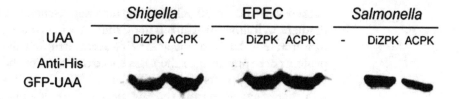

Fig. 2 Incorporation of DiZPK and ACPK into the model protein GFP-149TAG in EPEC, *Shigella* and *Salmonella*. Full-length protein is detected only after the addition of DiZPK or ACPK, indicating the successful genetic code expansion in these pathogens. Reprinted with permission from ref. 16. Copyright (2016) American Chemical Society

8. Centrifuge the samples at $13,000 \times g$ for 10 min and carefully transfer the supernatant to a new Eppendorf tube for further analysis.

9. Load 5 μL of each sample to 12% sodium dodecyl sulfate polyacrylamide gel electrophoresis (SDS-PAGE) and run the gel to separate the protein.

10. Transfer the protein to polyvinylidene difluoride (PVDF) membrane.

11. Immunoblot with 6×His antibody carrying horseradish peroxidase (HRP) fusion tag.

12. Detect the expression of full-length DiZPK or ACPK bearing GFP on ChemiDoc™ XRS⁺ System. The result is shown in Fig. 2.

3.6 Protein Purification and Mass Spectrometry Analysis

1. Expression of GFP-149DiZPK or GFP-149ACPK in 1 L EPEC cells are conducted according to the method in Subheading 3.4.

2. Lyse EPEC cell by sonication in buffer A (20 mM sodium phosphate, 0.5 M NaCl, 20 mM imidazole, and 8 M urea, pH 7.4).

3. Purify 6×His tag GFP protein with a Ni-nitrilotriacetic acid (NTA) agarose column (GE Healthcare) following the manufactured protocol.

4. Concentrate the eluted protein with Amicon Ultra-15 centrigual filter units (Millipore) with a molecular weight cutoff of 3kD to a final volume of 500–1000 μL.

5. Perform buffer exchange to the PBS buffer by using the HiTrap™ Desalting column (GE Healthcare).

6. Concentrate the eluted protein again to a final volume of 500–1000 μL.

7. Analyze GFP-149DiZPK and GFP-149ACPK with LC/MS with the following condition: Load protein samples (about 1 mg/mL) into an analytical capillary column (75 μm, 5 cm)

packed with Poros 20 R1 packing material. Generate HPLC gradient as follows: 0–100% B in 60 min (A = 0.1 M acetic acid in water; B = 0.1 M acetic acid/70% acetonitrile). The eluted proteins are sprayed into the mass spectrometer. Set the spray voltage at 2100 volts and acquire data in MS mode. The protein charge envelops from the raw spectrum are de-convoluted into non-charged form by the BioAnalyst software provided by the manufacturer.

4   Notes

1. Unless otherwise noted, sterilize all the buffers and media by autoclaving at 121 °C, 15 psi for 20 min for volumes less than 1 L; for larger batches, adjust autoclave time accordingly to ensure sterility.

2. Unless otherwise noted, all bacteria-related procedures are performed in laminar flow hood with high efficiency bacteria-retentive filter to protect from contamination.

3. Enteric bacterial pathogen can cause a variety of diseases including septicemia and diarrhea. All laboratory personnel should be trained in compliance with biosafety level 2 requirements. Every laboratory member should be notified before handing pathogen. In all the procedures, extreme precautions should be taken with pathogen contaminated items, including growth medium, tips, flasks, agar, groves, etc. All contaminated items should be autoclaved after experiments as soon as possible. Extra precautions should be taken with contaminated sharp items.

4. Choose appropriate plasmid backbone according to the target strain. Replication origin, antibiotics resistance marker, and species-specific expression condition are usually the most important thing to think out. We will discuss how to choose replication origin and species-specific expression in more detail in the following protocol.

5. The Schultz lab (TSRI) have previously constructed a series of pSupAR plasmids for the expression of ncAA containing proteins in laboratory *E. coli* strains [6, 19, 22]. These plasmids carry combination choices of replication origin (p15A, pBR322, and ColE1), PylRS expression promoter (arabinose inducible and IPTG inducible), and tRNA promoter (*E. coli*-proK and *E. coli*-lpp). Take EPEC for example, after choosing the appropriate origin, PylRS- tRNA$_{CUA}^{Pyl}$ pair promoter, the sequence closest pSupAR plasmid is used to construct new plasmid pSupAR-EPEC for expansion of the genetic code of EPEC.

6. Replication origin used for expression PylRS- tRNA$^{Pyl}_{CUA}$ is crucial. There are at least two plasmids needed for genetic code expansion, one for expression of PylRS- tRNA$^{Pyl}_{CUA}$ pair, and the other for expressing the POI with an in-frame amber codon. Thus, the plasmids compatibility needs to be considered carefully before choosing appropriate replication origin for PylRS- tRNA$^{Pyl}_{CUA}$ pair. Generally speaking, co-transformed plasmids with the same replication origin are incompatible as they will compete with each other for the replication machinery, resulting in loss of plasmids. Besides, co-transformed plasmids with different replication origins may not be compatible to each other if they are in the same incompatibility group. For example, the ColE1 origin is not compatible with pBR322 origin, because both of them are in the same incompatibility group A. (For more details about bacterial replication origin and compatibility, visit Addgene website: http://blog.add-gene.org/plasmid-101-origin-of-replication). When working on new target strain, always check available replication origins in the literature, and their compatibility for co-transformation.

7. When extending this strategy to other bacterial strains, make sure that the promoters of both PylRS and POI are compatible with your target strains. And optimize the promoter pair to obtain better ncAA incorporation efficiency if necessary. According to our experience, the arabinose inducible promoter works well for both PylRS and the POI. While stronger promoters like T7lac did not show better amber suppression efficiency in our case.

8. The tRNA promoters function as recruitment of RNA polymerase as well as a bunch of transcriptional factors to assure successful transcription initiation, thus is essential for tRNA transcription.

9. The tRNA transcription terminates after a stretch of several thymidines and the tRNAs are always carrying CCA 3' tail. Many enteric bacterial pathogens share similar mechanism of tRNA termination. Terminator rrnC works well in these strains in ours' study. However, more consideration needs to be taken on tRNA termination on other pathogen species such as *Pseudomonas aeruginosa*. To ensure correct transcription of tRNA$^{Pyl}_{CUA}$, northern blot could be carried out.

10. We have already known the sequence of proK tRNA promoter in *E. coli*, so we directly copied the corresponding sequence in these pathogens to align with *E. coli*-prok-promoter. However, if working on a completely unknown tRNA promoter, we suggest copying at least 100 nucleotides upstream of its tRNA gene for plasmid cloning.

11. We actually could not detect the expression of full-length ncAA incorporated protein in *Salmonella* using laboratory *E. coli* proK promoter, indicating the importance of tRNA promoter in controlling tRNA transcription in enteric bacterial pathogens.

12. We assume that both *salmonella*-prok-promoter and *salmonella*-lpp-promoter should work for transcription of $tRNA^{Pyl}_{CUA}$ gene. We decided to use lpp promoter because it is easy to make through our pSupAR template plasmids.

13. If the tRNA promoters of your target bacterial strain are quite different from any available tRNA promoter of laboratory *E. coli* strain, we suggest you to employ the prok-promoter or lpp-promoter of your target strain. Polymerase Incomplete Primer Extension (PIPE) or Gibson assembly could be used to construct genetic code expansion plasmids, since there is no restriction enzyme cutting site between the tRNA gene and its promoter.

14. The standard parameters for mutation primers are shown as follows: Melting temp (°C): 75–85, GC content (%): 40–60, Length (bp): 25–45, 5′ flanking region (bp): 11–21 and 3′ flanking region (bp): 11–21.

15. The annealing temperature of the PCR reaction is set according to primers' melting point. The extension time of the PCR reaction is set according to the vector size. For pSupAR series vector, the size is around 7 kbs. We use 4 min as a standard condition considering the rate of Phusion DNA Polymerase is 10–30 s/kb.

16. The EPEC strain in our lab has previously been selected with streptomycin, while laboratory *E. coli* cannot survive on the streptomycin medium. So Streptomycin is used to keep EPEC strain from contamination by other laboratory *E. coli* strains. EPEC strain evolves a streptomycin biosynthesis pathway and can be selected by streptomycin treatment, which is different from laboratory *E. coli* bacteria.

17. It usually takes 2.5–3.0 h to reach to desired OD_{600} value. Overgrown of EPEC strain will significantly affect the co-transformation efficiency.

18. For EPEC centrifugation, use new centrifugation tubes to avoid contamination from laboratory *E. coli* strains.

19. Perform all the procedures on ice. Over-warming of EPEC competent cells will significantly affect the co-transformation efficiency.

20. Always choose electro-transformation to co-transform plasmid into enteric bacterial pathogen due to the higher transformation efficiency than heat-shock transformation.

21. The plasmid pBAD-GFP-149TAG expresses the full-length GFP carrying a C-terminal 6xHis tag in the presence of aaRS-tRNA pair and ncAA. The immunoblotting analysis is performed with an anti-His antibody to detect the amber suppression efficiency in enteric bacterial pathogens. Full-length protein is only expressed in the presence of aaRS-tRNA pair and ncAA.

22. Avoid generating air bubbles in this process. Air bubble can be removed by flicking cuvette.

23. The EPEC strain after co-transformation grows slower than the WT strain. It takes at least 16 h incubation to be able to pick up positive colonies.

24. Protect the growth medium and EPEC cell from light after adding DiZPK, which carries photo sensitive functional group for protein photocrosslinking.

25. EPEC cell is much stickier than laboratory *E. coli*. Longer heat lysis time is usually needed.

Acknowledgment

This work was supported by the National Key Basic Research Foundation of China (2012CB917301), National Natural Science Foundation of China (21225206, 21521003 and 21432002). We thank the support from Peking University Principal Foundation.

References

1. Bielaszewska M, Mellmann A, Zhang W, Kock R, Fruth A, Bauwens A, Peters G, Karch H, Caprioli A (2011) Characterisation of the *Escherichia coli* strain associated with an outbreak of haemolytic uraemic syndrome in Germany, 2011: a microbiological study. Lancet Infect Dis 11(9):671–676

2. Scheutz F, Møller Nielsen E, Frimodt-Møller J, Boisen N, Morabito S, Tozzoli R, Nataro JP, Caprioli A (2011) Characteristics of the entero-aggregative Shiga toxin/ verotoxin-producing *Escherichia coli* O104:H4 strain causing the outbreak of haemolytic uraemic syndrome in Germany, May to June 2011. Euro Surveill 16(24):19889. http://www.eurosurveillance.org/ViewArticle.aspx?ArticleId=19889

3. Arbibe L, Kim DW, Batsche E, Pedron T, Mateescu B, Muchardt C, Parsot C, Sansonetti PJ (2007) An injected bacterial effector targets chromatin access for transcription factor NF-kappaB to alter transcription of host genes involved in immune responses. Nat Immunol 8(1):47–56

4. Quezada CM, Hicks SW, Galan JE, Stebbins CE (2009) A family of salmonella virulence factors functions as a distinct class of autoregulated E3 ubiquitin ligases. Proc Natl Acad Sci U S A 106(12):4864–4869

5. Pearson JS, Giogha C, Ong SY, Kennedy CL, Kelly M, Robinson KS, Lung TW, Mansell A, Riedmaier P, Oates CV, Zaid A, Muhlen S, Crepin VF, Marches O, Ang CS, Williamson NA, O'Reilly LA, Bankovacki A, Nachbur U, Infusini G, Webb AI, Silke J, Strasser A, Frankel G, Hartland EL (2013) A type III effector antagonizes death receptor signalling during bacterial gut infection. Nature 501(7466):247–251

6. Chin JW, Martin AB, King DS, Wang L, Schultz PG (2002) Addition of a photocrosslinking amino acid to the genetic code of *Escherichia coli*. Proc Natl Acad Sci U S A 99(17):11020–11024

7. Chin JW, Cropp TA, Anderson JC, Mukherji M, Zhang Z, Schultz PG (2003) An expanded eukaryotic genetic code. Science 301(5635): 964–967

8. Wang L, Xie J, Schultz PG (2006) Expanding the genetic code. Annu Rev Biophys Biomol Struct 35:225–249

9. Liu CC, Schultz PG (2010) Adding new chemistries to the genetic code. Annu Rev Biochem 79:413–444

10. Hao Z, Hong S, Chen X, Chen PR (2011) Introducing bioorthogonal functionalities into proteins in living cells. Acc Chem Res 44(9): 742–751

11. Greiss S, Chin JW (2011) Expanding the genetic code of an animal. J Am Chem Soc 133(36):14196–14199

12. Bianco A, Townsley FM, Greiss S, Lang K, Chin JW (2012) Expanding the genetic code of Drosophila melanogaster. Nat Chem Biol 8(9):748–750

13. Johnson JR (2000) Shigella and Escherichia coli at the crossroads: machiavellian masqueraders or taxonomic treachery? J Med Microbiol 49(7):583–585

14. Foster JW (2004) Escherichia coli acid resistance: tales of an amateur acidophile. Nat Rev Microbiol 2(11):898–907

15. Zurawski DV, Mumy KL, Faherty CS, McCormick BA, Maurelli AT (2009) Shigella flexneri type III secretion system effectors OspB and OspF target the nucleus to downregulate the host inflammatory response via interactions with retinoblastoma protein. Mol Microbiol 71(2):350–368

16. Lin S, Zhang Z, Xu H, Li L, Chen S, Li J, Hao Z, Chen PR (2011) Site-specific incorporation of photo-cross-linker and bioorthogonal amino acids into enteric bacterial pathogens. J Am Chem Soc 133(50):20581–20587

17. Zhang M, Lin S, Song X, Liu J, Fu Y, Ge X, Fu X, Chang Z, Chen PR (2011) A genetically incorporated crosslinker reveals chaperone cooperation in acid resistance. Nat Chem Biol 7(10):671–677

18. Hao Z, Song Y, Lin S, Yang M, Liang Y, Wang J, Chen PR (2011) A readily synthesized cyclic pyrrolysine analogue for site-specific protein "click" labeling. Chem Commun (Camb) 47(15):4502–4504

19. Chen PR, Groff D, Guo J, Ou W, Cellitti S, Geierstanger BH, Schultz PG (2009) A facile system for encoding unnatural amino acids in mammalian cells. Angew Chem Int Ed Engl 48(22):4052–4055

20. Ahmer BMM, van Reeuwijk J, Timmers CD, Valentine PJ, Heffron F (1998) Salmonella typhimurium encodes an SdiA homolog, a putative quorum sensor of the LuxR family, that regulates genes on the virulence plasmid. J Bacteriol 180(5):1185–1193

21. Newman JR, Fuqua C (1999) Broad-host-range expression vectors that carry the L-arabinose-inducible Escherichia coli araBAD promoter and the araC regulator. Gene 227(2):197–203

22. Ryu Y, Schultz PG (2006) Efficient incorporation of unnatural amino acids into proteins in Escherichia coli. Nat Methods 3(4):263–265

Chapter 7

Self-Directed in Cell Production of Methionine Analogue Azidohomoalanine by Synthetic Metabolism and Its Incorporation into Model Proteins

Ying Ma, Martino L. Di Salvo, and Nediljko Budisa

Abstract

Common protocols for the incorporation of noncanonical amino acids (ncAAs) into proteins require addition of the desired ncAA to the growth medium, its cellular uptake, and subsequent intracellular accumulation. This feeding scheme is generally suitable for small-scale proof-of-concept incorporation experiments. However, it is no general solution for orthogonal translation of ncAAs, as their chemical synthesis is generally tedious and expensive. Here, we describe a simple protocol that efficiently couples in situ semi-synthetic biosynthesis of L-azidohomoalanine and its incorporation into proteins at L-methionine (Met) positions. In our metabolically engineered Met-auxotrophic *Escherichia coli* strain, Aha is biosynthesized from externally added sodium azide and O-acetyl-L-homoserine as inexpensive precursors. This represents an efficient platform for expression of azide-containing proteins suitable for site-selective bioorthogonal strategies aimed at noninvasive protein modifications (Tornøe et al., J Org Chem 67:3057–3064, 2002; Kiick et al., Angew Chem Int Ed 39:2148–2152, 2000; Budisa, Angew Chem Int Ed Engl 47:6426–6463, 2004; van Hest, J Am Chem Soc 122:1282–1288, 2000).

Key words Azide, Azidohomoalanine, Methionine, Metabolic engineering, Orthogonal translation, Direct sulfhydrylation pathway

1 Introduction

In this chapter, a self-directed L-azidohomoalanine (Aha) biosynthesis system is described. The basic feature of this system is that the engineered host's methionine biosynthetic pathway is efficiently diverted toward the production of the desired ncAA by exploiting the broad reaction specificity of recombinant pyridoxal-5′-phosphate-dependent O-acetyl-L-homoserine sulfhydrylase from *C. glutamicum* (cgOAHSS; encoded by *metY*, GenBank: FJ483537.1) [1–3]. In the expression experiments herein described, an *E. coli* host strain equipped with genes for direct sulfhydrylation is used for the synthesis of Aha, a Met-analogue that will be incorporated in place of Met. The intracellular biosynthesis

Edward A. Lemke (ed.), *Noncanonical Amino Acids: Methods and Protocols*, Methods in Molecular Biology, vol. 1728, https://doi.org/10.1007/978-1-4939-7574-7_7, © Springer Science+Business Media, LLC 2018

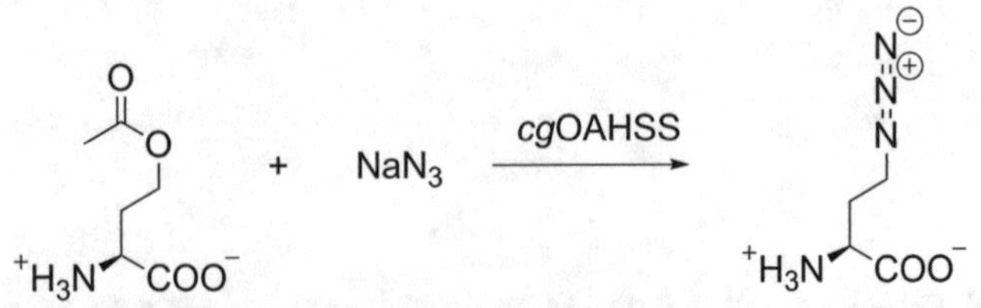

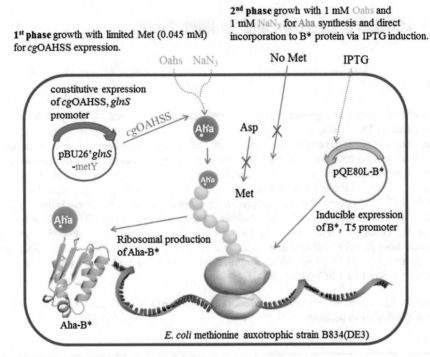

Fig. 1 Reaction scheme of L-azidohomoalanine (Aha) production by *C. glutamicum O*-acetyl-L-homoserine sulfhydrylase

Fig. 2 General scheme for intracellular L-azidohomoalanine (Aha) biosynthesis and incorporation into recombinant proteins

of Aha is carried out by external addition of sodium azide and *O*-acetyl-L-homoserine (Oahs) under the catalysis of the constitutively expressed *cg*OAHSS (Fig. 1). Aha is then charged onto tRNA[Met] and subsequently inserted into recombinant target protein barstar (B*) by reassignment of AUG codons (Fig. 2). The chosen target protein is a mutant pseudo-wild-type cysteine-free B* form (ψB*), which includes no other AUG codons than the N-terminal methionine position [4, 5]. This B* variant is an ideal model protein for bio-conjugation and secondary structure characterization studies. The system described in this chapter is of general use, and is suitable for expression of other modified target proteins. For instance, carried by the same vector, different target proteins such as green fluorescent protein (GFP) or *T. thermohydrosulfuricus* lipase can be expressed with this procedure.

2 Materials

2.1 Bacterial Strains

The bacterial host strain used in this system is the *ΔmetE* Met-auxotrophic *E. coli* B834(DE3). This is a commercial strain used for the production of IPTG-induced seleno-methionine recombinant proteins for X-ray crystallization trials [6]. The *metE* gene encodes for the essential enzyme methionine synthase (homocysteine methyltransferase). This deleted strain is therefore unable to produce endogenous Met from L-homocysteine.

2.2 Plasmids and Primers

See Tables 1 and 2.

2.3 Media and Reagent Solutions

Based on the *selective pressure incorporation* (SPI) method [6], the system (at flask-level expression procedure) requires new minimal media (NMM) with limited methionine amount. One liter medium-scale is enough for yielding ~5 mg of Aha-labeled B* protein. The concentration herein described for NMM represents, for each component, the final concentration.

Table 1
Plasmid list

Construct	Insert	Source	Plasmid backbone	Replication origin	Promoter	Resistant
pBU26'*glnS*-metY	*metY*	*Corynebacterium glutamicum*	pSEVA	p15A	*glnS'*	Kanamycin (50 mg/mL)
pQE80L-*psi*-barstar	Barstar	*Bacillus amyloliquefaciens*	pQE80L	ColEI	T5/LacO	Ampicillin (100 mg/mL)

Table 2
Primer list

Primer	Sequence[a]
metY-forward (*Nhe*I)	5'- CGCC<u>GCTAGC</u>CCAAAGTACGACAATTCCA-3'
metY-reverse (*Kas*I)	5'-AAAA<u>GGCGCCC</u>TAGATTGCAGCAAAGCCGCC-3'
T5-forward	5'-CCCGAAAAGTGCCACCTG-3'
T5-reverse	5'-GTTCTGAGGTCATTACTGG-3'

[a]Restriction sites are underlined

1. NMM: 7.5 mM $(NH_4)_2SO_4$, 50 mM K_2HPO_4, 22 mM KH_2PO_4, 8.5 mM NaCl, 1 mM $MgSO_4$, 20 mM D-glucose, 50 mg/L canonical amino acids (each) except Met, 1 μg/mL $FeCl_2$, 1 μg/mL $CaCl_2$, 10 μg/mL thiamine, 10 μg/mL biotin, 0.01 μg/mL trace elements ($CuSO_4$, $ZnCl_2$, $MnCl_2$ and $(NH_4)_2MoO_4$ mixture), all dissolved in deionized water (dH_2O), sterile (*see* **Note 1**).

2. Antibiotics: 100 μg/mL ampicillin and 50 μg/mL kanamycin (working concentration).

3. Met: 0.045 M L-methionine, dissolved in dH_2O as 1000× stock solution, sterile.

4. Oahs: For 1 L culture, measure 161 mg of Oahs, dissolve in 2 mL sterile dH_2O and add to the culture to reach the final concentration of 1 mM (*see* reference [1] for Oahs synthesis procedure).

5. Azide: 1 M NaN_3, dissolved in dH_2O as 1000× stock solution, sterile.

6. Induction: 1 M IPTG (isopropyl-β-D-1-thiogalactopyranoside), dissolved in dH_2O as 1000× stock, sterile.

7. LB medium: 0.5% yeast extract (w/v), 1% Tryptone (w/v), 1% NaCl (w/v), dissolved in dH_2O, sterile.

8. Lysozyme: 50 mg/mL in dH_2O as 1000× stock solution.

9. Protein purification: 5 mL HiTrap Q-Sepharose column for ion exchange chromatography.

10. Dialysis material: cellulose tubular membrane (nominal MWCO: 3500 Da).

2.4 Solutions

Solutions described here are used for recombinant target protein B* purification and SDS-PAGE analysis. Tris–HCl here refers to Tris(hydroxymethyl)aminomethane hydrochloride.

1. Lysis buffer (1 L): 50 mM Tris–HCl, pH 8.0, 100 mM NaCl, sterile.

2. Resuspension buffer (1 L): 50 mM Tris–HCl, pH 8, 7.5 M urea.

3. Dialysis buffer (5 L): 50 mM Tris–HCl, pH 8.0, 100 mM NaCl.

4. Milli-Q water (1 L), filter sterilized on a 0.22 μm membrane.

5. 20% Ethanol (v/v), diluted in Milli-Q water (1 L), filter sterilized on a 0.22 μm membrane.

6. 5× SDS loading buffer (1 mL): 80 mM Tris–HCl, pH 6.8, 4% β-mercaptoethanol (v/v), 0.2% bromophenol blue (w/v), 12.5% glycerol (v/v), 10% SDS (sodium dodecyl sulfate, w/v).

7. Wash buffer (1 L): 50 mM Tris–HCl, pH 8.0, 100 mM NaCl, filter sterilized on a 0.22 μm membrane.

8. Elution buffer (1 L): 50 mM Tris–HCl, pH 8.0, 1 M NaCl, filter sterilized on a 0.22 μm membrane (*see* **Note 2**).

9. Storage buffer (2 L): 50 mM Tris–HCl, pH 8.0, 100 mM NaCl, 20% glycerol (w/v).

---

3   Methods

3.1   Expression of Aha-Labeled B* in E. coli

3.1.1   Strain and Media Preparation

1. This step aims to introduce plasmid pBU26'*glnS*-metY together with pQE80L-*psi*-barstar into *E. coli* strain B834(DE3). To prepare the recombinant strain suitable for Aha production and incorporation, both plasmids need to be transformed into *E. coli* B834(DE3) chemically competent cells by heat shock method [7] and plated on LB agar with 100 μg/mL ampicillin and 50 μg/mL kanamycin (*see* **Note 3**).

2. From a transformation plate, pick up one single colony of recombinant strain and culture it in 5 mL LB media with 100 μg/mL ampicillin and 50 μg/mL kanamycin at 37 °C, 200 rpm, overnight.

3. Prepare 1 L of NMM containing 100 μg/mL ampicillin, 50 μg/mL kanamycin and 1 mL of Met 1000× stock solution, to the final Met concentration 0.045 mM.

4. Harvest the cells from overnight culture by centrifuging at 4 °C, 8000 × *g* for 10 min.

5. Carefully discard all the supernatant and gently rinse the cells with 2 mL of freshly prepared NMM medium, twice. Inoculate all the cells into 1 L of NMM.

3.1.2   Aha Incorporation

1. Incubate the bacterial culture at 37 °C, 200 rpm. For the first phase, grow until depletion of methionine. This step may take about 8 h, maximum overnight (*see* **Note 4**).

2. At this point, add 1 mL sodium azide 1000× stock solution and 161 mg of Oahs to the culture to reach the final concentration of 1 mM; continue shaking at 37 °C, 200 rpm for 1 h (*see* **Notes 5 and 6**).

3. Induce the target protein expression with 1 mL IPTG 1000× stock solution. Carry on the whole expression phase at 30 °C, 200 rpm, for 4 h (*see* **Note 7**).

4. In parallel, Met-incorporated B* protein should also be expressed and purified with the same recombinant strain and the same medium as "positive control." Instead of Oahs and NaN_3 in **step 3**, 0.3 mM Met is added to the culture and induced by 1 mL of IPTG stock solution.

3.2 Purification of Target Protein

3.2.1 Cell Harvest and Lysis

1. Harvest cells by centrifugation at $8000 \times g$, 4 °C for 20 min. Resuspend the pellet in 30 mL of lysis buffer. Add 30 μL of lysozyme stock solution, to a final concentration of 0.5 mg/mL. Incubate the lysate on ice for 1 h.

2. Break the lysate by sonication. The recommended working conditions are: 15 min, 70% amplitude at an interval of 2 s *on* and 3 s *off*, twice.

3. Centrifuge the lysate for 45 min at 4 °C, $18,000 \times g$.

4. Discard the supernatant and resuspend the pellet in 30 mL of resuspension buffer (*see* **Note 8**).

5. Centrifuge the sample for 30 min, $18,000 \times g$ at 4 °C.

6. Transfer the supernatant to the regenerated cellulose tubular membrane (Nominal MWCO: 3500 Da).

7. Dialyze the sample extensively at least three times against 5 L of dialysis buffer for 3 h, overnight and again for 3 h.

8. Centrifuge the dialyzed sample for 40 min, $18,000 \times g$, 4 °C.

9. Pass the sample through a 0.22 μm filter membrane.

3.2.2 B* Protein Purification

1. The purification of recombinantly expressed B* protein is carried out by ion exchange chromatography (e.g., using an ÄKTA protein purification chromatography system). For a 5 mL HiTrap Q-sepharose column, follow these steps to program the purification of target protein (Table 3):

Table 3
Ion exchange chromatography protocol

Step	Buffer	Injected amount	Elution position
Pre-wash	Milli-Q water	5 column volume(CV)s	Waste
Equilibration	Elution buffer	5 CVs	Waste
Lysate injection	Lysate	All	Collect as flow through
Wash	Wash buffer	3 CVs	Collect as flow through
Elution	Gradient elution buffer (from 0 to 100%)	10 CVs	Collect 50 fractions of 1 mL each
Cleaning	Elution buffer	2–3 CVs	Waste
Wash	Milli-Q water	5 CVs	Waste
Storage	20% ethanol	5–10 CVs	Waste

2. Take ~40 μL aliquots from each fraction, mix with 10 μL of 5× SDS loading buffer, and perform SDS-PAGE to determine which fraction contains the B* protein. The recommended acrylamide concentration of SDS-PAGE gel for B* analysis is 17%.

3. After electrophoresis analysis, combine fractions containing B* and dialyze against 2 L of storage buffer at 4 °C, overnight.

4. The protein can be used for next step analyses or stored at −80 °C. To confirm the Aha incorporation into B* protein, mass spectrometry analysis is recommended. The molecular weights of Met and Aha are 149.21 Da and 144.13 Da, respectively. Each Aha incorporation causes a -5.08 Da shift on the protein mass, compared to Met-incorporated B* (positive control). The yield of Aha- or Met-incorporated protein can be determined by Bradford protein assay. Normally, from 1 L culture, a yield of ~5 mg for Aha-B* and ~10 mg for Met-B* is obtained, respectively (*see* **Note 9**).

4 Notes

1. One optimal method to prepare NMM is to dissolve each of the components in a higher concentration sterile stock solution and then mix them together before use. To dissolve the 19 amino acids in water, tryptophan and tyrosine need to be first dissolved separately in a few drops of 32% hydrochloride and then mix together with other 17 amino acids solution. In addition, biotin can only be solved in water under basic condition (e.g., with the addition of KOH).

2. All the buffers for ion exchange chromatography (e.g., using an ÄKTA protein purification chromatography system), including the injected lysate, need to be filtered through a 0.22 μm membrane before use.

3. It is highly recommended to run a colony PCR to verify that recombinant *E. coli* B834(DE3) strain harbors both plasmids before expression growth. For *metY* (carried on pBU26'*glnS*-metY plasmid), the primers are *metY*-Forward and *metY*-Reverse; for B* (carried on pQE80L-*psi*-barstar plasmid), the primers are T5-Forward and T5-Reverse.

4. During the expression, Met and Aha are competitive to each other for being loaded onto methionyl-tRNA. It is absolutely necessary to deplete any Met from the culture before target protein induction. The depletion can be confirmed by measuring the optical density of the culture at 600 nm (OD_{600}) every half hour after 8 h of first phase growth. If the OD_{600} stays at the same level for at least half hour, the expression can be moved onto the next step.

5. For **step 2** in Subheading 3.1.2, it is important to give the cells enough time to produce in vivo sufficient amount of Aha for incorporation.

6. Due to cell toxicity of sodium azide, we determined that its concentration should be limited to not more than 1 mM.

7. For B* protein, the most efficient induction conditions are at 30 °C, 4 h.

8. The B* protein is expressed in the form of inclusion body. It can be dissolved in resuspension buffer thanks to the high concentration of urea. Therefore, the extensive dialysis is very important for removing all urea from the lysate. During the slow removal of urea in the dialysis step, B* protein is able to correctly refold and becomes soluble in lysis buffer (same composition as elution buffer).

9. Purified B* protein can be stored at −80 °C for at least 1 year. It is recommended keeping small aliquots (e.g., 100 μL) for further measurements or reactions, to avoid thawing and refreezing.

Acknowledgments

The authors acknowledge the financial support of the EU-funded SYNPEPTIDE (613981) consortium of FP7 and thank the Deutsche Forschungsgemeinschaft (DFG) for financial support within the research group FOR 1905. M.L.d.S. was also supported by a Research Stays Fellowship for University Academics and Scientists from Deutscher Akademischer Austauschdienst (DAAD). Y. M. acknowledges the financial support of the China Scholarship Council (CSC). We are very grateful to Dr. Hernán Biava for the preparation of Oahs.

References

1. Ma Y, Biava H, Contestabile R et al (2014) Coupling bioorthogonal chemistries with artificial metabolism: intracellular biosynthesis of azidohomoalanine and its incorporation into recombinant proteins. Molecules 19(1):1004–1022. https://doi.org/10.3390/molecules 19011004

2. Hwang B, Yeom H, Kim Y et al (2002) Corynebacterium glutamicum utilizes both Transsulfuration and direct Sulfhydrylation pathways for methionine biosynthesis. J Bacteriol 184(5):1277–1286. https://doi.org/10.1128/JB.184.5.1277-1286.2002

3. Di Salvo ML, Budisa N, Contestabile R (Dec. 2013) PLP-dependent Enzymes: a powerful tool for metabolic synthesis of non-canonical amino acids in molecular evolution and control (Molekulare Entwicklung und Kontrolle. Beilstein Symposium Ed

4. Nölting B, Golbik R, Fersht AR (1995) Submillisecond events in protein folding. Proc Natl Acad Sci U S A 92(23):10668–10672

5. Dong S, Moroder L, Budisa N (2009) Protein iodination by click chemistry. Chembiochem 10(7):1149–1151. https://doi.org/10.1002/cbic.200800816

6. Budisa N, Steipe B, Demange P et al (1995) High-level biosynthetic substitution of methionine in proteins by its analogs 2-aminohexanoic acid, selenomethionine, telluromethionine and ethionine in *Escherichia coli*. Eur J Biochem 230(2):788–796. https://doi.org/10.1111/j.1432-1033.1995.0788h.x

7. Seidman CE, Struhl K, Sheen J, Jessen T (1997) Chapter 1, Escherichia Coli, plasmids, and bacteriophages. Introduction of plasmid DNA into cells. Curr Protoc Mol Biol 1(8):1–1.8.10. https://doi.org/10.1002/0471142727

Chapter 8

Residue-Specific Incorporation of Noncanonical Amino Acids for Protein Engineering

Mark B. van Eldijk and Jan C.M. van Hest

Abstract

The incorporation of noncanonical amino acids has given protein chemists access to an expanded repertoire of amino acids. This methodology has significantly broadened the scope of protein engineering allowing introduction of amino acids with non-native functionalities, such as bioorthogonal reactive handles (azides and alkynes) and hydrophobic fluorinated side chains. Here, we describe the efficient residue-specific replacement of methionine by azidonorleucine in an engineered green fluorescent protein using a bacterial expression system to introduce a single reactive site for the strain-promoted azide-alkyne cycloaddition.

Key words Noncanonical amino acids, Residue-specific incorporation, Aminoacyl-tRNA synthetase, Azidonorleucine, Medium shift, Bioconjugation, Strain-promoted azide-alkyne cycloaddition, Unnatural amino acids

1 Introduction

In the field of protein engineering, scientists manipulate the amino acid sequence of proteins to alter or enhance their function. Originally, protein chemists only had access to the twenty proteinogenic amino acids. However, many efforts have been made to broaden the scope of protein engineering [1, 2]. Besides traditional targeted chemical or enzymatic posttranslational modifications, over the past 20 years additional functionality has been introduced in proteins via the cotranslational incorporation of noncanonical amino acids (ncAAs) with non-native chemical moieties [3, 4].

In translation of the genetic code to protein, codons are assigned to amino acids by base-pairing of the mRNA and charged tRNAs. Aminoacyl-tRNA synthetases (aaRSs), which catalyze the attachment of the appropriate amino acid to the correct tRNA, are the gatekeepers of the genetic code as they are responsible for the selection of both the amino acid and the tRNA. Manipulation of this aminoacylation step has allowed incorporation of a multitude of ncAAs via the residue-specific and the site-specific method. The

Edward A. Lemke (ed.), *Noncanonical Amino Acids: Methods and Protocols*, Methods in Molecular Biology, vol. 1728, https://doi.org/10.1007/978-1-4939-7574-7_8, © Springer Science+Business Media, LLC 2018

site-specific approach allows complementing the naturally-occurring amino acids with a new ncAA which makes it ideal for introducing a single modification [5, 6]. This method most often relies on reassignment of a stop codon, which requires genetic manipulation of the target protein and expression of an orthogonal (engineered) aaRS/tRNA pair.

The residue-specific method relies on replacement of one of the natural amino acids for a structural analogue [7]. This approach is ideal when one wishes to globally replace an amino acid by an unnatural derivative. An expression host, that is auxothropic for the amino acid that is being replaced, is often required to obtain efficient replacement. In 1957, one of the first examples of the residue-specific method was demonstrated when Cohen and coworkers showed replacement of methionine by selenomethionine, a contribution that later revolutionized X-ray crystallography of proteins [8, 9]. More recently, amino acids with bioorthogonal reactive side chains were introduced into proteins. These chemical handles, such as azides, alkynes, and aldehydes, can be selectively reacted through bioconjugation reactions. The amino acid repertoire of the residue-specific approach is not necessarily limited to the structural surrogates, because manipulation of the active site of the aaRS enables charging of an expanded set of ncAAs [10].

Often the replacement of methionine is the method of choice when a limited number of replacements with minimal perturbation are preferred, because methionine is of relatively low abundance in proteins [4]. If a higher level of selectivity is desired, for example when a reactive handle is introduced for protein labeling, the methionine codons at positions which should not be replaced can be mutated to other canonical amino acids. This is however not always necessary, as methionines are often positioned at inaccessible sites due to the hydrophobic nature of the residue [11]. In another example of residue-specific incorporation, multiple replacements of leucine by more hydrophobic surrogates result in better stabilized protein domains [12, 13]. More recently, the features of residue-specific incorporation of ncAAs have been optimally employed in the field of proteomics, which allows the broad labeling of newly-synthesized proteins in the proteome with temporal and spatial control [14].

In this chapter, we describe the residue-specific replacement of methionine by azidonorleucine (Anl, Fig. 1a) in an engineered green fluorescent protein (GFP) to introduce a single reactive site for bioconjugation using a bacterial expression system. We make use of an engineered genetic construct for a 6×His-tagged GFP mutant in which all methionine codons except for the N-terminal codon have been removed (Fig. 1b, c). Because Anl is not activated by the wild-type methionyl-tRNA synthetase (MetRS) we employ a mutant MetRS, which is able to efficiently charge the ncAA [10]. After expression the 6×His-tagged protein is purified via

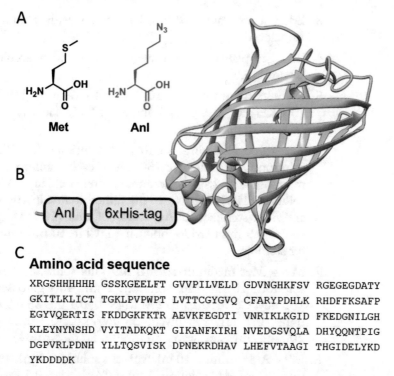

Fig. 1 (**a**) Structures of methionine (Met, 149.21 g/mol) and azidonorleucine (Anl, 172.19 g/mol). (**b**) Graphic representation of Anl-GFP. (**c**) Amino acid sequence of Anl-GFP. Anl is indicated by X

immobilized metal ion affinity chromatography and the purity is verified by SDS-PAGE. The incorporation of Anl is confirmed by mass spectrometry. Furthermore, we demonstrate the reactivity of the azido-modified protein via a strain-promoted azide-alkyne cycloaddition (SPAAC) reaction with a bicyclononyne-fluorophore [15]. This reaction is analyzed by in-gel fluorescence analysis and size exclusion chromatography. Elsewhere, we have demonstrated the site-specific immobilization of Anl-GFP onto a chip for surface plasmon resonance measurements [16].

2 Materials

2.1 Expression of Anl-GFP

All the solutions used for culturing should be made with ultrapure water and sterilized as indicated (by autoclaving or filter sterilization using filters with 0.2 μm pore size). Additionally, all glassware and disposables used during culturing should be sterile.

1. Competent methionine auxotrophic cells: M15MA[pREP4] (*see* **Notes 1** and **2**).

2. Expression plasmid encoding for engineered GFP and mutant MetRS (*see* **Note 3**).

3. Erlenmeyer flasks (300 mL) capped with aluminum foil, autoclaved.

4. LB agar plates and LB medium both with ampicillin and kanamycin.

5. M9 salts stock solution (10×): dissolve 64 g $Na_2HPO_4 \cdot 7H_2O$, 15 g KH_2PO_4, 2.5 g NaCl and 5.0 g NH_4Cl in 1 L water, autoclave to sterilize.

6. The three following amino acid solutions: (**a**) 19AA stock solution (25×): dissolve 250 mg of each L-amino acid except methionine in 250 mL water and filter sterilize; (**b**) Met stock solution (10×): dissolve 40 mg L-methionine in 100 mL water and filter sterilize; (**c**) Anl stock solution (10 mM, 10×): dissolve 20.9 mg L-azidonorleucine·HCl in 10 mL water and filter sterilize.

7. M9 + Met medium: 10 mL M9 salts solution, 4 mL 19AA solution, 10 mL Met solution, 1 mL 40% glucose, 0.1 mL 100 mM $CaCl_2$, 0.1 mL 2 M $MgSO_4$, 0.1 mL 35 mg/mL thiamine, 0.1 mL kanamycin 35 mg/mL, 0.1 mL ampicillin 200 mg/mL, add water up to 90 mL (*see* **Note 4**).

8. M9 + Anl medium: 10 mL M9 salts solution, 4 mL 19AA solution, 10 mL Anl solution, 1 mL 40% glucose, 0.1 mL 100 mM $CaCl_2$, 0.1 mL 2 M $MgSO_4$, 0.1 mL 35 mg/mL thiamine, 0.1 mL kanamycin 35 mg/mL, 0.1 mL ampicillin 200 mg/mL, add water up to 100 mL.

9. 0.9% NaCl solution (autoclaved and cooled at 4 °C before use).

10. 1 M isopropyl β-D-1-thiogalactopyranoside (IPTG) stock solution.

2.2 Purification of Anl-GFP Via Immobilized Metal Ion Affinity Chromatography

1. Lysis buffer: 50 mM NaH_2PO_4, 300 mM NaCl, 10 mM imidazole, pH 8.0. Autoclave to sterilize.

2. Wash buffer: 50 mM NaH_2PO_4, 300 mM NaCl, 20 mM imidazole, pH 8.0. Autoclave to sterilize.

3. Elution buffer: 50 mM NaH_2PO_4, 300 mM NaCl, 250 mM imidazole, pH 8.0. Autoclave to sterilize.

4. Phosphate-buffered saline (PBS) storage buffer: 10 mM Na_2HPO_4, 1.8 mM KH_2PO_4, 137 mM NaCl, 2.7 mM KCl, pH 7.4. Autoclave to sterilize.

5. Ni^{2+}-NTA agarose beads (washed with lysis buffer twice).

6. Disposable plastic columns.

7. Dialysis tubing (12–14 kDa MWCO) and clamps.

2.3 Strain-Promoted Azide-Alkyne Cycloaddition to Confirm Reactivity of Anl-GFP	1. Bicyclononyne-PEO$_3$-NH-lissamine rhodamine B conjugate (BCN-LisRhoB, Fig. 3a). Dissolve this compound in 50% DMSO/PBS at a concentration of 1.7 mM.
	2. Gel imaging system for the detection of fluorescence.
	3. Size exclusion chromatography system with Superdex 75 3.2/300 column with inline UV-VIS detector.

3 Methods

3.1 Expression of Anl-GFP	1. Transform pQE-80 L/MRGS6×His-GFPrm-FLAG/NLL-MetRS into competent M15MA[pREP4] cells and plate onto LB agar with 200 mg/L ampicillin and 35 mg/L kanamycin (*see* **Note 5**). Incubate overnight (16–24 h) at 37 °C.
	2. Pick a single colony to inoculate 25 mL LB medium with 200 mg/L ampicillin and 35 mg/L kanamycin. Grow overnight (16–24 h) at 37 °C while shaking vigorously (200–250 rpm) (*see* **Note 6**).
	3. Dilute 10 mL overnight culture in 90 mL M9 + Met medium and grow at 37 °C while shaking vigorously until a culture density of OD$_{600}$ = 0.6–0.8 has been reached (typically 2–3 h).
	4. Perform the medium shift: pellet the cells by centrifugation (5000 × g) at 4 °C. Then wash with 20 mL cold 0.9% NaCl twice to ensure complete removal of methionine. Then pellet the cells again and resuspend them in 100 mL M9 + Anl medium (use a clean Erlenmeyer flask) (*see* **Notes 7** and **8**).
	5. Induce expression of the engineered GFP by the addition of IPTG to a final concentration of 1 mM. Allow expression to proceed at 37 °C for 6 h while shaking vigorously (*see* **Note 9**).
	6. Harvest the cells by centrifugation (5000 × g) at 4 °C for 15 min. The cell pellets can be stored at −20 °C.
3.2 Purification of Anl-GFP Via Immobilized Metal Ion Affinity Chromatography	1. Thaw cell pellet on ice for 15 min and then resuspend in 5 mL lysis buffer.
	2. Lyse cells by ultrasonic disruption using a sonicator with a microtip (6 × 10 s pulse followed by 5 s pause to allow cooling).
	3. Clear lysate by centrifugation (15,000 × g at 4 °C for 30 min) and transfer the supernatant to a clean tube.
	4. Add 300 μL Ni^{2+}-NTA agarose beads to the supernatant. Incubate at 4 °C for 1 h while mixing gently to allow binding of 6×His-tagged protein to the Ni^{2+}-NTA agarose beads (for example on a rotator in a cold room).

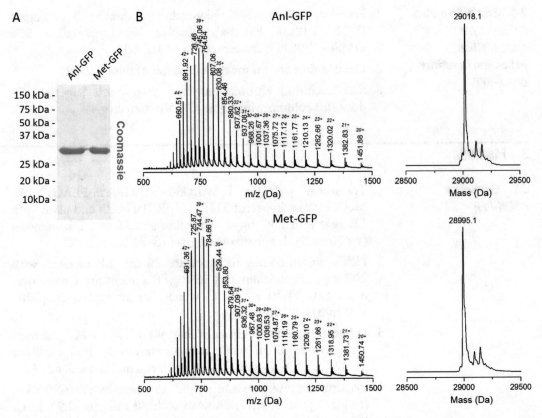

Fig. 2 (**a**) Confirmation of purity by SDS-PAGE showing both Anl-GFP and Met-GFP, as a single band around 30 kDa. (**b**) ESI-TOF mass spectrometry results of purified Anl-GFP and Met-GFP. Multiply-charged ion series (left) and deconvoluted total mass spectrum (right) (adjusted from [16] Published by The Royal Society of Chemistry)

5. Load the mixture into a column. Wash three times with 2 mL wash buffer and then elute Anl-GFP with 1 mL elution buffer.

6. Dialyze the protein solution to the PBS storage buffer. For dialysis use dialysis tubing with 12–14 kDa MWCO and refresh the buffer after 0.5 h, 1 h and 16–24 h and perform the buffer exchange at 4 °C. The protein solution can be stored at 4 °C.

7. Confirm the purity of the protein by SDS-PAGE and measure whole protein mass by electrospray ionization time-of-flight (ESI-TOF) mass spectrometry to validate expected molecular weight for the incorporation of Anl (Anl-GFP = 29018.4 Da and Met-GFP = 28995.4 Da). The expected results for SDS-PAGE and ESI-TOF mass spectrometry are shown in Fig. 2a, b, respectively (*see* **Note 10**).

8. Use the molar extinction coefficient to determine the yield. Typically, a yield of approximately 10–25 mg/L of bacterial culture is expected for Anl-GFP under these conditions (*see* **Note 11**).

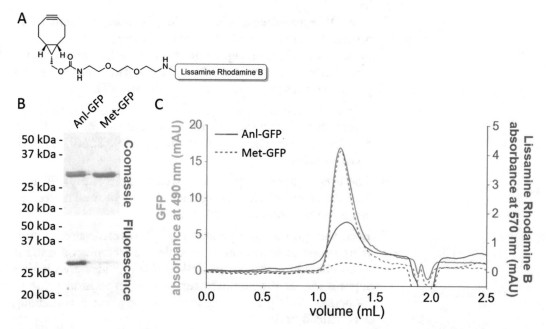

Fig. 3 (**a**) Structure of BCN-LisRhoB. (**b**) Analysis of reaction by Coomassie-stained and fluorescent SDS-PAGE and (**c**) by SEC of purified Anl-GFP (solid lines) and Met-GFP (dashed lines) with BCN-LisRhoB. The green lines correspond to the GFP absorbance at 490 nm and the red lines correspond to the lissamine rhodamine absorbance at 570 nm (adjusted from [16] Published by The Royal Society of Chemistry)

3.3 Strain-Promoted Azide-Alkyne Cycloaddition to Confirm Reactivity of Anl-GFP

1. Prepare 15 µL of 1.0 mg/mL (34 µM) for both Anl-GFP and Met-GFP solutions by diluting with PBS.

2. Add 15 µL of 1.7 mM BCN-LisRhoB to start the reaction (50-fold excess).

3. Incubate the reaction mixture at 20 °C for 20 h while shaking.

4. Analyze the reaction by SDS-PAGE analysis and fluorescence imaging (with a filter set for the detection of the fluorescent dye). Then stain the gel with Coomassie blue to validate equal loading. Expected results are shown in Fig. 3b.

5. Alternatively, the reaction can be monitored by size exclusion chromatography. Therefore, inject 10 µL of the reaction mixture onto the column and elute with PBS at 0.05 mL/min while measuring absorbance at both 490 nm and 570 nm for the detection of GFP and the fluorescent dye, respectively (Fig. 3c).

4 Notes

1. The M15MA strain is a methionine auxotroph, derived from the M15 strain after knockout of *metE*.

2. The QIAexpressionist is a very helpful resource with background information and protocols on cloning, expression, and purification of 6×His-tagged proteins (including protocols for preparation of competent cells and information about the T5 promoter transcription-translation system) [17].

3. The expression plasmid pQE-80L carries a copy of both the engineered GFP gene and the gene coding for the *E. coli* MetRS with the following mutations L13N, Y260L, H301L (NLL-MetRS, [10]).

4. The total volume of the M9 + Met medium is only 90 mL, because 10 mL of the overnight culture will be used to eventually bring the volume to 100 mL.

5. Even though the cis repressed pQE-80L vector already contains a copy of the repressor module, we recommend using it in combination with the pREP4 as this provides even higher suppression of the recombinant protein expression prior to induction. This prevents synthesis of the protein containing Met instead of Anl.

6. To avoid performing a transformation every time, it might be preferred to make a glycerol stock.

7. Some protocols describe an induction step prior to the medium shift to produce RNA polymerase. However, this is not required when the T5 promoter transcription-translation system is utilized, as this system uses the native *E. coli* RNA polymerase, which is already present. The advantage of the elimination of this pre-induction step is higher incorporation efficiency.

8. As a control, the expression can also be performed in M9 + Met medium. Met-GFP is useful for comparing protein molecular weights and expression yields, or as negative control for conjugation reactions.

9. Expression yields can be optimized by varying expression temperature (16–37 °C) and duration (2–16 h).

10. The theoretical molecular weight of the protein is calculated and adjusted for cyclization and chromophore formation (−20 Da). Also note that neither methionine nor the ncAA are cleaved by methionine aminopeptidase, probably because of the penultimate arginine residue.

11. Alternatively to improving target protein yields by optimizing expression parameters, the culture size can be scaled (for example to 1 L). Be aware that this will also require larger scale centrifugation during medium shift as well as harvesting.

Acknowledgment

This work was financially supported by Netherlands Organisation for Scientific Research (Rubicon grant 680-50-1407 and Gravitation program 024.001.035). The plasmids encoding the mutant MetRS and GFP were a kind gift from Prof. D. A. Tirrell (California Institute of Technology).

References

1. McKay CS, Finn MG (2014) Click chemistry in complex mixtures: bioorthogonal bioconjugation. Chem Biol 21:1075

2. Krall N, da Cruz FP, Boutureira O, Bernardes GJL (2016) Site-selective protein-modification chemistry for basic biology and drug development. Nat Chem 8:102

3. Link AJ, Mock ML, Tirrell DA (2003) Non-canonical amino acids in protein engineering. Curr Opin Biotechnol 14:603

4. Budisa N (2004) Prolegomena to future experimental efforts on genetic code engineering by expanding its amino acid repertoire. Angew Chem Int Ed 43:6426

5. Dumas A, Lercher L, Spicer CD, Davis BG (2015) Designing logical codon reassignment - expanding the chemistry in biology. Chem Sci 6:50

6. Young TS, Schultz PG (2010) Beyond the canonical 20 amino acids: expanding the genetic lexicon. J Biol Chem 285:11039

7. Johnson JA, Lu YY, Van Deventer JA, Tirrell DA (2010) Residue-specific incorporation of non-canonical amino acids into proteins: recent developments and applications. Curr Opin Chem Biol 14:774

8. Cowie DB, Cohen GN (1957) Biosynthesis by *Escherichia coli* of active altered proteins containing selenium instead of sulfur. Biochim Biophys Acta 26:252

9. Hendrickson WA, Horton JR, Lemaster DM (1990) Selenomethionyl proteins produced for analysis by multiwavelength anomalous diffraction (MAD) - a vehicle for direct determination of 3-dimensional structure. EMBO J 9:1665

10. Tanrikulu IC, Schmitt E, Mechulam Y, Goddard WA, Tirrell DA (2009) Discovery of *Escherichia coli* methionyl-tRNA synthetase mutants for efficient labeling of proteins with azidonorleucine in vivo. Proc Natl Acad Sci U S A 106:15285

11. Schoffelen S, Lambermon MHL, van Eldijk MB, van Hest JCM (2008) Site-specific modification of *Candida antarctica* lipase B via residue-specific incorporation of a non-canonical amino acid. Bioconjug Chem 19:1127

12. Montclare JK, Son S, Clark GA, Kumar K, Tirrell DA (2009) Biosynthesis and stability of coiled-coil peptides containing (2S,4R)-5,5,5-trifluoroleucine and (2S,4S)-5,5,5-trifluoroleucine. Chembiochem 10:84

13. Van Deventer JA, Fisk JD, Tirrell DA (2011) Homoisoleucine: a translationally active leucine surrogate of expanded hydrophobic surface area. Chembiochem 12:700

14. Ngo JT, Tirrell DA (2011) Noncanonical amino acids in the interrogation of cellular protein synthesis. Acc Chem Res 44:677

15. Dommerholt J, Schmidt S, Temming R, Hendriks LJA, Rutjes FPJT, van Hest JCM, Lefeber DJ, Friedl P, van Delft FL (2010) Readily accessible bicyclononynes for bioorthogonal labeling and three-dimensional imaging of living cells. Angew Chem Int Ed 49:9422

16. Wammes AEM, Fischer MJE, de Mol NJ, van Eldijk MB, Rutjes FPJT, van Hest JCM, van Delft FL (2013) Site-specific peptide and protein immobilization on surface plasmon resonance chips via strain-promoted cycloaddition. Lab Chip 13:1863

17. The QIAexpressionist, 2003, 5th edition

Chapter 9

Using Amber and Ochre Nonsense Codons to Code Two Different Noncanonical Amino Acids in One Protein Gene

Jeffery M. Tharp and Wenshe R. Liu

Abstract

In *Escherichia coli*, conventional amber and ochre stop codons can be separately targeted by engineered amber-suppressing *Methanocaldococcus jannaschii* tyrosyl-tRNA synthetase-tRNA^{Pyl} and ochre-suppressing *Methanosarcina maezi* pyrrolysyl-tRNA synthetase-tRNA^{Pyl} pairs for coding two different noncanonical amino acids in one protein gene. Here, we describe the application of this approach to produce a protein with two distinct chemical functionalites which can be selectively labeled with two fluorescent dyes.

Key words Double incorporation, Protein labeling, Amber codon, Ochre codon, MjTyrRS, PylRS, Noncanonical amino acid

1 Introduction

The genetic incorporation of noncanonical amino acids (ncAAs) via stop codon suppression has provided researchers with valuable tools for the study of protein structure and function. In general, ncAAs are incorporated into a target protein, in response to an amber codon, by an orthogonal aminoacyl-tRNA synthetase (aaRS) and its cognate tRNA which are concomitantly expressed with the target protein. Two such aaRS/tRNA pairs that have been widely used for this purpose are the tyrosyl-tRNA synthetase (*Mj*TyrRS)-*Mj*tRNA^{Pyl} pair from *M. jannaschii* and the pyrrolysyl-tRNA synthetase (*Mm*PylRS)-*Mm*tRNA^{Pyl} pair from *Methanosarcina mazei* [1–9]. We have previously demonstrated that these pairs are not only orthogonal to the natural aaRS/tRNA of *E. coli*, but that they are also orthogonal to each other—that is to say that the *Mj*TyrRS is incapable of acylating *Mj*tRNA^{Pyl} and vice versa [10]. Further, since *Mm*PylRS does not interact with the anticodon of tRNA^{Pyl}, this anticodon can be mutated from CUA to UUA, to allow for the suppression of ochre codons, without loss of enzyme function [10–12]. The orthogonality of *Mj*TyrRS and *Mm*PylRS combined with the flexibility at the anticodon position of *Mm*tRNA^{Pyl} has

Edward A. Lemke (ed.), *Noncanonical Amino Acids: Methods and Protocols*, Methods in Molecular Biology, vol. 1728,
https://doi.org/10.1007/978-1-4939-7574-7_9, © Springer Science+Business Media, LLC 2018

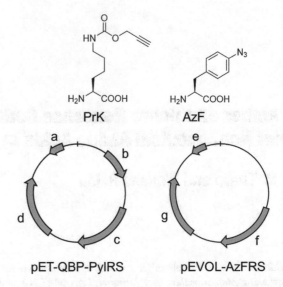

Fig. 1 (Top) Structures of N_ε-propargyloxycarbonyl-lysine (PrK) and 4-azidophenylalanine (AzF). (Bottom) Representative plasmid maps of pEVOL-AzFRS and pET-QBP-PylRS for incorporating AzF and PrK into glutamine binding protein. (a) $Mm\text{tRNA}_{UUA}^{Pyl}$, (b) QBP-3TAG-141TAA, (c) MmPylRS, (d) β-lactamase, (e) $Mj\text{tRNA}_{CUA}^{Tyr}$, (f) MjTyrRS, (g) chloramphenicol acetyltransferase

allowed us to use these two enzyme-tRNA pairs to incorporate two different ncAAs into one protein at amber and ochre codons. The ability to incorporate two different ncAAs has many applications including fluorescence labeling for Förster resonance energy transfer, studying cross-talk in posttranslational modifications, and the generation of non-natural protein and peptide libraries.

Here, we describe the use of MjTyrRS and MmPylRS for the incorporation of 4-azidophenylalanine (AzF) and N_ε-propargyloxycarbonyl-lysine (PrK) into the glutamine binding protein (QBP) of *E. coli* (Fig. 1). This general protocol requires a two plasmid system. The plasmid pEVOL-AzFRS encodes the MjTyrRS- $Mj\text{tRNA}_{CUA}^{Tyr}$ pair and the plasmid pET-QBP-PylRS encodes the MmPylRS- $Mm\text{tRNA}_{UUA}^{Pyl}$ pair along with the QBP with amber and ochre mutations at the desired locations (Fig. 1). This dual-plasmid system can be modified for use with other aaRS and target proteins.

2   Materials

Prepare all solutions and media using ultrapure water and store at room temperature unless otherwise indicated. Care should be taken to minimize exposure of recombinant microorganisms to the environment, including thoroughly decontaminating all waste before disposal and regularly cleaning the work area with 70% ethanol or 3% bleach.

2.1 Co-transformation of Expression Vectors

1. Petri dishes (100×15 mm).
2. Culture tubes (10 mL).
3. Cell spreaders.
4. Water bath set to 42 °C.
5. *E. coli* BL21(DE3) chemical competent cells (90 μ aliquots) (*see* **Note 1**).
6. Lysogeny broth (LB): 5 g tryptone, 2.5 g yeast extract, 5 g NaCl. Dissolve in about 400 mL water. Adjust pH to 7.5 with 5 M NaOH. Add water to a total volume of 500 mL and autoclave (*see* **Note 2**).
7. LB Agar: 5 g tryptone, 2.5 g yeast extract, 5 g NaCl, 7.5 g agarose. Dissolve in about 400 mL water. Adjust pH to 7.5 with 5 M NaOH. Add water to a total volume of 500 mL and autoclave (*see* **Notes 2** and **3**).
8. 1000× Ampicillin solution: 1.00 g ampicillin sodium salt dissolved in 10 mL water. Filter and store aliquots at −20 °C.
9. 1000× Chloramphenicol solution: 0.34 g chloramphenicol dissolved in 10 mL absolute ethanol. Filter and store aliquots at −20 °C.
10. Expression vectors: pET-QBP-PylRS and pEVOL-AzFRS (*see* Fig. 1).

2.2 Protein Expression

1. 2xYT Media: 8 g tryptone, 5 g yeast extract, 2.5 g NaCl. Dissolve in about 400 mL water. Adjust pH to 7.5 with 5 M NaOH. Add water to a total volume of 500 mL and autoclave (*see* **Notes 2** and **4**).
2. 4×1 L culture flasks. Autoclaved.
3. 1×3 L culture flask. Autoclaved.
4. 20% arabinose: 2.0 g L-arabinose dissolved in 10 mL water. Filter and store aliquots at −20 °C. Discard if mold growth is apparent.
5. 1 M isopropyl-β-D-1-thiogalactopyranoside (IPTG): 2.38 g IPTG dissolved in 10 mL water. Filter and store at −20 °C. Discard if mold growth is apparent.
6. 200 mM *para*-azido-L-phenylalanine (AzF): Add 0.206 g *para*-azido-L-phenylalanine to 3 mL of water and then add, dropwise, a minimal amount of 5 M NaOH to completely dissolve the solid. Dilute the solution to 5 mL with water. Store at −20 °C.

7. 200 mM *N*-propargyloxycarbonyl-L-lysine (PrK): Add 0.264 g *N*-propargyloxycarbonyl-L-lysine HCl salt to 3 mL of water and then add, dropwise, a minimal amount of 5 M NaOH to completely dissolve the solid. Dilute the solution to 5 mL with water. Store at −20 °C.

3 Methods

3.1 Co-transformation

1. Remove an aliquot of *E. coli* BL21(DE3) chemical competent cells from the freezer and allow it to completely thaw on ice.

2. Add 3 μL of pET-QBP-PylRS and 3 μL of pEVOL-AzFRS (*see* **Note 5**) to the competent cells. Mix the cells and plasmids by gently flicking the tube and incubate on ice for 30 min (*see* **Note 6**).

3. During this time, prepare an LB agar plate. Microwave the LB agar until it is completely melted. Pour 20 mL of melted LB agar into a sterile, 50 mL tube and let it cool until it can be comfortably handled. Add 20 μL of 1000× ampicillin and 20 μL of 1000× chloramphenicol and mix by inverting the tube several times (*see* **Note 7**). Carefully pour the molten LB agar into a sterile Petri dish, place the lid on top, and let it cool until completely solidified.

4. Place the tube containing BL21 and plasmids in a 42 °C water bath and incubate for exactly 1 min.

5. Place the tube on ice for 2 min.

6. Add the transformed cells to a 10 mL culture tube containing 900 μL LB and incubate at 37 °C, 250 rpm, for 45 min (*see* **Note 8**).

7. Transfer the culture to a 1.5 mL tube and centrifuge for 5 min at $16,000 \times g$.

8. Remove the upper 800 μL, and completely resuspend the pellet in the remaining 100 μL of media by gently pipetting.

9. Pipette the transformed cells onto the surface of the LB agar plate. Using a sterilized cell spreader, spread the liquid to completely cover the plate. Continue spreading until most of the liquid has been absorbed by the plate.

10. Recap the plate, and incubate inverted at 37 °C overnight.

3.2 Protein Expression

1. Inoculate 5 mL LB, containing 5 μL 1000× ampicillin and 5 μL 1000× chloramphenicol, with a single colony from the plated BL21(DE3) that had been co-transformed (*see* **Note 9**). Incubate the tube at 37 °C, 250 rpm, overnight.

2. The following day, use 4 mL of the overnight culture to inoculate 400 mL 2xYT in a 3 L culture flask containing 400 μL 1000× ampicillin and 400 μL 1000× chloramphenicol.

3. Incubate the culture at 37 °C, 250 rpm until the optical density at 600 nm reaches 0.7 (*see* **Note 10**).

4. Once the optical density at 600 nm has reached 0.7, evenly divide the culture into 4, 1 L, culture flasks. To one flask, add 1 mL of 200 mM AzF and 1 mL of 200 mM PrK. To a second flask, add 1 mL of 200 mM AzF. To a third flask add 1 mL of 200 mM PrK. The final concentration of each ncAA should be 2 mM. Label each of the flasks accordingly (*see* **Note 11**).

5. Induce the expression of the QBP and aaRS by adding 100 μL of 1 M IPTG and 1 mL of 20% arabinose to each of the 4 flaks. Incubate the flasks at 37 °C, 250 rpm, overnight (*see* **Note 12**).

6. The following day, harvest the cells by centrifugation at 1700 × g for 20 min.

7. Purify the target protein, and analyze via SDS-PAGE. The presence of a band corresponding to the target protein that is present in the sample from the +Azf/+PrK culture, but absent in the other culture samples is indicative of successful dual incorporation (Fig. 2). ncAAs with reactive functional groups can also be verified by dual labeling of the protein (Fig. 3). In either case, the successful incorporation of both ncAAs should be further verified by high accuracy mass spectrometry (*see* **Note 13**).

4 Notes

1. Alternative *E. coli* expression cell strains can be used however, some *E. coli* strains are able to suppress amber codons (e.g., *E. coli* ER2738), and these strains cannot be used for non-natural amino acid incorporation via amber codon suppression. The strain should also be compatible with the expression vector.

2. Use aseptic techniques when working with any culture media. Always use fresh, sterile pipette tips. Work next to a Bunsen burner flame. Keep work area disinfected by cleaning surfaces and pipettes with 70% ethanol or 3% bleach before and after use. Discard media if mold or bacterial contamination is evident.

3. Agarose may not completely dissolve before autoclaving. After autoclaving, and while media is still melted, thoroughly mix by inverting the bottle several times.

4. Alternative media may be necessary depending on which aaRS is being used. Some aaRS mutants with lower fidelity require expression in minimal media to reduce background expression from the incorporation of canonical amino acids. Another alternative is to use synthetic media in which the concentration of the various amino acids can be carefully controlled (*see* ref. 13).

5. Depending on the concentration of plasmid DNA obtained from the miniprep, more or less may be required.

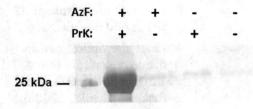

Fig. 2 SDS-PAGE analysis of purified QBP-3TAG-141TAA expressed in the presence and absence of 2 mM PrK and AzF

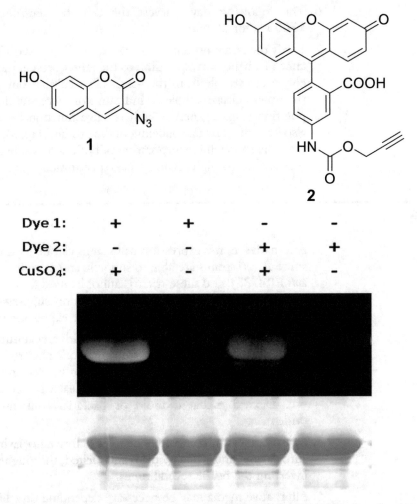

Fig. 3 (Top) Structure of dyes **1** and **2**. (Bottom) Labeling of QBP-3AzF-141PrK with dyes **1** and **2**. In-gel fluorescence and Coomassie stained gels are shown

6. Chemical competent *E. coli* are susceptible to mechanical stress, therefore, it is better to mix the cells by flicking the tube rather than pipetting. Incubation on ice can generally be skipped at the cost of transformation efficiency.

7. If bubbles are present after pouring into the plate, they can usually be removed by quickly passing the flame of a Bunsen burner over the surface of the molten agar.

8. Agarose plates should be pre-warmed and dried 15 min before plating the transformed cells. To do this, place the plates in a 37 °C incubator with the lids slightly canted. This will allow some of the excess liquid to evaporate and prevent smearing when plating.

9. To pick a colony, gently touch a well-isolated colony with a 10 μL pipette tip, dip the tip just below the surface of the LB, and swirl it in the media several times.

10. Under these conditions it generally takes 3–4 h for the OD_{600} to reach 0.7, however, the time will vary depending on the cell strain and media being used. The OD_{600} at which the protein expression is induced may need to be optimized for the protein of interest.

11. The ncAA concentration may also need to be optimized for each aaRS mutant and ncAA substrate as mutants have varying affinities for their substrates.

12. Expression conditions may vary depending on target protein. We have successfully used this protocol over a range of expression times and temperatures.

13. The absence of a band corresponding to the target protein can be caused by many factors. Because the yield is often low during dual stop codon suppression, it can be difficult to see the desired band on SDS-PAGE of whole-cell lysates before purifying and concentrating the protein. Therefore, we recommend blotting the lysates and checking for protein expression with an appropriate antibody.

References

1. Wang L, Brock A, Herberich B, Schultz PG (2001) Expanding the genetic code of *Escherichia coli*. Science 292(5516):498–500. https://doi.org/10.1126/science.1060077 292/5516/498 [pii]

2. Wang L, Schultz PG (2004) Expanding the genetic code. Angew Chem 44(1):34–66. https://doi.org/10.1002/anie.200460627

3. Xie J, Schultz PG (2005) An expanding genetic code. Methods 36(3):227–238. https://doi.org/10.1016/j.ymeth.2005.04.010

4. Wang L, Xie J, Schultz PG (2006) Expanding the genetic code. Annu Rev Biophys Biomol Struct 35:225–249. https://doi.org/10.1146/annurev.biophys.35.101105.121507

5. Liu CC, Schultz PG (2010) Adding new chemistries to the genetic code. Annu Rev Biochem 79:413–444. https://doi.org/10.1146/annurev.biochem.052308.105824

6. Srinivasan G, James CM, Krzycki JA (2002) Pyrrolysine encoded by UAG in Archaea: charging of a UAG-decoding specialized

tRNA. Science 296(5572):1459–1462. https://doi.org/10.1126/science.1069588 296/5572/1459 [pii]

7. Blight SK, Larue RC, Mahapatra A, Longstaff DG, Chang E, Zhao G, Kang PT, Green-Church KB, Chan MK, Krzycki JA (2004) Direct charging of tRNA(CUA) with pyrrolysine in vitro and in vivo. Nature 431(7006):333–335. https://doi.org/10.1038/nature02895 nature02895 [pii]

8. Neumann H, Peak-Chew SY, Chin JW (2008) Genetically encoding N(epsilon)-acetyllysine in recombinant proteins. Nat Chem Biol 4(4):232–234. https://doi.org/10.1038/nchembio.73

9. Wang YS, Fang X, Wallace AL, Wu B, Liu WR (2012) A rationally designed pyrrolysyl-tRNA synthetase mutant with a broad substrate spectrum. J Am Chem Soc 134(6):2950–2953. https://doi.org/10.1021/ja211972x

10. Wan W, Huang Y, Wang Z, Russell WK, Pai PJ, Russell DH, Liu WR (2010) A facile system for genetic incorporation of two different noncanonical amino acids into one protein in Escherichia coli. Angew Chem Int Ed 49: 3211–3214

11. Wu B, Wang Z, Huang Y, Liu WR (2012) Catalyst-free and site-specific one-pot dual-labeling of a protein directed by two genetically incorporated noncanonical amino acids. Chembiochem 13:1405–1408

12. Odoi KA, Huang Y, Rezenom YH, Liu WR (2013) Nonsense and sense suppression abilities of original and derivative Methanosarcina mazei pyrrolysyl-tRNA synthetase-tRNA^{Pyl} pairs in the Escherichia coli Bl21(DE3) cell strain. PLoS One 8:e57035

13. Hammill JT, Miyake-Stoner S, Hazen JL, Jackson JC, Mehl RA (2007) Preparation of site-specifically labeled fluorinated proteins for ^{19}F-NMR structural characterization. Nat Protoc 2:2601–2607

Chapter 10

Genetic Incorporation of Unnatural Amino Acids into Proteins of Interest in *Streptomyces venezuelae* ATCC 15439

Jingxuan He and Charles E. Melançon III

Abstract

Site-specific, genetic incorporation of unnatural amino acids (UAAs) into proteins in living cells using engineered orthogonal aminoacyl-tRNA synthetase (aaRS)/tRNA pairs is a powerful tool for studying and manipulating protein structure and function. To date, UAA incorporation systems have been developed for several bacterial and eukaryotic model hosts. Due to the importance of *Streptomyces* as prolific producers of bioactive natural products and as model hosts for natural product biosynthesis and bioengineering studies, we have developed systems for the incorporation of the UAAs p-iodo-L-phenylalanine (pIPhe) and p-azido-L-phenylalanine (pAzPhe) into green fluorescent protein (GFP) in *Streptomyces venezuelae* ATCC 15439. Here, we describe the procedure for using this system to site-specifically incorporate pIPhe or pAzPhe into proteins of interest in *S. venezuelae*. The modular design of plasmids harboring UAA incorporation systems enables use of other aaRS or aaRS/tRNA pairs for the incorporation of other UAAs; and the vector backbone used allows the system to be transferred to diverse *Streptomyces* species via both protoplast transformation and conjugation.

Key words Unnatural amino acids, *Streptomyces*, p-iodo-L-phenylalanine, p-azido-L-phenylalanine, Green fluorescent protein, Natural products

1 Introduction

Techniques for the site-specific, genetic incorporation of unnatural amino acids (UAAs) into proteins of interest have revolutionized research at the chemistry-biology interface, providing an additional level of control over protein structure and function that is useful in a wide range of applications [1–3]. This technology relies on co-expression of three components: an engineered aminoacyl-tRNA synthetase (aaRS), cognate suppressor tRNA, and a mutant gene encoding the protein in which the UAA is to be installed. To date, engineered aaRS capable of accepting over 100 unique UAAs have been developed [1–3]. Among aaRS/tRNA pairs developed to date, those derived from the *M. jannaschii*

Edward A. Lemke (ed.), *Noncanonical Amino Acids: Methods and Protocols*, Methods in Molecular Biology, vol. 1728, https://doi.org/10.1007/978-1-4939-7574-7_10, © Springer Science+Business Media, LLC 2018

tyrosyl tRNA synthetase/amber suppressor tRNA pair (MjTyrRS/MjtRNA$^{Tyr}_{CUA}$) and the $M.$ $barkeri$ and $M.$ $mazei$ pyr-rolysyl-tRNA synthetase/tRNA pairs (MbPylRS/MbtRNA$^{Pyl}_{CUA}$, MmPylRS/MmtRNA$^{Pyl}_{CUA}$) are the most widely used [1–3].

High GC Gram-positive Actinobacteria of the genus *Streptomyces* are prolific producers of natural products—complex organic molecules that often possess useful bioactivities. Natural products are typically produced by complex, multi-step biosynthetic pathways that often rely on large, multi-domain enzymes and multi-component enzyme complexes [4, 5]. Due to the complexity of natural product pathways and marked differences in codon usage and gene regulation between Actinobacteria and commonly used bacterial hosts such as *E. coli*, model *Streptomyces* are preferred hosts for heterologous expression of actinobacterial natural product biosynthetic enzymes and pathways [6–8].

We envisioned that site-specific genetic incorporation of UAAs, particularly photo-crosslinking amino acids such as *p*-azido-L-phenylalanine (pAzPhe) [9] and *p*-benzoyl-L-phenylalanine (pBpa) [10], into target proteins in *Streptomyces* would be a valuable tool to study protein-protein interactions within natural product biosynthetic enzyme complexes, and to produce analogs of ribosomally-synthesized, posttranslationally modified peptides (RiPPs) for drug development applications and mechanism of action studies. Toward these ends, we have developed single plasmid systems harboring engineered variants of the MjTyrRS/MjtRNA$^{Tyr}_{CUA}$ pair that are capable of site-specifically incorporating a single UAA (*p*-iodo-L-phenylalanine, pIPhe; or *p*-azido-L-phenylalanine, pAzPhe; Fig. 1) into superfolder green fluorescent protein (sfGFP) [11] in the rapidly growing model host *Streptomyces venezuelae* ATCC 15439 [12]. Here, we provide a detailed protocol for using these systems to site-specifically incorporate pIPhe or pAzPhe into proteins of interest in *S. venezuelae*. The modular vector design, which employs unique restriction sites at the flanks of each functional element (Fig. 2), facilitates replacement of the aaRS, gene of interest, or tRNA for development of other UAA incorporation systems. To maximize system versatility, both plasmids (pSUA1-pAzPheRS,

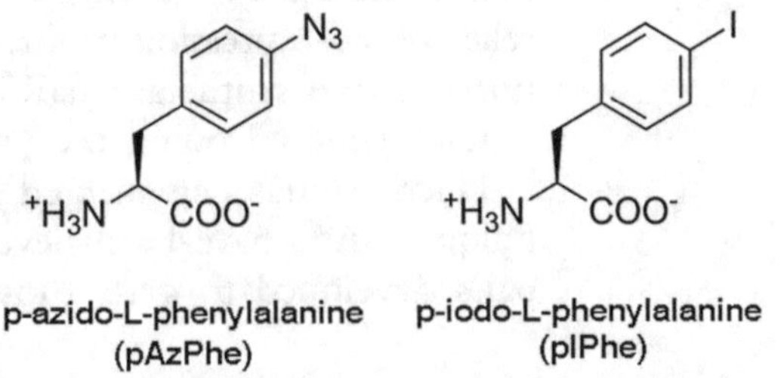

p-azido-L-phenylalanine   p-iodo-L-phenylalanine
(pAzPhe)   (pIPhe)

Fig. 1 Structures of UAAs *p*-azido-L-phenylalanine (pAzPhe) and *p*-iodo-L-phenylalanine (pIPhe) used in this study

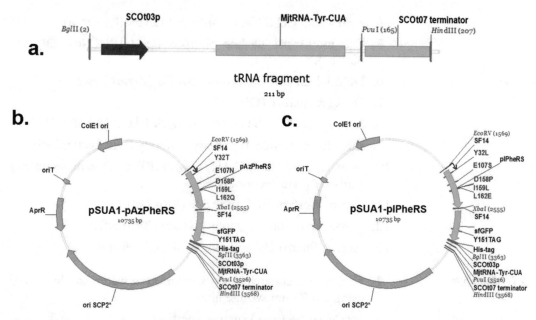

Fig. 2 Diagrams of the tRNA fragment, pSUA1-pAzPheRS, and pSUA1-pIPheRS plasmids. (**a**) SCOt03p and SCOt07t are the promoter and terminator flanking MjtRNA$^{Tyr}_{CUA}$. (**b**) pSUA1-pAzPheRS plasmid for pAzPhe incorporation. Expressions of the pAzPheRS and sfGFP-151TAG genes are both controlled by the SF14 promoter. The sfGFP-151TAG gene can be replaced by the gene of interest. (**c**) pSUA1-pIPheRS plasmid for pIPhe incorporation. pSUA1-pIPheRS is identical to pSUA1-pAzPheRS except for the aaRS sequence

and pSUA1-pIPheRS) are derived from the *E. coli-Streptomyces* shuttle vector pCM1d [13], which harbors the SCP2* origin of replication and the oriT conjugal transfer element. The SCP2* origin allows the vector to replicate in diverse *Streptomyces* hosts; and the oriT element allows the vector to be introduced into hosts from donor *E. coli* via conjugal transfer.

2   Materials

Prepare all solution with ultrapure water and use analytical grade reagents unless otherwise indicated. Carry out routine *E. coli* molecular biology work (plasmid DNA transformation and isolation, *E. coli* growth, PCR amplification, restriction enzyme and calf intestinal phosphatase (CIP) treatments, gel purification, DNA cleanup, and ligation—*see* Subheadings 3 and 3.1) according to standard procedures and/or the enzyme/kit manufacturer's protocol unless otherwise indicated.

1. QIAprep Spin Miniprep Kit (Qiagen).

2. Phusion DNA polymerase (ThermoFisher Scientific).

3. XbaI and BglII restriction enzymes (New England Biolabs).

4. Zymoclean Gel DNA recovery Kit (Zymo Research).

5. Calf intenstinal alkaline phosphatase (CIP) (New England Biolabs).

6. DNA Clean and Concentrator-5 Kit (Zymo Research).

7. Ni-NTA agarose (Qiagen).

8. Mini-PROTEAN TGX pre-cast SDS-PAGE gels (Bio-Rad).

9. anti-His (C-terminal) primary antibody (Life Technologies).

10. Goat anti-mouse IgG (H + L)/HRP conjugate secondary antibody (Life Technologies).

11. Clarity Western ECL substrate (Bio-Rad).

12. Poly-Prep chromatography column (Bio-Rad).

13. HiTrap Phenyl HP column, 1 mL (GE Healthcare).

2.1 Solutions

1. 50 mg/mL apramycin stock solution in water, sterile filtered using a 0.22 μm syringe filter.

2. Sterile 10.3% (w/v) sucrose solution.

3. P buffer for *Streptomyces* protoplast preparation [14]. To prepare P buffer, add 103 g sucrose, 0.25 g K_2SO_4, 2.02 g $MgCl_2 \bullet 6H_2O$, and 2 mL trace element solution (40 mg/L $ZnCl_2$, 200 mg/L $FeCl_3 \bullet 6H_2O$, 10 mg/L $CuCl_2 \bullet 2H_2O$, 10 mg/L $MnCl_2 \bullet 4H_2O$, 10 mg/L $Na_2B_4O_7 \bullet 10H_2O$, 10 mg $(NH_4)_6Mo_7O_{24} \bullet 4H_2O$) to 800 mL distilled water. Dispense in 80 mL aliquots, autoclave. Before each use, add to each flask, in order: 1 mL 0.5% (w/v) KH_2PO_4, 10 mL 3.68% (w/v) $CaCl_2 \bullet 2H_2O$, and 10 mL 5.73% (w/v) TES buffer (pH 7.2).

4. 100 mg/mL lysozyme solution in sterile water.

5. Sterile 20% (v/v) glycerol.

6. 100 mM pAzPhe (10 mL): add 206 mg of pAzPhe (Chem-Impex) to 8 mL of water and mix. Add 1 M NaOH dropwise with intermittent mild vortexing until the solution becomes clear. Add water to 10 mL total volume. Sterilize the solution by filtration using a 0.22 μm syringe filter.

7. 100 mM pIPhe (10 mL): add 291 mg pIPhe (Sigma-Aldrich) to 8 mL water and mix. Add 1 M NaOH dropwise with intermittent mild vortexing until the solution becomes clear. Add water to 10 mL total volume. Sterilize the solution by filtration using a 0.22 μm syringe filter.

8. Ni-NTA lysis buffer: 50 mM NaH_2PO_4, 300 mM NaCl, 10 mM imidazole, pH 8.0.

9. Ni-NTA wash buffer: 50 mM NaH_2PO_4, 300 mM NaCl, 20 mM imidazole, pH 8.0.

10. Ni-NTA elution buffer: 50 mM NaH_2PO_4, 300 mM NaCl, 250 mM imidazole, pH 8.0.

11. Solutions and reagents for Western blotting and chemiluminescent detection [15].

12. 100 mM phenylmethylsulfonyl fluoride (PMSF, Sigma-Aldrich) stock solution in isopropanol.

13. HIC Loading buffer: 0.7 M $(NH_4)_2SO_4$, 50 mM NaH_2PO_4, pH 8.0.

14. 1.7 M ammonium sulfate buffer: 1.7 M $(NH_4)_2SO_4$, 50 mM NaH_2PO_4, pH 8.0.

2.2 Media

1. Luria-Bertani (LB) broth (Difco).

2. Tryptic Soy Broth (TSB) (Cole Parmer).

3. YEME medium, prepared according to [14]. To prepare YEME medium, add 0.3 g yeast extract (Difco), 0.5 g Bacto-peptone (Difco), 0.3 g malt extract (BD Biosciences), 1 g glucose, and 34 g sucrose to 100 mL water, and autoclave. After autoclaving, add 0.2 mL of filter sterilized 2.5 M $MgCl_2 \cdot 6H_2O$ (5 mM final concentration) and 2.5 mL of filter sterilized 20% (w/v) glycine (0.5% final concentration) to media and mix thoroughly.

4. R5 agar, prepared according to [14]. To prepare R5 agar, add 103 g sucrose, 10 g glucose, 5 g yeast extract (Difco), 100 mg casamino acids (Fisher), 5.73 g TES, 250 mg K_2SO_4, 10.12 g $MgCl_2 \cdot 6H_2O$, 2 mL trace element solution (40 mg/L $ZnCl_2$, 200 mg/L $FeCl_3 \cdot 6H_2O$, 10 mg/L $CuCl_2 \cdot 2H_2O$, 10 mg/L $MnCl_2 \cdot 4H_2O$, 10 mg/L $Na_2B_4O_7 \cdot 10H_2O$, 10 mg $(NH_4)_6Mo_7O_{24} \cdot 4H_2O$), and 3 g NaOH to 1 L of deionized water, adjust pH to 7.2 with HCl, add 22 g agar, and autoclave. After autoclaving, add 15 mL of 20% (w/v) sodium glutamate, 15 mL of 2% (w/v) $NaNO_3$, 10 mL of 0.5% (w/v) KH_2PO_4, and 4 mL of 5 M $CaCl_2$ (all sterile) to media, mix thoroughly, and pour 20–25 mL agar plates.

3 Methods

3.1 Plasmid Construction

1. Transform competent *E. coli* with pSUA1-pIPheRS or pSUA1-pAzPheRS (*see* **Note 1**), select on Luria-Bertani (LB) agar containing 50 µg/mL apramycin. Pick one or more colonies using a pipette tip or sterile toothpick, transfer to culture tubes with 2–5 mL LB liquid media containing 50 µg/mL apramycin, incubate at 37 °C, 250 rpm, for 12–16 h. Harvest cells, isolate and quantify plasmid DNA.

2. Amplify the gene of interest containing the amber (TAG) stop codon at the desired site (*see* **Note 2**). Append the sequence containing the XbaI restriction site, SF14 promoter, and 5′-untranslated region (5′-UTR) from the pSUA1 vector to the 5′ end of the gene of interest; and the sequence con-

taining linker, hexa-histidine tag, stop codon, and BglII restriction site to the 3′ end of the gene of interest using PCR (*see* **Note 3**). Gel purify the resulting fragment.

3. Digest the pSUA1-pIPheRS/pSUA1-pAzPheRS vector with XbaI and BglII. Gel purify the resulting fragment, treat with CIP, purify the resulting linearized vector using a PCR cleanup kit.

4. Digest the amplified gene of interest containing amber stop codon and correct 5′ and 3′ ends appended (*see* **step 2**, above) with XbaI and BglII, purify the resulting insert using a PCR cleanup kit.

5. Ligate the linearized vector and insert from **steps 3** and **4**, transform competent *E. coli*, screen the resulting clones by restriction enzyme digestion or colony PCR, verify the insert sequence by Sanger sequencing (*see* **Note 4**).

6. Transform competent *E. coli* with the sequence verified vector and carry out culture and plasmid DNA isolation as in **step 1**.

3.2 Protoplast Preparation

Prepare *S. venezuelae* ATCC 15439 protoplasts according to the standard protocol used for *S. lividans* 66 and *S. coelicolor* ([14], p. 56–58) with minor modifications as follows.

1. Inoculate 5 mL of TSB with 50 μL of *S. venezuelae* spores in a 25 × 150 mm glass culture tube with around fifteen 4 mm glass beads, grow at 28 °C, 250 rpm, for 72 h.

2. Use 1 mL of the resulting culture to inoculate 50 mL of YEME medium in a 250 mL Erlenmeyer flask without beads. Grow at 28 °C, 250 rpm, until the cells reach stationary phase (7–10 d). *S. venezuelae* grows slowly in YEME medium due to osmotic stress from the high sucrose concentration.

3. Harvest cells by centrifugation ($1000 \times g$, 10 min) in a sterile 50 mL conical tube, discard the supernatant.

4. Resuspend the resulting cell pellet in 30 mL of sterile 10.3% (w/v) sucrose in a sterile 50 mL conical tube, collect cells by centrifugation ($1000 \times g$, 10 min), carefully decant the supernatant.

5. Resuspend cells in 30 mL of P buffer containing 2 mg/mL lysozyme in a sterile 50 mL conical tube, incubate at 37 °C for 1 h on a mechanical rotator. Check for the formation of protoplasts (spherical cell-derived structures formed by enzymatic removal of the peptidoglycan layer from *Streptomyces* cells) using a light microscope.

6. Filter the resulting cell suspension through sterile 1 mL pipette tips packed loosely with cotton, into a sterile 50 mL conical tube to remove cell debris and collect protoplasts.

7. Collect protoplasts by centrifugation ($1000 \times g$, 10 min), carefully decant supernatant.

8. Resuspend protoplasts in 4 mL of P buffer, transfer in 300 µL aliquots to sterile 1.5 mL eppendorf tubes, freeze aliquots at −20 °C overnight, and transfer to −80 °C on the following day. One 300 µL aliquot is sufficient for six transformations (*see* **Note 5**).

3.3 Protoplast Transformation

Transform *S. venezuelae* protoplasts using the pSUA1-pIPheRS or pSUA1-pAzPheRS derived plasmid harboring the gene of interest according to the standard protocol for rapid small-scale polyethylene glycol (PEG)-mediated protoplast transformation ([14], p. 234–235) with minor modifications as follows.

1. Thaw one 300 µL aliquot of protoplasts by placing in a 37 °C water bath with gentle shaking.

2. Transfer thawed protoplasts in 50 µL aliquots into sterile Eppendorf tubes.

3. Add 200 ng of plasmid DNA (comprising not more than 5 µL total volume) to one 50 µL aliquot of protoplasts and mix by gently flicking the tube.

4. Add 200 µL P buffer containing 25% v/v polyethylene glycol (PEG) and mix by pipetting up and down several times.

5. Gently spread the entire solution on a 20 mL R5 agar plate, and incubate at 28 °C for 16 h.

6. After 16 h incubation, overlay the plate with 1.5 mL of 700 µg/mL sterile aqueous apramycin solution, dry the plate by leaving open in the hood for ~30 min, incubate at 28 °C for 3–5 days, or until colonies appear (*see* **Note 6**).

3.4 Transformant Recovery and Storage

1. Pick one or more colonies using a pipette tip or sterile toothpick, transfer to a glass tube with glass beads and 5 mL of TSB medium containing 50 µg/mL apramycin, incubate at 28 °C, 250 rpm, for 72 h.

2. Collect 1 mL of cells by centrifugation, remove supernatant, resuspend cells in 0.5 mL of sterile 20% glycerol, and store at −80 °C (*see* **Note 7**).

3.5 Small-Scale Protein Expression and Batch Purification

The following is a variant of the standard protocol for Ni-NTA affinity purification of proteins using polyhistidine tags [16] that has been optimized for efficient lysis of *Streptomyces* cells and small-scale batch purification of protein for Western blot analysis.

1. Transfer 50 µL of glycerol stock from the transformed *S. venezuelae* mutant to a glass tube with glass beads containing 10 mL of TSB medium, 50 µg/mL apramycin, and either 1 mM pIPhe (for mutants harboring pSUA1-pIPheRS derived constructs), 5 mM pAzPhe (for mutants harboring

pSUA1-pAzPheRS derived constructs), or no UAA (as a negative control) and incubate at 28 °C, 250 rpm, for 72 h (*see* **Note 8**).

2. Transfer the cells to a 15 mL conical tube, centrifuge (4000 × *g*, 10 min, room temperature), decant the supernatant.

3. Resuspend the cell pellet in 1 mL lysis buffer containing 2 mg/mL lysozyme, transfer to a 1.5 mL Eppendorf tube, and incubate at 37 °C for 1 h on a mechanical rotator.

4. Sonicate the mixture on ice for 2 min (20% amplitude, 24 × 5 s pulses with 10 s pauses), and pellet the cell debris by centrifugation (24,000 × *g*, 10 min, room temperature).

5. Transfer the supernatant to an Eppendorf tube containing 50 µL of Ni-NTA slurry that has been pre-equilibrated with lysis buffer, incubate at 4 °C for 1 h on a mechanical rotator (*see* **Note 9**).

6. Pellet Ni-NTA by centrifugation at 300 × *g* for 2 min, remove the supernatant.

7. Wash Ni-NTA resin with 1 mL wash buffer, pellet by centrifugation (300 × *g*, 2 min), remove the supernatant. Repeat wash and centrifugation one time, and completely remove the supernatant.

8. Elute the bound protein from Ni-NTA resin by adding 35 µL of elution buffer, gently flicking the tube several times, centrifuging (300 × *g*, 2 min) to collect the resin, and transferring the supernatant to an Eppendorf tube.

3.6 Western Blot Analysis

Separate and image UAA-containing protein of interest by SDS-PAGE and anti-hexahistidine tag Western blotting according to the standard protocol [15] with the following specifics. Load 13 µL of Ni-NTA eluate from each sample onto one lane of the SDS-PAGE gel. Use a 1:5000 dilution of anti-His (C-terminal) primary antibody and a 1:2000 dilution of the goat anti-mouse IgG (H + L)/HRP conjugate secondary antibody. Develop the membrane by incubation with chemiluminescent substrate solution according to the manufacturer's protocol, and image it using a Gel Documentation system capable of chemiluminescent detection. An example of Western blot analysis of sfGFP proteins produced in *S. venezuelae* using pSUA1-pIPheRS and pSUA1-pAzPheRS is shown in Fig. 3). The production of sfGFP protein by *S. venezuelae* harboring pSUA1-pAzPheRS grown in the absence of pAzPhe (Fig. 3, lane 2) is due to the fact that pAzPheRS retains significant catalytic activity toward endogenous amino acids tyrosine and phenylalanine [12].

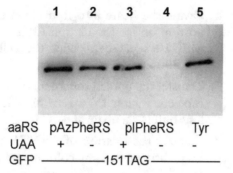

Fig. 3 Analysis of pIPhe- and pAzPhe-containing sfGFP variants in *Streptomyces*. Anti-His Western blot of Ni-NTA affinity purified sfGFP proteins obtained from *S. venezuelae* harboring pSUA1-pAzPheRS grown in the presence (lane 1) and the absence (lane 2) of 1 mM pAzPhe, pSUA1-pIPheRS grown in the presence (lane 3) and the absence (lane 4) of 1 mM pIPheRS, and opt-pSUA2-sfGFP (which harbors MjTyrRS) (lane 5). The blot was imaged for 165 s

3.7 Large-Scale Expression and Purification of UAA-Containing sfGFP Variants

The following is a protocol specifically developed for purification of small amounts of high purity sfGFP variants from *S. venezuelae* for mass spectrometric characterization. While the *S. venezuelae* culturing (**steps 1–3**) and initial Ni-NTA purification procedures (**steps 4–10**) are generally useful for purification of His-tagged UAA-containing proteins of interest, the hydrophobic interaction chromatography (HIC) procedure (**steps 11–18**) is primarily useful for purification of hydrophobic proteins such as sfGFP. sfGFP proteins obtained using this method are >90% pure, with expressed yields of 200–250 μg/L culture and isolated yields of 15–20 μg/L culture.

1. Transfer 50 μL of *S. venezuelae* glycerol stock to 40 mL of TSB medium containing 50 μg/mL apramycin and either 1 mM pIPhe or 5 mM pAzPhe in a 250 mL Erlenmeyer flask containing about one hundred 4 mm glass beads, incubate at 28 °C, 250 rpm, for 72 h.

2. Transfer 8 mL of the resulting seed culture to each of 4 2 L Erlenmeyer flasks containing 400 mL of TSB medium (1.6 L in total) containing 50 μg/mL apramycin and either 1 mM pIPhe or 5 mM pAzPhe with no glass beads, incubate at 28 °C, 200 rpm, for 96 h (*see* **Note 10**).

3. Harvest the cells by centrifugation (6,000 × *g*, 20 min, room temperature) and remove the supernatant (*see* **Note 11**).

4. Resuspend the cells in 24 mL of lysis buffer containing 2 mg/mL lysozyme, incubate at 37 °C for 1 h on a mechanical rotator.

5. Force the resulting mixture through a 26-gauge needle to disrupt mycelia.

6. Divide the lysate into four equal portions and sonicate each portion on ice for 12 min (30% amplitude, 72×10 s pulses with 10 s pauses), pellet the cell debris by centrifugation ($35,000 \times g$, 45 min, 4 °C).

7. Transfer the pooled supernatants to a prechilled 50 mL conical tube containing 1.6 mL of Ni-NTA slurry pre-equilibrated in lysis buffer, incubate at 4 °C for 2 h on a mechanical rotator.

8. Transfer the mixture of Ni-NTA and supernatant to a Poly-Prep chromatography column and drain the flow-through.

9. Wash the column with 20 mL of lysis buffer, followed by 10 mL of wash buffer.

10. Elute the protein with 5 mL of elution buffer.

11. Mix the Ni-NTA eluate with the appropriate amount of 1.7 M ammonium sulfate buffer to give a final concentration of 0.7 M $(NH_4)_2SO_4$.

12. Remove the insoluble proteins by centrifugation ($24,000 \times g$, 1 min, room temperature).

13. Apply the supernatant by syringe to aHiTrap Phenyl HP HIC column pre-equilibrated with loading buffer. Wash the column with 8 mL of loading buffer.

14. Elute the proteins by step gradient (addition of 7×1.5 mL portions of loading buffer containing decreasing concentrations of $(NH_4)_2SO_4$—0.6 M, 0.5 M, 0.4 M, 0.3 M, 0.2 M, 0.1 M, 0 M), collect 1.5 mL fractions.

15. Check the purity of the fractions by SDS-PAGE.

16. Mix the purest fractions with the appropriate amount of ammonium sulfate buffer to give a final concentration of 0.7 M $(NH_4)_2SO_4$, remove the insoluble proteins by centrifugation ($24,000 \times g$, 1 min, room temperature).

17. Employ a similar procedure to that used for the first HIC column for the isolation of pure sfGFP. Use the same wash and step gradient, but collect 1 mL fractions.

18. Analyze fractions by SDS-PAGE, pool pure sfGFP-containing fractions for further characterization.

4   Notes

1. Plasmids pSUA1-pIPheRS, pSUA1-pAzPheRS, positive control vectors opt-pSUA2 and pSUA5, and negative control vectors pSUA3 and pSUA4 [12] can be requested from Charles E. Melançon III. Because expression of UAA-containing sfGFP variants in *S. venezuelae* using pSUA1-pIPheRS and pSUA1-pAz-PheRS has been previously demonstrated [12], these plasmids can also be used as positive controls.

2. If the codon corresponding to the desired site of UAA incorporation is more than 50 bp from either end of the gene of interest, we typically amplify the gene as two fragments, each with an 18–25 bp overlapping region containing the TAG codon mutation. These two fragments are then joined by overlap extension PCR. If the codon corresponding to the desired site of UAA incorporation is within 50 bp of either end of the gene of interest, we typically amplify the gene as a single fragment, and install the TAG codon mutation using an upstream or downstream primer. If the gene of interest is from a high GC Gram positive Actinobacterium, codon optimization is not typically necessary. However, if the gene of interest is from an organism with significantly different GC content or codon usage, codon optimization via gene synthesis may be required. Codon usage statistics from the *S. coelicolor* A3(2) genome are used as the basis for codon optimization. If the gene of interest contains one or more natural internal XbaI or BglII sites, these sites must be mutated prior to gene cloning. We typically accomplish this by installing a silent mutation via overlap extension PCR to eliminate each restriction site while maintaining a high frequency codon at the mutated position.

3. The sequence to be installed at the 5′ end of the gene of interest is as follows:
   NNNNNN<u>TCTAGA</u>GG<u>TTGACCTTGATGAGGCG GCGTGAGCTACAATCAAT</u>AGTCGATTAGAGGAGA AATTAGAA<u>ATG</u> where NNNNNN (hereafter referred to as N$_6$) is a 6 bp sequence that ensures efficient digestion by XbaI (hereafter referred to as the digestion spacer), the doubly underlined sequence is the XbaI site, the dotted underlined sequence is the SF14 promoter, and the underlined ATG represents the native start codon of the gene of interest (can also be GTG). This 70 bp sequence can be appended to the 5′ end of the gene of interest by two sequential PCR amplifications using two primers.

   The sequence to be installed at the 3′ end of the gene of interest is as follows:
   <u>N N N</u> G G C T C G **C A C C A C C A T C A CCACCAC**<u>TGAAGATCT</u>NNNNNN where the underlined NNN represents the last sense codon of the gene of interest, the dotted underlined sequence encodes a Gly-Ser linker, the bold sequence encodes the C-terminal hexa-histidine tag, the underlined TGA codon is the stop codon, the doubly underlined sequence is the BglII site, and N$_6$ is the digestion spacer. This 39 bp sequence can be appended to the 3′ end of the gene of interest by PCR amplification using a single primer that contains the reverse complementary sequence.

4. A similar cloning strategy to that described for the gene of interest can be employed to replace the aaRS and/or tRNA genes in pSUA1-pIPheRS or pSUA1-pAzPheRS. If introducing an aaRS variant derived from *Mj*TyrRS, the codon optimized *Mj*TyrRS gene in opt-pSUA2 [12] can be used as a starting point for the construction of the desired *Mj*TyrRS variant by one pot overlap extension PCR. The unique restriction sites EcoRV and XbaI are used to replace the aaRS. Note that the aaRS gene should be codon optimized for expression in *Streptomyces* and must contain the appropriate codon changes to allow UAA recognition by the aaRS; the sequence containing the digestion spacer, EcoRV site, SF14 promoter, and 5′ UTR must be appended to the 5′ end of the aaRS gene; and the sequence containing the XbaI site and digestion spacer must be appended to the 3′ end of the aaRS gene immediately following the aaRS stop codon. The sequence to be installed at the 5′ end of the aaRS gene is as follows:

NNNNNNGATATCGGTTGACCTTGATGAGGCGG CGTGAGCTACAATCAATAGTCGATTAGAGGAGAA ATTAGAAATG where N_6 is the digestion spacer, the doubly underlined sequence is the EcoRV site, the dotted underlined sequence is the SF14 promoter, and the underlined ATG represents the native start codon of the gene of interest (can also be GTG). Similarly, BglII and PvuI are used to replace the tRNA. Note that the sequence containing the digestion spacer, the BglII site, and the SCOt03p tRNA promoter/5′ spacer sequence must be appended to the 5′ end of the tRNA gene; and the sequence containing the 3′ spacer sequence, the PvuI site, and the digestion spacer must be appended to the 3′ end of the tRNA gene.

The sequence to be installed at the 5′ end of the tRNA gene is as follows:

NNNNNNAGATCTAATTGTGCTGGCCCCGG AATCCGCTAGAGTTCATCACGTCGCCGGGC CGCGCAAGCGGACCGAACCGACAGNNN where N_6 is the digestion spacer, the doubly underlined sequence is the BglII site, the dotted underlined sequence is the SCOt03p promoter, and the underlined NNN represents the first three nucleotides of the tRNA gene.

The sequence to be installed at the 3′ end of the tRNA gene is as follows:

NNNCACTCACGATCGNNNNNN where the underlined NNN represents the last three nucleotides of the tRNA gene, the doubly underlined sequence is the PvuI site, and N_6 is the digestion spacer.

5. After preparing *Streptomyces* protoplasts, aliquot the remaining protoplast buffer (P buffer) into sterile tubes and store

them at −80 °C along with the protoplasts for use in the subsequent protoplast transformation step.

6. The transformation efficiency of protoplasts is dependent on the quality of each individual protoplast preparation. The amount of plasmid DNA can be increased to potentially overcome the problem of inefficient protoplast transformation.

7. Spore suspensions [14] are superior to glycerol stocks for long-term storage of *Streptomyces*.

8. The typical UAA concentration used 1 mM. However in the case of pAzPhe, 5 mM is required to obtain a high level of UAA incorporation due to the fact that pAzPheRS retains significant catalytic activity toward endogenous amino acid tyrosine in the presence of 1 mM pAzPhe [12].

9. To avoid blocking pipettes tips when transferring Ni-NTA slurry, remove the end of the pipette tip with a razor blade to increase the inlet diameter to 4 mm.

10. The length of time required for optimal expression of soluble protein of interest should be determined empirically. *S. venezuelae* harboring pSUA1-pIPheRS/pSUA1-pAzPheRS typically reach stationary phase after 48–72 h.

11. Buffers used in all subsequent steps contain 1 mM PMSF to prevent protease degradation of the protein of interest.

Acknowledgments

We gratefully acknowledge Peter G. Schultz of the Scripps Research Institute for providing the plasmid harboring MjtRNA$^{Tyr}_{CUA}$ and Hung-wen Liu at The University of Texas at Austin for providing *S. venezuelae* ATCC 15439. This work is supported by NIH NM-INBRE grant P20 GM103451 (C.E.M.) and by the University of New Mexico.

References

1. Xiao H, Schultz PG (2016) At the Interface of chemical and biological synthesis: an expanded genetic code. Cold Spring Harb Perspect Biol 8(9):a023945

2. Liu CC, Schultz PG (2010) Adding new chemistries to the genetic code. Annu Rev Biochem 79:413–444

3. Chin JW (2014) Expanding and reprogramming the genetic code of cells and animals. Annu Rev Biochem 83:379–408

4. Weissman KJ (2015) The structural biology of biosynthetic megaenzymes. Nat Chem Biol 11:660–670

5. Xu W, Qiao K, Tang Y (2013) Structural analysis of protein-protein interactions in type I polyketide synthases. Crit Rev Biochem Mol Biol 48:98–122

6. Ongley SE, Bian X, Neilan BA, Müller R (2013) Recent advances in the heterologous expression of microbial natural product biosynthetic pathways. Nat Prod Rep 30:1121–1138

7. Zhang H, Boghigian BA, Armando J, Pfeifer BA (2011) Methods and options for heterologous production of complex natural products. Nat Prod Rep 28:125–151

8. Baltz RH (2010) *Streptomyces* and *Saccharopolyspora* hosts for heterologous expression of secondary metabolite gene clusters. J Ind Microbiol Biotechnol 37:759–772

9. Chin J, Santoro S, Martin A, King D, Wang L, Schultz P (2002) Addition of *p*-Azido-L-phenylalanine to the genetic code of *Escherichia coli*. J Am Chem Soc 124:9026–9027

10. Chin JW, Martin AB, King DS, Wang L, Schultz PG (2002) Addition of a photocrosslinking amino acid to the genetic code of *Escherichia coli*. Proc Natl Acad Sci U S A 99:11020–11024

11. Pédelacq J, Cabantous S, Tran T, Terwilliger T, Waldo G (2005) Engineering and characterization of a superfolder green fluorescent protein. Nat Biotechnol 24:79–88

12. He J, Van Treeck B, Nguyen HB, Melançon CE III (2016) Development of an unnatural amino acid incorporation system in the Actinobacterial natural product producer *Streptomyces venezuelae* ATCC 15439. ACS Synth Biol 5:125–132

13. Melançon C, Liu H (2007) Engineered biosynthesis of macrolide derivatives bearing the non-natural Deoxysugars 4-*epi*-D-Mycaminose and 3-*N*-Monomethylamino-3-Deoxy-D-Fucose. J Am Chem Soc 129:4896–4897

14. Kieser T, Bibb MJ, Buttner MJ, Chater KF, Hopwood DA (2000) Practical *Streptomyces* genetics. The John Innes Foundation, Norwich

15. Mahmood T, Yang P-C (2012) Western blot: technique, theory, and trouble shooting. North American J Med Sci 4:429–434

16. Bornhorst JA, Falke JJ (2000) Purification of proteins using Polyhistidine affinity tags. Methods Enzymol 326:245–254

Chapter 11

Expression and Purification of Site-Specifically Lysine-Acetylated and Natively-Folded Proteins for Biophysical Investigations

Michael Lammers

Abstract

N-(ε)-lysine-acetylation (short: lysine-acetylation) is a dynamic and powerful posttranslational modification to regulate protein function. Mutational approaches are often poor to access the real mechanistic impact of lysine-acetylation at the molecular level. Therefore, the ability to site-specifically incorporate N-(ε)-acetyl-L-lysine (short: AcK) into proteins dramatically increased our understanding how lysine-acetylation regulates protein function by using diverse molecular mechanisms going far beyond neutralizing a positive charge at the lysine-side chain. Genetically encoding AcK is a powerful way to introduce AcK into proteins, resulting in homogenously, quantitatively, and site-specifically lysine-acetylated proteins. Thereby, lysine-acetylated proteins can be produced in their natively-folded state in a high quality and in a yield sufficient to perform biophysical studies, including X-ray crystallography. This protocol describes the expression and purification of site-specifically lysine-acetylated proteins in *Escherichia coli* using the genetic-code expansion concept (GCEC) and subsequent steps to assess the successful incorporation of AcK by immunoblotting and mass-spectrometry.

Key words Acetyl-L-lysine, Amber stop codon, Genetic-code expansion, Aminoacyl-tRNA-snthetase, Acetyl-L-lysyl-tRNA-synthetase, *Methanosarcina barkeri*, PylRS, PylT, MbtRNA$_{CUA}$.

1 Introduction

Most functional studies of lysine-acetylation were performed by either studying the glutamine mutants as a mimic for lysine-acetylation or the arginine mutants to conserve the non-acetylated state, respectively. However, studies performed by others and our lab showed that these mutational approaches are often misleading as glutamine often constitutes a poor molecular mimic for a lysine-acetylation. In fact, arginine is a closer mimic of the acetylated state, if lysine-acetylation mediates steric rather than electrostatic effects [1–3].

A challenge to study the impact of lysine-acetylation on protein function was the production of quantitatively, homogeneously, and site-specifically lysine-acetylated proteins in quality and yield

Edward A. Lemke (ed.), *Noncanonical Amino Acids: Methods and Protocols*, Methods in Molecular Biology, vol. 1728, https://doi.org/10.1007/978-1-4939-7574-7_11, © Springer Science+Business Media, LLC 2018

sufficient to perform biophysical studies [2, 4, 5]. For biophysical techniques such as isothermal-titration calorimetry, stopped-flow, microscale-thermophoresis, and X-ray crystallography highly pure material in multi-miligramm quantities is required. Furthermore, the proteins need to be highly stable during the course of the experiments and should be modifiable by fluorescence labeling, lipid modifications etc. [1, 6–8].

Attempts to lysine-acetylate proteins by either using purified lysine-acetyltransferases (KATs) or semisynthetic, chemical approaches have been reported previously [9–11]. However, each of those has several limitations. Semisynthetic, chemical approaches are often restricted to the termini of proteins and result in low yields. Enzymatic acetylation in vitro using purified KATs often yields a non-quantitative and heterogenic products since the reactions rarely proceed until completion and are not site-specific. Furthermore, it is challenging to purify catalytically-active KAT-complexes. A powerful solution to these problems was developed in the Chin lab by genetically encoding N-(ε)-acetyl-L-lysine (AcK) in recombinant proteins [2, 4, 5], enabling the study of the real molecular impact of lysine-acetylation to control protein function.

The pyrrolysyl-tRNA-synthetase (PylRS)/amber suppressor tRNA (tRNA$_{CUA}$) pair, present in *Methanosarcina* species and some bacteria, is the basis for this system. Methanogenic archaea use PylRS/tRNA$_{CUA}$ to incorporate pyrrolysine as the 22nd proteinogenic amino acid into monomethylamine methyltransferase, an enzyme used in methanogenesis (Fig. 1a) [12]. PylRS has been used as a powerful synthetic biology tool to incorporate more than 100 different noncanonical amino acids (ncAA) into proteins. This is due to some remarkable features of PylRS: (a) it is orthogonal in most organisms such as *Escherichia coli (E. coli)*, *Caenorabditis elegans*, *Drosophila melanogaster* and mammals, (b) it has a no editing domain, (c) it has a high promiscuity for substrate α-substituents and can adopt different side-chains as it binds to the pyrrolysyl-side chain mostly via hydrophobic interactions and (d) it does not recognize the tRNA's anticodon (Fig. 1b) [13].

In an elegant iterative-saturation mutagenesis (ISM) strategy six active-site residues (L266, L270, Y271, L274, C313, and W382) were selected for the directed evolution of *Methanosarcina barkeri* MS PylRS to finally result in an enzyme, AcKRS-3 (L266M, L270I, Y271F, L274A, C313F, i.e., W382 was not altered) that charges the amber-suppressor tRNA (*Mbt*RNA$_{CUA}$) with acetyl-L-lysine (AcK) (Fig. 1c) [5]. Using the amber stop codon (TAG) in the POI$_{TAG}$ creates the site-specificity of AcK incorporation. Notably, the amber codon is the least frequently used stop-codon in *E. coli* (TAA$_{ochre}$: 64%; TGA$_{umber/opal}$: 29%; TAG$_{amber}$: 7%), a prerequisite for the functionality of this approach.

We used this system in *E. coli* to produce site-specifically, lysine-acetylated and natively-folded proteins [1–3, 6, 7, 14]. This

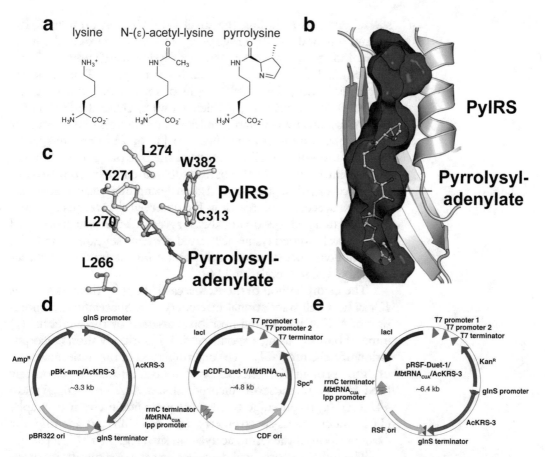

Fig. 1 A synthetically evolved acetyl-lysyl-tRNA-synthetase (AcKRS-3)/*Mbt*RNA$_{CUA}$ pair from *Methanosarcina barkeri* is used to site-specifically incorporate *N*-(ε)-acetyl-L-lysine (AcK) into proteins. (**a**) Structures of L-lysine, AcK, and pyrrolysine (Pyl). (**b**) The pyrrolysine-binding pocket of the pyrrolysyl-tRNA-synthetase (PylRS) of *M. mazei* (PDB: 2ZIM). The structure shows PylRS in complex with pyrrolysyl-adenylate. The pocket extends far beyond the pyrroline ring, which is an excellent prerequisite to evolve PylRS to genetically encode highly extended and bulky ncAA. (**c**) The active site residues selected for mutagenesis to all 20 aa to yield AcKRS-3 (L266M, L270I, Y271F, L274A, C313F). The residues shown were conserved between *M. mazei* and *M. barkeri*. (**d**) Two-plasmid system to incorporate AcK into proteins. AcKRS-3 is encoded on pBK-amp, a plasmid derived from pBR322 with Amp^R and pBR322 origin of replication (ori). The POI$_{TAG}$ is cloned into pCDF-Duet-1 (CDF ori and Spc^R) using restriction enzymes of multiple cloning site 1 (MCS-1) and/or MCS-2, 3′ from the T7 promoters 1 and 2, respectively. The suppressor tRNA$_{CUA}$ of *M. barkeri* (*Mbt*RNA$_{CUA}$) including its regulatory elements is cloned into the *Xba*I-restriction site. Both plasmids need to be compatible regarding its antibiotic resistance and its ori. (**e**) The one-plasmid system is based on pRSF-Duet-1 (RSF ori and Kan^R). It contains the coding regions for *Mbt*RNA$_{CUA}$, AcKRS-3 and the POI$_{TAG}$

enabled us to solve the first crystal structures of a protein, Cyclophilin A (CypA), carrying a posttranslationally relevant lysine-acetylation, and complexes of acetylated CypA with HIV1-CA^N and cyclosporine [2]. Recently, we were able to solve the first crystal structure of a protein, RhoGDIα, carrying two acetylated lysine-side chains and performed extensive biophysical studies on several proteins including the small GTP-binding proteins K-Ras 4B and Ran and the Rho-regulator RhoGDIα regarding the functional and structural consequences of lysine-acetylation [1, 7, 8]. We designed a one-plasmid system based on pRSF-Duet-1 (kanamycin resistance, Kan^R; RSF origin of replication) encoding the AcKRS-3 and MbtRNA$_{CUA}$, which allows to clone the POI$_{TAG}$ (or two POIs) into two available multiple-cloning sites (MCS) and/or to co-express two proteins (Fig. 1e) [1, 6, 7]. The site-specific incorporation of AcK allows studying the role of lysine-acetylation on protein function in the natively-folded context, compared to lysine-acetylated peptides and also analyzing how deacetylases catalyze the deacetylation reaction. We observed that results reported for acetylated peptides do not always agree with the results obtained with natively-folded lysine-acetylated proteins suggesting that the three-dimensional structure is important for the functional consequence of lysine acetylation [14].

The incorporation of the AcK is carried out by the endogenous *E. coli* host cell translational machinery. The successful incorporation of AcK at the desired position is assessed by three criteria: (a) immunoblotting using a specific anti-AcK antibody shows incorporation of AcK into POI$_{AcK}$, (b) determination of the molecular mass of the protein by electrospray ionization-mass-spectrometry (ESI-MS) shows successful incorporation of AcK into POI$_{AcK}$ (mass: +42 Da), (c) tryptic-digest and subsequent liquid chromatography (LC)-tandem mass-spectrometry (LC-MS/MS) to identify the exact position of the lysine-acetylation site (Fig. 3a–c) [1, 2].

The following protocol describes first the site-specific incorporation of acetyl-L-lysine into a protein of interest (POI$_{AcK}$) taking the small GTP-binding protein Ran, His$_6$-Ran acetylated at position K37 (His$_6$-Ran AcK37), as a model [1] and second the verification of site specific AcK incorporation.

2 Materials

2.1 Cloning of Constructs and Protein Expression

1. *E. coli* DH5α chemically competent or electrocompetent cells (Thermo Fisher Scientific Inc.; or other company). Any other *E. coli* strain for efficient cloning might be used.

2. *E. coli* BL21 (DE3) or *E. coli* C41 (DE3) (Thermo Fisher Scientific Inc.; Sigma-Aldrich Co. LLC). Other *E. coli* expression strains might be used but the strain should contain the λDE3 lysogen as expression is initiated using a T7-RNA-polymerase promoter.

3. Luria–Bertani (LB) growth medium and Terrific Broth (TB), twofold Yeast Extract and Trypton (2-YT) growth medium.

4. LB agar plates with 100 μg/L ampicillin (Amp).

5. LB agar plates with 100 μg/L Amp and 50 μg/L spectinomycin (Spc) (for the two-plasmid system with pBK-amp/AcKRS-3 and pCDF-Duet1/MbtRNA$_{CUA}$/POI$_{TAG}$; available from Dr. J. Chin, MRC-LMB, Cambridge/UK).

6. LB agar plates with 25 μg/L kanamycin (Kan) (for the one-plasmid system pRSF-Duet-1 AcKRS-3 MbtRNA$_{CUA}$/POI$_{TAG}$).

7. QuikChange® site-directed mutagenesis kit (Agilent Technologies, Santa Clara, CA) or other site-directed mutagenesis kits.

2.2 Protein Expression, Site-Specific Incorporation of N-(ε)-acetyl_L-lysine, and Protein Purification

1. Branson Sonifier 250 for cell lysis (or any other sonifier).

2. Resuspension/lysis buffer: 50 mM Tris/HCl, pH 7.4, 100 mM NaCl, 5 mM MgCl$_2$, 2 mM nicotinamide, 100 μM PMSF, 1 mM Pefabloc SC® (Sigma-Aldrich Co. LCC), 2 mM β-mercaptoethanol.

3. Ni Sepharose 6 Fast Flow/Ni Sepharose High Performance resin packed in XK 26/20 column (or other columns) (GE Healthcare Europe GmbH, Freiburg, Germany).

4. Equilibration buffer (His I): 25 mM Tris/HCl, pH 8.0, 500 mM NaCl, 5 mM MgCl$_2$, 2 mM nicotinamide, 10 mM imidazole, 2 mM β-mercaptoethanol, 100 μM PMSF.

5. High salt wash buffer (His II): 25 mM Tris/HCl, pH 8.0, 1 M NaCl, 5 mM MgCl$_2$, 10 mM imidazole, 2 mM β-mercaptoethanol.

6. Low salt wash buffer (His III): 25 mM Tris/HCl, pH 8.0, 300 mM NaCl, 5 mM MgCl$_2$, 10 mM imidazole, 2 mM β-mercaptoethanol.

7. Elution buffer (His IV): 25 mM Tris/HCl, pH 8.0, 300 mM NaCl, 5 mM MgCl$_2$, 500 mM imidazole, 2 mM β-mercaptoethanol.

8. HiLoad 26/600 or 16/600 Superdex 200 pg/75 pg size exclusion chromatography columns (depending on the molecular size and oligomeric state of the protein and the protein yield) (GE Healthcare Europe GmbH, Freiburg, Germany).

9. Gelfiltration buffer: 50 mM Tris/HCl, pH 7.5, 100 mM NaCl, 5 mM MgCl$_2$, 2 mM β-mercaptoethanol.

10. Fast protein liquid chromatographic (FPLC) equipment including Unicorn software (GE Healthcare, Europe GmbH, Freiburg, Germany).

11. N-(ε)-acetyl-L-lysine (Chem-Impex International, Inc., U.S.A; or other company).

12. 0.22 μm syringe filter (Carl Roth GmbH & Co. KG, Karlsruhe, Germany; or other company).

13. Nicotinamide (NAM, Sigma-Aldrich Co. LCC).

14. β-mercaptoethanol, dithiothreitol (DTT), dithioerythritol (DTE) (Sigma-Aldrich Co. LCC).

15. SDS-PAGE running buffer: 25 mM Tris/HCl, pH 8.3, 192 mM glycine, 2% (w/v) SDS.

16. SDS-PAGE sample buffer: 50 mM Tris/HCl, pH 6.8, 50% (v/v) Glycerol, 500 mM DTE, 10% (w/v) SDS, 0.5% (w/v) bromphenolblue.

17. SDS-PAGE staining solution: 40% (v/v) methanol, 10% (v/v) acetic acid, 0.4% (w/v) Coomassie-brilliant blue R-250, 0.4% (w/v) Coomassie-brilliant blue G-250.

18. SDS-PAGE destaining solution: 20% (v/v) ethanol, 10% (v/v) acetic acid.

19. Amicon® Ultra centrifugal filter device dependent on the molecular weight of the POI$_{AcK}$ (Merck Millipore, Darmstadt, Germany).

20. UV–visible spectrophotometer (Eppendorf, Hamburg, Germany; or other company).

2.3 Quality Control of Site-Specifically Lysine-Acetylated Protein by Immunoblotting, ESI-MS, and LC-MS/MS

1. Anti-acetyl-L-lysine antibody (e.g., ab21623; 1:1500 in 3% (w/v) milk/PBST; Abcam plc, Cambridge/UK).

2. Goat HRP-coupled anti-rabbit IgG (ab6271; 1:1000 in PBST; Abcam plc, Cambridge/UK).

3. Equipment for immunoblotting (semi-dry blotter, nitrocellulose/PVDF membrane, Roti-Lumin (Carl Roth GmbH & Co. KG), Super RX Fujifilms (Fuji), CCD camera-based Vilber Fusion Express detection system, ImageJ software).

4. Transfer buffer for semi-dry blotting: 25 mM Tris base, 150 mM Glycine, 10% (v/v) methanol.

5. PBST buffer: 10 mM di-sodium hydrogen phosphate, 1.8 mM potassium di-hydrogen phosphate, 137 mM NaCl, 2.7 mM KCl, 0.1% (v/v) Tween-20, pH 7.4.

6. ESI-MS equipment (nano-electrospray ionization source (Thermo Fisher Scientific Inc.), QExactive Plus mass spectrometer (Thermo Fisher Scientific Inc.), MagTran software).

7. ESI-buffer: 20% (v/v) acetonitrile (ACN), 1% (v/v) acetic acid (AA).

8. ESI-measure buffer: 25% (v/v) acetonitrile (ACN), 1% (v/v) acetic acid (AA).

9. Direct Detect® (Merck Millipore, Darmstadt, Germany).

10. LC-MS/MS equipment (Easy nano-flow LC1000 system (Thermo Fisher Scientific Inc.), nano-electrospray ionization source (Thermo Fisher Scientific Inc.), quadrupole orbitrap mass spectrometer (QExactive Plus, Thermo Fisher Scientific Inc.), MaxQuant software).

11. ESI-MS and LC-MS/MS buffer A: 80% (v/v) acetonitrile (ACN), 0.1% (v/v) formic acid (FA).

12. ESI-MS and LC-MS/MS buffer B: 0.1% (v/v) formic acid (FA) in 80% (v/v) acetonitrile (ACN).

13. In-solution digest: iodacetamide, trypsin (Sigma-Aldrich Co. LLC), Lys-C (Wako), triethylammoniumbicarbonate (TEAB).

3 Methods

3.1 Cloning of the Gene Encoding the Protein-of-Interest (POI$_{TAG}$) into pCDF-Duet-1/MbtRNA$_{CUA}$ (Two-Plasmid-System) or pRSF-Duet-1/AcKRS-3/MbtRNA$_{CUA}$ (One-Plasmid-System)

1. Clone the coding sequence of the POI into either the vector pCDF-Duet-1/MbtRNA$_{CUA}$ (if using the two-plasmid system together with pBK-amp/AcKRS-3) or the vector RSF-Duet-1/AcKRS-3/MbtRNA$_{CUA}$ (if using the one-plasmid-system) using the restriction enzymes available in both multiple cloning sites or alternatively by Gibson assembly (Fig. 1d, e) [15].

2. For the expression of His$_6$-tagged K37-acetylated Ran full-length (His$_6$-Ran AcK37), the gene encoding AcKRS-3 including the glutaminyl-tRNA-synthetase promoter and terminator is inserted into the pRSF-Duet-1 vector using the *Sph*I restriction site, the gene encoding the amber-suppressor tRNA, MbtRNA$_{CUA}$, is inserted via the *Xba*I restriction site, creating a one-plasmid system for expression of a site-specifically lysine-acetylated POI$_{AcK}$ (Fig. 1e; *see* **Note 1**).

3. Perform site-directed mutagenesis, via QuikChange®, or alternative protocols, to alter the position of acetyl-L-lysine incorporation to TAG (POI$_{TAG}$) (*see* **Note 2**).

3.2 Preparation of Chemically Competent BL21 (DE3) pBK-amp/AcKRS-3 Cells (If Using Two-Plasmid System)

1. Transform competent *E. coli* BL21 (DE3) cells with pBK-amp/AcKRS-3 and spread cells on LB agar plates with 100 µg/L ampicillin (Amp). Grow cells overnight at 37 °C. Pick a single colony and transfer it into LB medium containing 100 µg/L Amp and grow cells overnight at 37 °C. Use this culture as a pre-culture to incoculate a culture to prepare chemically competent BL21 (DE3) pBK-AcKRS-3 cells.

2. Inoculate 400 mL LB medium with 5 mL of the pre-culture and grow culture until an OD$_{600}$ of 0.3. During the growth of the culture add Amp at half the usual concentration, e.g., 50 µg/mL. Centrifuge the culture at 500 × g for 10 min at 4 °C in eight prechilled 50 mL centrifuge tubes. Resuspend each pellet in 25 mL sterile ice-cold 0.1 M CaCl$_2$ and incubate for 20 min on ice. Centrifuge the cell suspension as described before (500 × g, 10 min, 4 °C). Resuspend each cell pellet in 2 mL ice-cold 0.1 M CaCl$_2$/15% (v/v) glycerol and incubate the suspension on ice for 1–4 h. Afterward, snap-freeze the

competent *E. coli* cells in 200 µL aliquots liquid N_2 and stored at −80 °C (*see* **Note 3**).

3. Use these cells to transform with pCDF-Duet-1/MbtRNA$_{CUA}$/POI$_{TAG}$.

3.3 Transformation of E. coli BL21 (DE3) and Starting the Culture

1. Transform competent *E. coli* BL21 (DE3)/pBK-amp/AcKRS-1 cells (two-plasmid-system) with pCDF-Duet-1/MbtRNA$_{CUA}$/POI$_{TAG}$ or alternatively competent *E. coli* BL21 (DE3) cells (one-plasmid-system) with pRSF-Duet-1/AcKRS-3/MbtRNA$_{CUA}$/POI$_{TAG}$ (*see* **Note 4**).

2. Plate cells on either LB agar plates containing 37.5 µg/L Spc and 50 µg/L Amp (antibiotic resistance Spc^R/Amp^R) (two-plasmid system) or on 25 µg/L Kan (antibiotic resistance Kan^R) (one-plasmid system).

3. Pick an individual colony to grow 100 mL pre-culture in LB Spc/Amp (two-plasmid system) or LB Kan (one-plasmid system) medium in 500 mL flask overnight at 37 °C (*see* **Note 4**).

3.4 Protein Expression

1. Inoculate eight 5 L baffled flasks, each containing 1.25 L liquid LB medium with appropriate antibiotics (50 µg/mL Spc/100 µg/mL Amp or 25 µg/mL Kan), with 12.5 mL of the overnight pre-culture (1:100). Allow cells to grow under constant shaking (160 rpm) at 37 °C (*see* **step 3** from Subheading 3.3).

2. When the cells reach the exponential/log growth phase, at an optical density OD$_{600}$ of 0.6, add acetyl-L-lysine and nicotinamide to a final concentration of 10 mM and 20 mM, respectively (*see* **Notes 5** and **6**).

3. Let the cells grow further for 30 min while adjusting the temperature to 18 or 20 °C (*see* **Note 6**).

4. Induce expression of site-specifically lysine-acetylated POI (POI$_{AcK}$) by the addition of 100–400 µM IPTG.

5. Conduct POI$_{AcK}$ expression overnight at 18 or 20 °C at 160 rpm for 12–16 h (*see* **Notes 7–12**).

6. Analyze the successful IPTG-induced POI$_{AcK}$ expression by SDS-PAGE (*see* **Note 8**).

7. Harvest the cells by centrifugation 4000 × g for 20 min at 4 °C. Discard the supernatant, resuspend the cells in 3 volumes of resuspension buffer, and freeze the suspension at −80 °C (*see* **Notes 9–12**).

3.5 Protein Purification (Part I)

1. Thaw the cell suspension and lyse the cells by sonication (2 × 3 min, micro tip limit 8, duty cycle 60%). The freeze-thawing before sonication improves cell lysis (Fig. 2a).

2. Centrifuge the lysate at 50,000 × g for 45 min. Collect the supernatant containing the POI$_{AcK}$ and discard the pellet. Filter

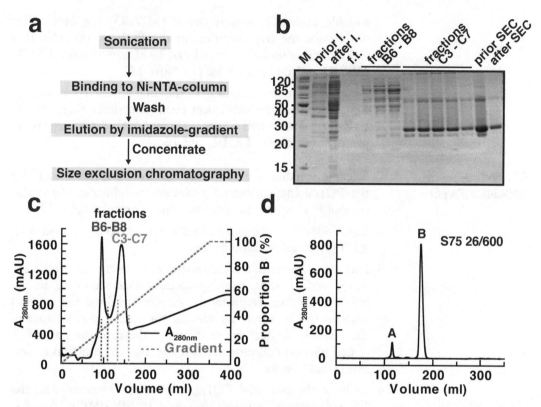

Fig. 2 Purification of lysine-acetylated full-length His_6-Ran AcK37. (**a**) Workflow of the purification including a Ni^{2+}-NTA-chromatography and a size-exclusion chromatography. (**b**) SDS-PAGE gel analysis of the purification process. I: lysis, f.t. flow-through of Ni^{2+}-NTA-column, SEC: size-exclusion chromatography. (**c**) Elution profile of His_6-Ran AcK37 from the Ni^{2+}-NTA-column during the imidazole gradient. (**d**) Chromatogram of the final size-exclusion chromatography on a HiLoad 26/600 Superdex 75 pg column. Peak B contains pure His_6-Ran Ack37 protein

the supernatant using a bottle-top filter with 0.45 μm pore size to remove unsoluble reminiscent cell debris.

3. Load the supernatant onto an equilibrated (≥3 CV (column volumes) of His I buffer) Ni^{2+}-NTA column using an automated FPLC system (at 4 °C, flowrate 1 mL/min) (Fig. 2c).

4. Wash the column with 5–10 CV of His II buffer followed by 2 CV buffer His III (4 °C, flowrate 1 mL/min), until the absorption at 280 nm, A_{280}, reaches a stable minimum.

5. Elute the POI_{AcK} by applying a gradient from 10 mM imidazole (buffer His III) to 500 mM imidazole (buffer His IV) over 10 CV at a flowrate of 1 mL/min. This gradient is followed by 1 CV of 100% buffer His IV to ensure complete elution of the target protein. Collect the eluate in fractions of 4 mL (Fig. 2c).

6. Run an SDS-PAGE gel to screen for the POI_{AcK} containing fractions (Fig. 2b, *see* **Note 8**).

7. Pool the fractions containing the POI_{AcK} and concentrate the protein using an Amicon® Ultra centrifugal filter device with a

suitable molecular weight cutoff (MWCO) (or other filter devices for the concentration of the POI_{AcK}). For His_6-Ran AcK37 with a molecular weight of 26 kDa, we used a 10 kDa MWCO amicon filter device (*see* **Note 13**).

8. Determine the concentration of the total protein by measuring the A_{280}, using the extinction coefficient determined by the ExPASy, ProtParam tool (http://web.expasy.org/protparam/) (*see* **Notes 12, 14**, and **15**).

3.6 Protein Purification (Part II)

1. Select a suitable size-exclusion chromatography column for the POI_{AcK} (dependent on molecular weight and oligomeric state of POI_{AcK} and the total amount of protein) (Fig. 2d).

2. Equilibrate the size-exclusion chromatography column with ≥ 1.5 CV of gelfiltration buffer.

3. Perform size-exclusion chromatography using an automated FPLC system using running specifications according to the selected column. Importantly, do not overload the column with protein to ensure an optimal purification performance of the column. Collect the fractions of the run using an automated fraction collector. Record the A_{280} as a function of elution volume/time.

4. Analyze the potential POI_{AcK} containing fractions, i.e., the fractions showing an increased A_{280}, by SDS-PAGE (the A_{280} depends on extinction coefficient and the amount of the POI_{AcK}, etc.) (Fig. 2c; *see* **Notes 8** and **16**).

5. Pool the POI_{AcK} containing and pure fractions. Importantly, size-exclusion chromatography also separates different oligomeric states enabling to pool these accordingly.

6. Calculate the apparent molecular weight of the POI_{AcK} by the determination of the elution volume at the A_{280} maximum for the POI_{AcK} on a calibrated size-exclusion chromatography column.

7. Concentrate the target protein containing fractions using an Amicon® Ultra centrifugal filter device with a suitable MWCO (or other filter devices for the concentration of the POI_{AcK}) (*see* **Note 13**).

8. Determine the protein concentration and shock freeze concentrated POI_{AcK} in liquid nitrogen and store in suitable aliquots at $-80\ ^{\circ}C$ to avoid repeated freeze-thaw cycles (*see* **Notes 14, 15**, and **16**).

3.7 Detection of Incorporated Acetyl-L-lysine by Immunoblotting

1. Analyze 1–5 μg of the lysine-acetylated POI_{AcK}, a non-acetylated protein (here: Ran wild-type) as a negative control and another lysine-acetylated protein as a positive control by SDS-PAGE (Fig. 3a; *see* **Notes 8, 14, 15** and **17**).

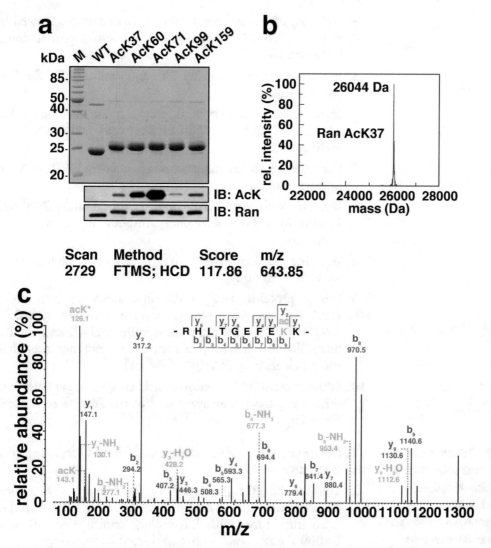

Fig. 3 Verification of site-specific, quantitative, and homogeneous incorporation of *N*-(ε)-acetyl-ʟ-lysine into His$_6$-Ran AcK37. (**a**) The lysine-acetylation of His$_6$-Ran is assessed by immunoblotting using a specific anti-acetyl-ʟ-lysine antibody. Upper panel: Coomassie-brilliant blue (CMB) staining of purified Ran proteins (5 μg per lane). Middle panel: immunoblotting using acetyl-ʟ-lysine antibody shows that the wild-type (WT) Ran protein is not lysine-acetylated. The other proteins show different signal intensities suggesting that the detection by the antibody depends on residues next to the acetylated lysine side chain. Lower panel: immunoblotting using an anti-Ran antibody shows equal loading in all lanes [1]. (**b**) The acetylation of His$_6$-Ran at K37 is confirmed by electrospray ionization-mass-spectrometry (ESI-MS) showing the expected molecular mass [1]. (**c**) LC-MS/MS is performed to determine the site of acetylation. His$_6$-Ran is acetylated at K37 as shown by the MS/MS fragmentation spectrum

2. Blot 100–250 ng of POI$_{AcK}$ on a PVDF or nitrocellulose membrane by semi-dry transfer in transfer buffer at 150 mA for 45 min. Prior to blotting, activate the PVDF membrane in methanol for 1 min (Fig. 3a).

3. Stain the membrane for 1 min with Ponceau S followed by a washing with ddH_2O to monitor successful protein transfer (*see* **Note 18**).

4. Block the membrane using 3% (w/v) milk/PBST for 1 h at room temperature.

5. Incubate with primary anti-acetyl-L-lysine antibody (abcam 21623; anti-rabbit; 1:1500 in 3% (w/v) milk/PBST) overnight at 4 °C.

6. Wash the membrane three times, 10 min with PBST, to remove primary antibody.

7. Incubate with goat HRP-coupled anti-rabbit IgG (ab6271; 1:1000 in PBST) secondary antibody for 1 h at room temperature.

8. Wash the membrane three times, 10 min with PBST, to remove secondary antibody.

9. Detect proteins using a chemiluminescence substrate for HRP. For long-term storage, strip membranes with 0.2 M NaOH for 20 min, rinse them with ddH_2O, and then dry them. For re-activation, incubate the membranes in methanol and block again as described (Fig. 3a).

10. Perform quantitative densitometric analysis of gel band intensities using the gel analyzer tool of the ImageJ software (*see* **Note 19**).

3.8 Determination of the Molecular Mass of Lysine-Acetylated Protein by Electrospray-Ionization Mass-Spectrometry (ESI-MS)

1. Determine the protein concentration with Direct Detect®, or alternative methods.

2. Transfer the POI_{AcK} into ESI-buffer. Use a 3 kDa Amicon® Ultra centrifugal filter device to remove any detergent and salt. Wash three times with ESI-buffer, centrifuge for 30 min at 14,000 × g after each step, and discard flow-through.

3. Transfer the app. 40 μL POI_{AcK} solution into a new tube and determine the protein concentration again using Direct Detect®. Adjust the protein concentration to 2 μg/μL (or less) and store on ice until measurement.

4. Connect the static electrospray ionization source (ESI) to the orbitrap mass spectrometer (MS) (ESI-MS). Use manual injection.

5. Fill 25 μL Hamilton syringe with ESI-measure buffer as a blank, and with 25 μL of prepared POI_{AcK} in ESI-measure buffer and start the ESI-MS measurement.

6. Analyze data using the MagTran software (or alternative software), which performs a charge state deconvolution of the mass-to-charge ratio spectra resulting in the molecular mass of the POI_{AcK} (Fig. 3b; *see* **Note 20**).

3.9 In Solution Digest and Peptide Purification for LC-MS/MS Analysis

1. Take 1–3 μg of POI$_{AcK}$ (1–3 μg) in solution and add urea buffer to reach a concentration of ≥6 M urea (*see* **Note 21**).

2. To reduce the POI$_{AcK}$, add dithiothreitol (DTT, 500 mM stock) to a final concentration of 5 mM and incubate at 37 °C for 1 h. Afterward add iodoacetamide (IAA, 500 mM stock) to a final concentration of 10 mM, vortex and incubate in the dark for 30 min.

3. Add endoproteinase Lys-C at an enzyme:substrate ratio of 1:100 and incubate at 37 °C for 4 h. Then, dilute the sample with 50 mM triethylammoniumbicarbonate (TEAB) to obtain a final urea concentration of 2 M. Add trypsin at an enzyme:substrate ratio of 1:100 and incubate at 37 °C overnight (*see* **Note 22**).

4. Stop digestion by acidifying by the addition of formic acid to a final concentration of 1% (*see* **Note 20**).

5. Perform stop-and-go-extraction-tip (StageTip) purification of the peptides obtained. Prepare one stage tip for each sample by stacking two layers of C18 material in a 0.2 mL pipette tip (=StageTip). Wash the StageTips with 20 μL methanol and centrifugation at 700 × *g* for 2 min, with 20 μL buffer B and centrifugation at 700 × *g* for 2 min and twice with 20 μL buffer A and centrifuge at 700 × *g* for 2 min. Avoid that the disk gets completely dry after the last centrifugation step and keep the C18 material always in buffer A.

6. Load the samples, acidified with formic acid (*see* **step 4**), onto the StageTips. Centrifuge at 700 × *g* for 4 min. Wash StageTips with 20 μL buffer A and spin at 700 × *g* for 2 min. Dry StageTip using a syringe and keep the StageTips at 4 °C.

3.10 Confirmation of Site-Specificity of Acetyl-L-lysine Incorporation Using LC-MS/MS

1. Peptides were separated using a buffer system containing buffer A and buffer B on an Easy nano-flow LC1000 UHPLC system. Execute a linear gradient from 4% (v/v) to 30% (v/v) buffer B for 40 min, followed by 95% (v/v) buffer B for 10 min and reequilibrate to 5% (v/v) buffer B in 5 min on a 50 cm column (75 μm ID), packed with C18 1.8 μm-diameter resin. Set flowrate to 250 nl/min and the column temperature to 45 °C.

2. Couple the UHPLC via a nano-electrospray ionization source to the quadrupole-Orbitrap mass spectrometer. Acquire MS spectra using 3×10^6 as an AGC (automatic gain control) target at a resolution of 70,000 (200 m/z) in a mass range of 350–1650 m/z.

3. Set the maximum injection time to 60 ms for ion accumulation. Measure MS/MS events for the 10 most abundant peaks (TOP 10 method) in the data-dependent mode in the high mass accuracy Orbitrap after HCD (Higher energy C-Trap

Dissociation) fragmentation in a 100–1650 m/z mass range. Set resolution to 35,000 at 200 m/z combined with an injection time of 120 ms and an AGC target of 5×10^5.

4. Process raw files with MaxQuant (version 1.5.2.8) using the implemented Andromeda search engine [16, 17]. For protein assignment, ESI-MS/MS fragmentation spectra were correlated with the Uniprot *E. coli* database including the respective POI$_{AcK}$ sequence.

5. Perform searches with trypsin specificity allowing two missed cleavages and a mass tolerance of 4.5 ppm for MS and 6 ppm for MS/MS spectra. Set carbamidomethyl at cysteine residues as a fixed modification. Define acetylation at lysine residues, oxidation at methionine, and acetylation at the N terminus as variable modifications. Set the minimal peptide length to 7, and the false discovery rate (FDR) on proteins and peptides level is 1%.

6. Analyze the obtained MS/MS spectra for the peptide containing the acetylated lysine (quality of obtained b- and y-ion series, peptide-score, m/z for peptide) to verify the exact lysine-acceptor sites. Analyze if the acetyl-acceptor lysine can be found in the non-acetylated form in any peptide (should not be the case if the acetylated-lysine is not deacetylated during the expression in *E. coli* or during the sample preparation process) (Fig. 3c; *see* **Note 23**).

4   Notes

1. If cloning the POI$_{TAG}$ encoding sequence into pCDF-Duet-1/*Mbt*RNA$_{CUA}$ or pRSF-Duet-1/AcKRS-3/*Mbt*RNA$_{CUA}$ using restriction digest and ligation rather than Gibson assembly, use a 5′ restriction enzyme from multiple-cloning site (MCS)-1 and a 3′ restriction enzyme from MCS-2. Otherwise, expression will be initiated from both T7 promoters resulting in the expression of an unwanted small peptide. If using enzymes solely from MCS-2 for cloning of POI$_{TAG}$, a His$_6$-tagged peptide will be expressed from T7 promoter 1, which will be co-purified with the His$_6$-tagged POI$_{AcK}$. This might affect the expression and subsequent purification of the POI$_{AcK}$. *Bam*HI in MCS-1 is the preferred 5′-restriction site compared to restriction sites further 3′, as *Bam*HI creates a fusion protein with the smallest possible linker between the His$_6$-tag and the POI$_{AcK}$. Notably, the *Bam*HI restriction site is not in frame in pRSF-Duet-1/pCDF-Duet-1. The plasmid pBK-amp/AcKRS-3 is a plasmid derived from pBR322, i.e., it has a pBR322 origin replication (oriV, from pMB1), and contains an ampicillin resistance gene (Amp^R) (Fig. 1d, e) [2, 18].

2. Importantly, the stop codon used for the termination of the POI_{TAG} should not be TAG (amber), but either TAA (ochre) or TGA (opal/umber).

3. During the growth of the culture use half the usual Amp concentration to reach a concentration of 50 μg/mL, (a) to minimize a negative effect of using the antibiotic on competence development, and, (b) to ensure plasmid propagation.

4. Both the one-plasmid system and two-plasmid system described here are compatible with most *E. coli* strains for cloning of the constructs. Instead of the pBK expressing AcKRS-3 and pCDF-Duet-1 and pRSF-Duet-1 expressing $MbtRNA_{CUA}$ and the POI_{TAG}, other plasmids can be used as well (Fig. 1d, e). However, if using the two-plasmid system, the plasmid expressing the $MbtRNA_{CUA}$ and the POI_{TAG} should be compatible to the plasmid expressing AcKRS-3 regarding the antibiotic resistance (must be different) and the origin of replication (must belong to different incompatibility groups). A well-established system to express the acetylated POI_{AcK} is based on expression from a T7-RNA-polymerase promoter (T7 promoter). This does imply that the *E. coli* strain used, harbors a genomic integration of the λT7 lysogen expressing T7-RNA-polymerase. Several *E. coli* (DE3) strains are commercially available and might be selected concerning the needs for expression of the POI_{AcK}. For example, if observing leaky expression of the prematurely terminated POI_{AcK} prior to induction with IPTG, it is advisable to use *E. coli* (DE3) pLys strains.

5. To prepare 100 mL of an 1 M N-(ε)-acetyl-L-lysine (AcK; MW: 188.22 g/mol) stock solution suitable to obtain a concentration of 10 mM AcK in a 10 L culture (add 12.5 mL of 1 M AcK to 1.25 L of culture), dissolve 18.82 g of AcK in app. 80 mL of 1 M HCl and neutralize with 1 M NaOH. Prepare fresh before the cultivation. Filter-sterilize AcK-solution using a 0.22 μm-syringe sterile filter. AcK is also soluble in water. Using 10 mM of AcK ensures that incorporation of AcK into the POI_{AcK} is highly preferred over other molecules present endogenously within the *E. coli* cell, which might be loaded onto $MbtRNA_{CUA}$ by AcKRS-3 if binding to the active-site. A high level of homogeneous and quantitative incorporation is important considering that the POI_{AcK} is used for biophysical studies including X-ray crystallography. ESI-mass-spectrometry can be used to analyze the homogeneity of AcK incorporation (Fig. 3b). To prepare 100 mL of a 2 M nicotinamide (NAM; MW: 122.12 g/mol) stock solution, dissolve 24.4 g of NAM in 100 mL of ddH_2O. Filter-sterilize NAM-solution using a 0.22 μm-syringe sterile filter. Add 12.5 mL of 2 M NAM per 1.25 L culture to yield a final concentration of 20 mM NAM for efficient inhibition of CobB deacetylase.

6. The AcK and the NAM is added to the culture 30 min prior to induction of expression. This is advisable to adjust the system to obtain an efficient incorporation of this ncAA. This allows to charge the MbtRNA$_{CUA}$ with AcK and to effectively inhibit the *E. coli* deacetylase CobB. Notably, *E. coli* DE3 strains with genetic deletion of CobB are available, which makes the addition of NAM needless.

7. Expression was done for 12–16 h at 18–20 °C. This turned out to be optimal for most proteins. However, other protocols might be optimal for some POI$_{AcK}$, such as expressing the POI$_{AcK}$ for 3–4 h at 37 °C. This has to be optimized for each individual POI$_{AcK}$. The medium used for expression does mainly affect the yield of the POI$_{AcK}$ obtained. LB medium can be used for expression. However, using media such as 2YT- or TB-medium does strongly increase the biomass obtained after expression by $\geq$2–3-fold. If observing a high amount of prematurely terminated POI$_{AcK}$, this can be due to several reasons. Available AcK might be limited during expression or alternatively, the expression of the AcKRS-3 or of MbtRNA$_{CUA}$ was not sufficient or comprised. It is advisable to check the sequences of AcKRS-3 and MbtRNA$_{CUA}$ and its regulatory parts for correctness (Fig. 1d, e). It might be helpful to include additional copies of AcKRS-3 and MbtRNA$_{CUA}$ genes. The expression might be conducted at a lower temperature, for a shorter period of time to slow the folding kinetics of the POI$_{AcK}$ and to lower the overall amount of expressed POI$_{AcK}$.

8. SDS-PAGE analysis is used to confirm the expression of the POI$_{AcK}$ and to analyze the collected fractions during purification (Fig. 3a). Some reports describe a massive effect of migration behavior of lysine-acetylated proteins versus the non-acetylated. The effect depends on the position of the acetylated lysine in the protein and on the number of lysines being acetylated. It is often not observable if using a standard discontinuous, linear SDS-PAGE gel, i.e., with a stacking and a separation gel with a given constant percentage. It is only observable if running a SDS-PAGE gel in a percentage suitable for the size of the POI$_{AcK}$ or alternatively a gradient gel (4–20% SDS-PAGE) to get the most possible separation of the bands. The acetylated protein runs at a slightly higher molecular weight compared to the non-acetylated counterpart due to the neutralization of the positive charge upon acetylation. If analyzing a POI$_{AcK}$ that is acetylated at multiple sites, this might be additive and might affect the running behavior in an SDS-PAGE gel.

9. Sometimes, the co-translational incorporation of AcK is not possible, i.e., no soluble protein is obtained. This might be due to co-translational incorporation of AcK, in contrast to the

otherwise posttranslational nature of this modification, interfering with protein folding. This is affected by the position of the amber suppressor site. Moreover, the solubility and other physicochemical properties of the protein might be comprised upon lysine-acetylation.

10. As the amber-suppression is sometimes not 100% efficient, a prematurely terminated product of the protein of interest (POI) can be obtained. However, dependent on the position of the TAG-codon in the open-reading frame, this truncated product is often not properly folded and therefore insoluble or aggregated and it is therefore removed or can be separated from the non-truncated POI_{AcK} during the expression/purification process. However, particularly if the acetyl-L-lysine (AcK) target site is at the far termini of the POI_{AcK}, a prematurely terminated product might be soluble and might behave similar to the non-terminated POI_{AcK} in the expression/purification process. If this is the case, the position of the purification tag can be selected to allow for an efficient removal of translational truncation product.

11. The conditions used for protein expression might vary and need to be adapted to the POI_{AcK}. However, the expression conditions described here were successfully implemented for full-length lysine-acetylated Ran. Other POI_{AcK} might require the development of a different purification strategy including other purification steps to yield pure protein. As an example, the purification can be done with various affinity tags (GST, His_6, MBP, etc.), including removal of the tags by including a protease (Tabacco-etch virus (Tev), thrombin, precision, etc.) cleavage site. Ion exchange chromatography often constitutes a powerful purification step, strongly improving the purity. For some POI_{AcK}, we observed an altered solubility compared to the non-acetylated proteins showing that lysine-acetylation interferes with general physicochemical properties, such as the electrostatics affecting the protein's isoelectric point, pI (pI CypA: 7.68; CypA AcK125: 6.9; K-Ras 4B: 8.24, K-Ras 4B AcK104: 7.64). This might require altering the buffer composition and the buffer pH accordingly to increase solubility (buffer pH 0.5 to 1 pH units lower or above the pI value of the POI_{AcK}). As an example, for CypA AcK125 and K-Ras 4B AcK104, we observed that purification in a mild acidic 50 mM K_2HPO_4/KH_2PO_4, pH 6.4 buffer compared to a 50 mM Tris/HCl, pH 7.4 buffer did significantly increase the stability and as a result the yield. The pI of a POI_{AcK} can be obtained using the ProtParam tool on the ExPASy webpage (http://web.expasy.org/protparam.html). To assess the impact of lysine-acetylation on the pI, replace the acetylated lysine by an uncharged amino acid such as alanine or glutamine and run ProtParam.

12. Several labs developed *E. coli* strains which carry a genomic deletion of the amber-stop codon (ΔTAG) and/or a deletion/modification of the ribosomal release-factor 1 (RF1), which endogenously terminates the translation at amber and ochre stop-codons [19–21]. Both strategies increase the efficiency of the incorporation of ncAA into proteins using genetic-code expansion, particularly when simultaneously incorporating ncAA at two positions. For the efficient incorporation of AcK into a POI_{TAG} at one site, the usage of the ΔRF1 and/or ΔTAG strains is often dispensable. However, for strategies to genetically encode more than one ncAA incorporated at amber, ochre, and/or even quadruplet codons, using these improved tools is beneficial for the yield of product [13, 22–28].

13. For concentration using Amicon® Ultra centrifugal filter device (or other filter device) use a molecular weight cutoff (MWOC), at least twofold lower than the molecular weight of the POI_{AcK}. Otherwise, the POI_{AcK} might be lost during filtration into the flow-through.

14. The molecular mass of the POI_{AcK}, its extinction coefficient (EC), and its isoelectric point (pI) are calculated using programmes such as ProtParam from the ExPASy bioinformatics resource portal (http://web.expasy.org/protparam.html). As these programmes do not allow for the calculation of a mass of a POI_{AcK} with incorporated ncAA, the mass for the acetylation of the lysine-side chain (+42 Da) has to be added. Determine the concentration of the POI_{AcK} using absorption at near UV (280 nm). The measurement is done using either a quartz cuvette or a cuvette suitable for UV measurements, i.e., being transparent at this wavelength. The absorption by the POI_{AcK} depends on the content of Tyr and Trp residues (to a smaller extent also on Phe and disulfide bonds). A suitable dilution is made using the storage buffer of the POI_{AcK}, which should not contain chromophores absorbing at 280 nm (A_{280}) in a UV-spectrophotometer (protein concentration range: 0.024–6.25 mg/mL) [29]. The protein-concentration is adjusted to the dynamic range of the assay and the instrument. Measurement of the absorbance A_{280} must be performed within the linear range of 0.05–1.0, while a value of 0.3 is most accurate. If the A_{280} is below this range, use a higher concentrated solution, if the A_{280} is above this range, use a higher dilution of the POI_{AcK}. The concentration of the POI_{AcK} is calculated using the EC determined for the POI_{AcK} using ProtParam (http://web.expasy.org/protparam.html). As the acetylation at a lysine side chain does not interfere with the A_{280}, the EC corresponds to the value obtained for the non-acetylated protein. The EC_{POI} can be calculated using the following formula: $EC_{POI} = n \times 5690_{Trp} + n \times 1280_{Tyr} + n \times 120_{Cys}$ (n: number of residues; Trp: tryptophan, Tyr: tyrosine, Cys: cysteine). The

concentration is determined by the law of Lambert-Beer: $c_{POI} = A_{280}/\varepsilon_{POI} \times l \times df$ (c: concentration in mol/L; A_{280}: absorption at 280 nm; ε: extinction coefficient; l: optical path-length in cm; df: dilution factor). For small GTP-binding proteins of the Ras superfamily include also the absorption of the bound guanine-nucleotide (either GDP or GTP) for protein concentration determination by adding the EC of 7765 M^{-1} cm^{-1} to the EC of the protein. For full-length Ran AcK37 the EC would be 29,910 M^{-1} cm^{-1} plus 7756 M^{-1} cm^{-1} to result in a total EC of 37,666 M^{-1} cm^{-1} [30].

15. Also follow the scattering in the sample by determining the absorption at 340 nm during the concentration process using the Amicon® Ultra centrifugal filter device (some instruments measure A_{310} or A_{320}), A_{340} as this shows any precipitation occurring in the sample and is therefore an indicator for the stability of the POI_{AcK} in the buffer of choice. This is particularly important if the POI_{AcK} is purified for the first time. Correct the A_{280nm} for this value. If there is substantial A_{340nm}, spin the sample at 15,000 × g for 15 min at 4 °C to remove precipitated protein. Importantly, the lysine-acetylated POI_{AcK} might behave differently compared to the non-acetylated protein (*see* **Note 9**). If the A_{340} increases, stop the concentration process and optimize the buffer composition used for purification, if the desired concentration for the planned application cannot be obtained. This includes adding buffer additives such as 5–10% (v/v) glycerol, adding reductants (β-mercaptoethanol, DTT or DTE, increasing/decreasing the ionic strength, altering the buffer system and/or the buffer pH, etc.).

16. The intensity of A_{280} depends on the amount of protein loaded on the chromatography column and also on the specifications of the FPLC system used, such as the path-length of the UV cell and the intensity of the Xenon flash lamp. Furthermore, it depends on the level of absorption of the POI_{AcK} at 280 nm, which is represented by the EC of the POI_{AcK}. If the EC should be very low, also record the absorption at 220 nm, A_{220}, which detects the peptide bonds. For Ran and other GTP-binding proteins (or for ATP/ADP-binding proteins) also record the absorption at 254 nm, A_{254}, which detects the bound nucleotide (*see* **Note 12**).

17. Always analyze a positive (site-specifically acetylated protein) and a negative (non-acetylated counterpart of POI_{AcK}) control in immunoblottings using an anti-AcK antibody (Fig. 3a). The amino acids surrounding the acetylation site do interfere with the detection by the anti-AcK antibody (Fig. 2a). This shows that the antibodies available commercially do not recognize the acetylated-lysine as its sole antigen as they show preferences for some sequences. As an example, full-length Ran

AcK37 protein was strongly recognized by the anti-AcK antibody (sequence: 30-LTGEFE-(AcK)-KYVATL-43), in contrast to full-length Ran AcK71 (sequence: 65-DTAGQE-(AcK)-FGGLRD-77) (Fig. 2a). For certain applications such as mass-spectrometry it is therefore recommendable to use a mixture of different anti-AcK antibodies from diverse suppliers. As loading control for immunoblottings, perform a Ponceau S staining of the blotting membrane, stain the SDS-PAGE gel with Coomassie-brilliant blue, and perform an immunoblotting after stripping off the anti-AcK antibody using a protein-specific antibody or an antibody recognizing a tag (His$_6$-tag, GST-tag, etc.). The acetylated POI$_{AcK}$ should overlay with the Ponceau S stained/Coomassie-brilliant blue-stained/immunostained POI$_{AcK}$ band. This will show if there is prematurely terminated POI$_{AcK}$.

18. The Ponceau S washing step of the nitrocellulose or PVDF membrane should not exceed 1 min as Ponceau S staining is reversible and longer washing would remove all staining.

19. Depending on signal strength, detection is carried out either with X-ray films or with a CCD camera-based detection system.

20. There are many protocols available for the determination of the molecular mass of a protein by ESI-MS and the analysis of the data including the software. Along this line, also for the in-solution protease digest of proteins, the purification of the obtained peptides and the subsequent analysis by LC-MS/MS and for the data analysis, diverse protocols are available and different software can be used. The settings used are described for His$_6$-Ran AcK37 described in this protocol but the settings used depend on the instrumentation available and also on the POI$_{AcK}$ to be analyzed and might be adjusted accordingly.

21. If your POI$_{AcK}$ sample contains detergents (e.g., Triton, NP40, Tween, Glycerol, PEG, SDS etc.), these compounds need to be removed prior to analysis, e.g., by acetone precipitation before conducting the in-solution protocol. If using purified proteins, 1–3 μg are sufficient, if using a cell lysate, at least 10 μg of protein after lysis should be used. Importantly, StageTip purification removes salts, but no detergents and other compounds such as glycerol, PEG, Triton, Tween, SDS, NP40, etc.

22. Reconstitute 1 mg of the lyophilized trypsin in 2 mL 1 mM HCl and the lyophilized Lys-C in ddH$_2$O to obtain a final concentration of 0.5 μg/μL.

23. The successful incorporation of AcK is analyzed by (a) immunoblotting using an anti-AcK antibody, (b) ESI-MS to analyze the mass of the POI$_{AcK}$ and to assess the homogeneity of the

product, and (c) tryptic digest of the recombinantly expressed and site-specifically lysine-acetylated and purified POI$_{AcK}$ and following LC-MS/MS analysis, to assess the site of AcK incorporation (Fig. 3). In the case of a low Andromeda score for the POI$_{AcK}$ a manual inspection of MS/MS fragment spectra is recommended.

Acknowledgments

I thank Christian Frese from the CECAD proteomics facility for support in writing the mass-spectrometry section and Marcus Krüger critical reading of the mass-spectrometry part of this manuscript. Furthermore, I thank CECAD for support and all members of my laboratory, particularly Dr. Susanne de Boor, for the preparation of some figures and discussions, and the Deutsche Forschungsgemeinschaft (DFG) for funding of my position by the Heisenberg Programme (LA 2984/3-1).

References

1. de Boor S et al (2015) Small GTP-binding protein Ran is regulated by posttranslational lysine acetylation. Proc Natl Acad Sci U S A 112(28):E3679–E3688

2. Lammers M, Neumann H, Chin JW, James LC (2010) Acetylation regulates cyclophilin A catalysis, immunosuppression and HIV isomerization. Nat Chem Biol 6(5):331–337

3. Knyphausen P, Kuhlmann N, de Boor S, Lammers M (2015) Lysine-acetylation as a fundamental regulator of Ran function: implications for signaling of proteins of the Ras-superfamily. Small GTPases 6(4):189–195

4. Neumann H et al (2009) A method for genetically installing site-specific acetylation in recombinant histones defines the effects of H3 K56 acetylation. Mol Cell 36(1):153–163

5. Neumann H, Peak-Chew SY, Chin JW (2008) Genetically encoding N(epsilon)-acetyllysine in recombinant proteins. Nat Chem Biol 4(4):232–234

6. Kuhlmann N et al (2016) Structural and mechanistic insights into the regulation of the fundamental rho regulator RhoGDIalpha by lysine acetylation. J Biol Chem 291(11):5484–5499

7. Kuhlmann N, Wroblowski S, Scislowski L, Lammers M (2016) RhoGDIalpha acetylation at K127 and K141 affects binding toward non-prenylated RhoA. Biochemistry 55(2):304–312

8. Knyphausen P, Lang F, Baldus L, Extra A, Lammers M (2016) Insights into K-Ras 4B regulation by post-translational lysine acetylation. Biol Chem 397(10):1071–1085

9. Shogren-Knaak M et al (2006) Histone H4-K16 acetylation controls chromatin structure and protein interactions. Science 311(5762):844–847

10. Robinson PJ et al (2008) 30 nm chromatin fibre decompaction requires both H4-K16 acetylation and linker histone eviction. J Mol Biol 381(4):816–825

11. Gertz M et al (2012) A molecular mechanism for direct sirtuin activation by resveratrol. PLoS One 7(11):e49761

12. Burke SA, Lo SL, Krzycki JA (1998) Clustered genes encoding the methyltransferases of methanogenesis from monomethylamine. J Bacteriol 180(13):3432–3440

13. Wan W, Tharp JM, Liu WR (2014) Pyrrolysyl-tRNA synthetase: an ordinary enzyme but an outstanding genetic code expansion tool. Biochim Biophys Acta 1844(6):1059–1070

14. Knyphausen P et al (2016) Insights into lysine deacetylation of natively folded substrate proteins by sirtuins. J Biol Chem 291(28): 14677–14694

15. Gibson DG et al (2009) Enzymatic assembly of DNA molecules up to several hundred kilobases. Nat Methods 6(5):343–345

16. Cox J et al (2011) Andromeda: a peptide search engine integrated into the MaxQuant environment. J Proteome Res 10(4):1794–1805

17. Cox J, Mann M (2008) MaxQuant enables high peptide identification rates, individualized p.p.b.-range mass accuracies and proteome-wide protein quantification. Nat Biotechnol 26(12):1367–1372

18. Chin JW, Martin AB, King DS, Wang L, Schultz PG (2002) Addition of a photocross-linking amino acid to the genetic code of Escherichiacoli. Proc Natl Acad Sci U S A 99(17):11020–11024

19. Johnson DB et al (2012) Release factor one is nonessential in Escherichia coli. ACS Chem Biol 7(8):1337–1344

20. Johnson DB et al (2011) RF1 knockout allows ribosomal incorporation of unnatural amino acids at multiple sites. Nat Chem Biol 7(11): 779–786

21. Mukai T et al (2015) Highly reproductive Escherichia coli cells with no specific assignment to the UAG codon. Sci Rep 5:9699

22. Schmied WH, Elsasser SJ, Uttamapinant C, Chin JW (2014) Efficient multisite unnatural amino acid incorporation in mammalian cells via optimized pyrrolysyl tRNA synthetase/tRNA expression and engineered eRF1. J Am Chem Soc 136(44):15577–15583

23. O'Donoghue P et al (2012) Near-cognate suppression of amber, opal and quadruplet codons competes with aminoacyl-tRNAPyl for genetic code expansion. FEBS Lett 586(21):3931–3937

24. Wan W et al (2010) A facile system for genetic incorporation of two different noncanonical amino acids into one protein in Escherichia coli. Angew Chem Int Ed Engl 49(18):3211–3214

25. Odoi KA, Huang Y, Rezenom YH, Liu WR (2013) Nonsense and sense suppression abilities of original and derivative Methanosarcina mazei pyrrolysyl-tRNA synthetase-tRNA(Pyl) pairs in the Escherichia coli BL21(DE3) cell strain. PLoS One 8(3):e57035

26. Niu W, Schultz PG, Guo J (2013) An expanded genetic code in mammalian cells with a functional quadruplet codon. ACS Chem Biol 8(7):1640–1645

27. Neumann H, Wang K, Davis L, Garcia-Alai M, Chin JW (2010) Encoding multiple unnatural amino acids via evolution of a quadruplet-decoding ribosome. Nature 464(7287):441–444

28. Chatterjee A, Sun SB, Furman JL, Xiao H, Schultz PG (2013) A versatile platform for single- and multiple-unnatural amino acid mutagenesis in Escherichia coli. Biochemistry 52(10):1828–1837

29. Chutipongtanate S, Watcharatanyatip K, Homvises T, Jaturongkakul K, Thongboonkerd V (2012) Systematic comparisons of various spectrophotometric and colorimetric methods to measure concentrations of protein, peptide and amino acid: detectable limits, linear dynamic ranges, interferences, practicality and unit costs. Talanta 98:123–129

30. Smith SJ, Rittinger K (2002) Preparation of GTPases for structural and biophysical analysis. Methods Mol Biol 189:13–24

Chapter 12

Site-Specific Incorporation of Sulfotyrosine Using an Expanded Genetic Code

Xiang Li and Chang C. Liu

Abstract

Tyrosine sulfation is an important posttranslational modification found in bacteria and higher eukaryotes. However, the chemical synthesis or expression of homogenously sulfated proteins is particularly difficult, limiting our study and application of tyrosine-sulfated proteins. With the recent development of genomically recoded organisms and orthogonal translation components, we can often treat otherwise posttranslationally-modified amino acids as noncanonical amino acids (ncAAs) encoded by an expanded genetic code. Here, we describe methods for the co-translational incorporation of one or multiple sulfotyrosines into proteins using standard or genomically recoded *Escherichia coli* stains, thereby achieving the direct expression of site-specifically tyrosine sulfated proteins in vivo.

Key words Sulfotyrosine, Expanded genetic code, Noncanonical amino acid, Protein expression, Genetically recoded organism, Green fluorescent protein

1 Introduction

Sulfated proteins play a key role in chemotaxis [1, 2], HIV and malarial infection [3, 4], coagulation [5], and plant immunity [6, 7]. In nature, proteins are sulfated by tyrosylprotein sulfotransferases (TPSTs), enzymes that posttranslationally modify a target tyrosine with a sulfate group [8]. However, when TPSTs modify a protein that contains multiple target tyrosines in a short sequence window, the resulting products may contain heterogenous sulfation patterns [9, 10] that are problematic for mass spectroscopy analysis and purification of minor sulfated products. In addition, TPSTs only modify specific target tyrosines, limiting our ability to dissect the role of non-native sulfation patterns or single sulfates in a stretch of multiple sulfated tyrosines.

We developed an expanded genetic code system to express homogenously sulfated proteins where the sulfated tyrosines can be installed at any location in a protein [11]. This system relies on the expression of a *Methanococcus jannaschii* (*Mj*) aminoacyl-tRNA

Edward A. Lemke (ed.), *Noncanonical Amino Acids: Methods and Protocols*, Methods in Molecular Biology, vol. 1728,
https://doi.org/10.1007/978-1-4939-7574-7_12, © Springer Science+Business Media, LLC 2018

synthetase (aaRS) and tRNA$_{CUA}$ pair that is engineered to incorporate sulfotyrosine (sY, Fig. 1a) at amber (UAG) stop codons. sY incorporation at a UAG codon is highly site-specific, as the engineered Mj aaRS and tRNA$_{CUA}$ do not cross-react with endogenous amino acids, aaRSs, or tRNAs [11]. While various homogenously sulfated proteins were successfully expressed using this expanded code for sY [12–14], our initial systems were limited to the incorporation of one or two sYs due to two main sources of inefficiency: suboptimal expression of the Mj aaRS for sY (STyrRS) and competition with the endogenous $E.$ $coli$ release factor 1 (RF-1) for UAG codons.

To improve the incorporation of multiple sYs, we implemented our expanded genetic code system for sY in a genomically recoded $E.$ $coli$ strain, C321.ΔA, whose genome lacks UAG codons and the gene that encodes RF-1 [15]. The removal of RF-1 competition in C321.ΔA should improve the incorporation of ncAAs at UAG codons (Fig. 1b). In addition, we used an upgraded expression vector, pUltra [16], to express STyrRS and tRNA$_{CUA}$. Here, we outline protocols for incorporating one or multiple sYs into green fluorescent protein (GFP) in SS320 cells, a standard $E.$ $coli$ strain, or in C321.ΔA cells. STyrRS is encoded in the pUltra-sY plasmid (Fig. 2a). Wild-type GFP or GFP containing one or three permissive UAG codons are expressed under the AraBAD promoter in the pGLO expression vector (Fig. 2b). After pUltra-sY and pGLO are co-transformed into SS320 or C321.ΔA cells, the expression of STyrRS and GFP are co-induced by IPTG and L-arabinose, respectively. Fluorescence of cultures was analyzed using a fluorescent plate reader. We found that the expression of GFP containing zero or one sY is higher in SS320 cells while the expression of GFP containing three sYs is higher in C321.ΔA cells (Fig. 3). This should serve to guide any effort to heterologously produce any tyrosine-sulfated protein.

2   Materials

Solutions are made with sterilized DI water and stored at room temperature unless otherwise specified.

2.1 E. coli Strains

1. SS320: purchased from Lucigen.
2. C321.ΔA: available on Addgene (ID # 48998).

2.2 Plasmids

1. pUltra-sY: available on Addgene (ID # 82417).
2. pGLO-GFP-WT: purchased from Bio-Rad.
3. pGLO-GFP-1UAG: available on Addgene (ID # 82500).
4. pGLO-GFP-3UAG: available on Addgene (ID # 82501).
5. pUC19: purchased from NEB.

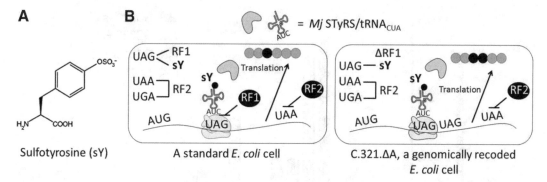

Fig. 1 Expanded genetic code for sY in a standard or a genomically recoded *E. coli* cell. (**a**) The chemical structure of sY. (**b**) A cartoon of the incorporation of sY in a standard *E. coli* cell or C321.ΔA, a genomically recoded *E. coli* cell. Genome-wide reassignment of UAG to UAA codons and the deletion of RF-1 in C321.ΔA cells enable maximum incorporation of sY at UAG codons via *Mj* STyrRS/tRNA$_{CUA}$

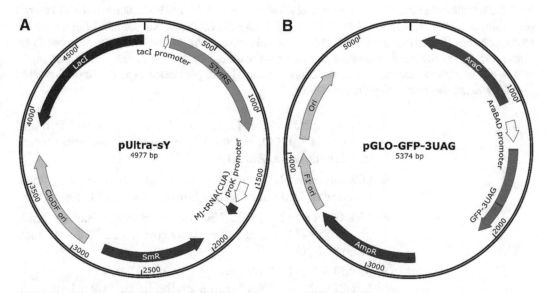

Fig. 2 Plasmid maps. (**a**) pUltra-sY, the expression plasmid for the orthogonal STyrRS/tRNA$_{CUA}$ pair. Expression of STyrRS is driven by an IPTG-inducible tacI promoter and *Mj* tRNA$_{CUA}$ is constitutively expressed from the proK promoter. pUltra-sY contains the CloDF origin and spectinomycin resistance gene (SmR), both of which are compatible with common expression vectors. (**b**) pGLO, the expression plasmid for GFP variants. pGLO-GFP-3UAG encodes GFP with three UAG codons at amino acid positions 3, 151, and 153. pGLO-GFP-1UAG and pGLO-GFP-WT (plasmid maps are not shown) encode GFP with one UAG codon at amino acid position 3 and the wild-type GFP gene without UAG codons, respectively. Expression of GFP variants is driven by an ʟ-arabinose-inducible AraBAD promoter. Plasmid maps were created using SnapGene

2.3 Stock Solutions

1. 10× PBS (1 L, pH 7.4): dissolve 25.6 g Na$_2$HPO$_4$ · 7H$_2$O, 80 g NaCl, 2 g KCl, and 2 g KH$_2$PO$_4$ in 900 mL of water. Adjust pH to 7.4. Fill the final volume to 1 L with water. Autoclave at 121 °C for 45 min.

2. 20% ᴅ-glucose (1 L): dissolve 200 g of ᴅ-glucose in 900 mL of water. Fill the final volume to 1 L with water. Autoclave at 121 °C for 15 min.

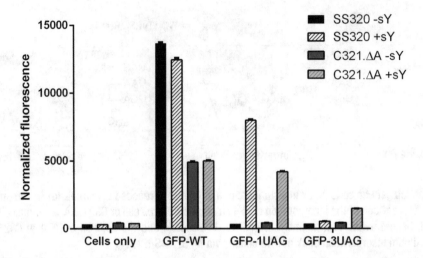

Fig. 3 Incorporation of sY into GFP. Fluorescence of wild-type GFP (GFP-WT), GFP with one or three UAG codons (GFP-1UAG or GFP-3UAG) expressed in SS320 or C321.ΔA cells without or with 20 mM sY in the growth media. Only the cells that express GFP-1UAG or GFP-3UAG contain pUltra-sY. "Cells only" are cells that do not contain expression vectors. Fluorescence signals were normalized by dividing by OD_{600}. Error bars represent $\pm$ s.d. of triplicate samples. Data was adapted from [7] with permission from The Royal Society of Chemistry. Graph was made using Graphpad Prism

3. 10× Phosphate buffer (1 L): dissolve 23.1 g of KH_2PO_4 and 125.4 g of K_2HPO_4 in 900 mL of water. Fill the final volume to 1 L with water. Autoclave at 121 °C for 45 min.

4. 80% Glycerol (1 L): add 800 mL of molecular biology grade Glycerol to 200 mL of water. Autoclave at 121 °C for 45 min.

5. 15% Glycerol (1 L): add 150 mL of molecular biology grade glycerol to 850 mL of water. Autoclave at 121 °C for 45 min. Store at 4 °C.

6. 5× M9 Salts (1 L): dissolve 64 g of $Na_2HPO_4 \cdot 7H_2O$, 15 g of KH_2PO_4, 2.5 g of NaCl, and 5 g of NH_4Cl in 900 mL of water. Fill the final volume to 1 L with water. Autoclave at 121 °C for 45 min.

7. 20% L-arabinose (250 mL): dissolve 50 g of L-arabinose in 200 mL of water. Fill the final volume to 250 mL with water. Filter the solution using a 0.22 μm filter.

8. 100 mM IPTG (250 mL): dissolve 5.96 g of IPTG in 200 mL of water. Fill the final volume to 250 mL with water. Filter the solution using a 0.22 μm filter.

9. 1 M $MgSO_4$ (40 mL): dissolve 4.8 g of $MgSO_4$ in 40 mL of water. Mix with caution as the reaction will release heat. Filter the solution using a 0.22 μm filter.

10. 1 M $CaCl_2$ (40 mL): dissolve 4.4 g of $CaCl_2$ in 40 mL of water and vortex to mix. Filter the solution using a 0.22 μm filter.

11. 2 M $MgCl_2$ (40 mL): dissolve 7.6 g of $MgCl_2$ in 40 mL of water and vortex to mix. Filter the solution using a 0.22 μm filter.

12. 0.02% D-biotin (250 mL): dissolve 50 mg of D-biotin in 250 mL of water and vortex to mix. Filter the solution using a 0.22 μm filter.

13. Sulfotyrosine (250 mL): synthesize sY following a previously reported protocol [17]. Resuspend lyophilized sY in water to make a sY solution of at least 100 mM (*see* **Note 1**). Filter the solution using a 0.22 μm filter. Store at 4 °C.

2.4 Media

1. SOB - Mg (1 L): dissolve 20 g Tryptone, 5 g Yeast extract, 0.584 g NaCl, 0.186 g KCl in 900 mL of water. Fill the final volume to 1 L with water. Autoclave at 121 °C for 45 min.

2. SOC Medium (40 mL): vortex mix 200 μL of 2 M $MgCl_2$, 721 μL of 20% Glucose, and 39 mL of SOB - Mg under sterile conditions.

3. M9T Medium (1 L): dissolve 10 g of tryptone and 5 g of NaCl in 200 mL of 5× M9 salts. Fill the final volume to 1 L with water. Autoclave at 121 °C for 45 min. When the autoclaved M9T Medium reaches room temperature, add 20 mL of 20% D-glucose, 2 mL of 1 M $MgSO_4$, 100 μL of $CaCl_2$, and 1 mL of 0.02% D-biotin (*see* **Note 2**). Stir until all of the $CaCl_2$ precipitants have fully dissolved.

4. Terrific Broth (TB) Medium (1 L): dissolve 12 g Tryptone and 24 g Yeast extract in 800 mL of water. Add 5 mL of 80% Glycerol and then fill the final volume to 900 mL with water. Autoclave at 121 °C for 45 min. Once the medium is cooled to <60 °C, add 100 mL of 10× Phosphate buffer under sterile conditions.

2.5 Agar Plates

1. LB Agar with antibiotics (500 mL): dissolve 18.5 g of LB Agar Mix in water. Autoclave at 121 °C for 20 min. Store the solution at 60 °C. When ready, adjust the solution to ~45 °C, add antibiotics of choice, swirl to mix, and pour plates. Store plates at 4 °C.

2.6 Equipment

1. Tecan Infinite M200 PRO fluorescence plate reader.

---

3 Methods

3.1 Construction of Plasmids

1. pUltra-sY can be obtained from Addgene. Alternatively, amplify STyrRS from pSup-STyrRS [11] using primers aaRStopUltraPCR1 (5′-CATCGCGGCCGC<u>AT GGACGAATTTGAAATGATAAAGAGA</u>-3′) and aaRStopUltra-

PCR2 (5'-CATC GCGGCCGCTTATAATCTCTTTCTAATTGG CTCTAAAATC-3'), and subsequently insert STyrRS into pUltra [16] using NotI cloning sites.

2. pGLO-GFP-1UAG can be obtained from Addgene. Alternatively, construct pGLO-GFP-1UAG via site-directed mutagenesis PCR from pGLO-GFP-WT using pGLO-GFP-1TAG-F (5'-CATATGGCTTAGAGCAAAGGAGAAGA ACTTTT-3') and pGLO-GFP-1TAG-R (5'-TCCTTTGCTCTAAGCCATATGTATAT CTCCTT-3').

3. pGLO-GFP-3UAG can be obtained from Addgene. Alternatively, construct pGLO-GFP-3UAG via site-directed mutagenesis PCR from pGLO-GFP-1UAG using pGLO-GFP-3TAG-F (5'-TCACACTAGGTATAGATCACGGCA GACAAACAAAA-3') and pGLO-GFP-3TAG-R (5'-GTGATCTATACCTAGTGTGA GTTATAGTTGTACTC-3').

3.2 Preparation of Electrocompetent Bacterial Cells

The following protocol is applicable to SS320 or C321.ΔA cells, or cells that contain pUltra-sY (*see* **Notes 3** and **4**).

1. Day 1: Streak the frozen glycerol stock of SS320 cells onto an LB Agar plate or C321.ΔA cells onto an LB Agar plate containing Zeocin.

2. Incubate the plate streaked with cells in a 37 °C incubator for 12–14 h.

3. Day 2: Pick a single colony into 10 mL of SOB - Mg media in a glass tube.

4. Incubate in a shaker at 37 °C, 200 rpm overnight.

5. Chill 15% glycerol, 250 mL plastic centrifuge bottles, 50 mL conical bottles, microcentrifuge tubes, and serological pipette tips at 4 °C overnight.

6. Day 3: Transfer 5 mL of the saturated culture into 500 mL of SOB - Mg media in a 2 L Erlenmeyer flask.

7. Incubate in a shaker at 37 °C, 200 rpm for ~2.5 h until the cells have reached mid-log phase (OD_{600} = 0.6).

8. Immediately chill the cells in an ice water bath for 10 min by constantly swirling the culture to evenly chill the cells.

The following steps should be performed at 4 °C.

9. Evenly split the chilled cell cultures into four 250 mL plastic centrifuge bottles.

10. Centrifuge the cells down at 2500 × g for 10 min.

11. Immediately discard the supernatant and then place bottles on ice.

12. Resuspend the cells in chilled 250 mL 15% glycerol without inducing air bubbles.

13. Repeat **steps 10–12**.

14. Centrifuge the cells down at $2500 \times g$ for 10 min.

15. Discard as much supernatant as possible by dumping supernatant and then aspirating supernatant at the screw cap with a pipette. Maximum removal of supernatant ensures appropriate cell density in **step 18**. Immediately place bottles on ice.

16. Resuspend cells in the residual 15% glycerol and transfer to a chilled 50 mL conical bottle. The total volume should be ~1.5 mL.

17. Dilute 5 μL of the resuspended cells into 1 mL of ddH$_2$O.

18. Measure the OD$_{600}$ of this diluted resuspension. If OD$_{600}$ is between 0.85 and 1.0, then the cells are at the appropriate density for transformation. If OD$_{600}$ is <0.85, then centrifuge the cells down at $2500 \times g$ for 10 min and remove enough supernatant to achieve the right cell density.

19. Aliquot 50 μL of the resuspended cells into chilled microcentrifuge tubes.

20. Directly store the aliquots at −80 °C.

3.3 Transformation via Electroporation

1× (100 mg/L) Ampicilin is required to select for cells that were transformed with pGLO expression vectors, and 1× (100 mg/L) Spectinomycin is required to select for cells that were transformed with pUltra-sY.

1. Thaw a 50 μL aliquot of electrocompetent cells on ice for 5 min.

2. Chill an electroporation cuvette on ice.

3. Add up to 5 μL of plasmids suspended in ddH$_2$O (*see* **Note 4**) into the cells and incubate on ice for 5 min.

4. Transfer the cell suspension into a chilled electroporation cuvette without inducing air bubbles.

5. Electroporate using the Biorad MicroPulser at 1.6–1.8 kV.

6. Immediately rescue the cells in 1 mL of warm SOC. Carefully pipette up and down two times to mix without inducing air bubbles.

7. Incubate in a shaker at 37 °C, 200 rpm for 1 h.

8. Dilute the cells as appropriate and then spread 50–200 μL of the cells onto selective, warm LB Agar plates.

9. Incubate the plates at 37 °C for 12–16 h.

3.4 Protein Expression

1. Pick a colony (*see* **Note 5**) from **step 9** into 1 mL of TB containing the appropriate antibiotics (*see* **Note 6**).

2. Incubate the culture in a shaker at 37 °C, 200 rpm for 16 h.

3. Add 20 μL of the saturated TB culture into 2 mL of M9T containing the appropriate antibiotics.

4. Incubate the culture in a shaker at 37 °C, 200 rpm for 12 h.

5. Add 10 μL of the saturated M9T culture into 1 mL of M9T containing the appropriate antibiotics with or without 20 mM sulfotyrosine (*see* **Notes** 7 and **8**).

6. Incubate the culture in a shaker at 37 °C, 200 rpm for approximately 3 h or until the culture has reached log phase (OD_{600} = 0.6–0.8) (*see* **Note 9**).

7. Induce protein expression by adding 10 μL of 20% L-Arabinose and 10 μL of 100 mM IPTG to the culture.

8. Incubate the culture in a shaker at 30 °C, 175 rpm for 18 h.

9. Transfer 100 μL of the culture into a UV transparent 96-well plate.

10. Measure the GFP fluorescence (excitation: 395 nm, emission: 509 nm) and the optical density of the culture by measuring its absorbance at 600 nm. Then divide the fluorescence signal by the optical density to obtain normalized fluorescence signal. *See* Fig. 3 for typical results (*see* **Notes 10** and **11**).

4   Notes

1. Ideally, the concentration of sY stock solution should be higher than 100 mM so that the addition of sY to cell culture media will not significantly change the concentrations of media components.

2. Do not use amber suppressor *E. coli* strains for incorporating sY to avoid incorporating natural amino acids at UAG codons.

3. Alternative to co-transformation, one can transform pGLO into chemically competent or electrocompetent cells that already contain pUltra-sY. To prepare electrocompetent cells containing pUltra-sY, simply grow the cells in SOB - Mg supplemented with Spectinomycin and then follow the same preparation protocol in Subheading 3.2.

4. To achieve high transformation efficiency, add a volume of plasmids that is ≤1/10th volume of the competent cells. 1 μg of DNA is recommended.

5. Avoid picking abnormally large colonies of cells that were transformed with pUltra-sY from LB Agar plates, as large colonies are typically mutants that have disabled components for sY incorporation such as the tRNA.

6. The initial growth of cells in TB before transferring to M9T increases protein yield in comparison to a full growth in M9T.

7. The volumes of cultures and solutions used for protein expression can be scaled up proportionally, as we routinely express sulfated proteins at 200 mL and 1 L culture volumes.

8. Incorporation of sY using low concentrations of sY in rich media has been difficult presumably due to transport of the negatively charged sY (Fig. 1a). We and others found that the incorporation of one sY into proteins can be achieved at 3 mM in M9T [6, 7]. Incorporation of multiple sYs will require higher sY concentration in the expression media even when using C321.ΔA cells. We have used concentrations of sY up to 25 mM.

9. C.321.ΔA cells have a slower doubling time than SS320 cells (56 min versus 46 min when the cells are grown in M9T media at 37 °C and 200 rpm). Take this factor into account when growing cells to their mid-log phase. For example, prepare the mid-log phase C321.ΔA cells ~30 min before SS320 cells to induce protein expression in both strains at the same time.

10. Expression of different proteins requires further optimization by altering sY concentration, expression temperature, induction culture OD, IPTG induction concentration, and protein expression duration. As a rule of thumb, incorporation of three or more sYs is more efficient in C321.ΔA cells than in standard protein expression cells.

11. When a coding region contains a UAG codon, minimal protein should be produced without sY. The protein expression difference in the presence and absence of sY is a good quick check on sY being incorporated into the desired protein. To further ensure that sYs are properly incorporated into the proteins of interest, analyze the purified proteins via mass spectroscopy (MS). Unfortunately, MS positive ion mode, especially in MALDI, deprotonates sulfate modifications [7, 18]. One can use selected reaction monitoring mass spectrometry to assess the relative loss of sulfates during MS based on the difference in retention times of the sulfated versus unsulfated tryptic peptides [7]. In our experience, the presence of tyrosine where sY is expected in mass spectra is almost always due to the loss of sulfate during mass spectrometry rather than the undesired incorporation of tyrosine during protein expression.

References

1. Colvin RA, Campanella GSV, Manice LA et al (2006) CXCR3 requires tyrosine sulfation for ligand binding and a second extracellular loop arginine residue for ligand-induced chemotaxis. Mol Cell Biol 26:5838–5849

2. Tan JHY, Ludeman JP, Wedderburn J et al (2013) Tyrosine sulfation of chemokine receptor CCR2 enhances interactions with both monomeric and dimeric forms of the chemokine monocyte chemoattractant protein-1 (MCP-1). J Biol Chem 288:10024–10034

3. Choe H, Moore MJ, Owens CM et al (2005) Sulphated tyrosines mediate association of chemokines and Plasmodium vivax Duffy binding protein with the Duffy antigen/receptor for chemokines (DARC). Mol Microbiol 55:1413–1422

4. Farzan M, Mirzabekov T, Kolchinsky P et al (1999) Tyrosine sulfation of the amino terminus of CCR5 facilitates HIV-1 entry. Cell 96:667–676

5. Priestle JP, Rahuel J, Rink H et al (1993) Changes in interactions in complexes of hirudin derivatives and human a-thrombin due to different crystal forms. Protein Sci 2:1630–1642

6. Pruitt RN, Schwessinger B, Joe A et al (2015) The rice immune receptor XA21 recognizes a tyrosine-sulfated protein from a Gram-negative bacterium. Sci Adv 1:e1500245

7. Schwessinger B, Li X, Ellinghaus TL et al (2016) A second-generation expression system for tyrosine-sulfated proteins and its application in crop protection. Integr Biol (Camb) 8:542

8. Choe H, Farzan M (2009) Chapter 7. Tyrosine sulfation of HIV-1 coreceptors and other chemokine receptors. Methods Enzymol 461:147–170

9. Mikkelsen J, Thomsen J, Ezban M (1991) Heterogeneity in the tyrosine sulfation of Chinese hamster ovary cell produced recombinant FVIII. Biochemistry 30:1533–1537

10. Seibert C, Veldkamp CT, Peterson FC et al (2008) Sequential tyrosine sulfation of CXCR4 by tyrosylprotein sulfotransferases. Biochemistry 47:11251–11262

11. Liu CC, Schultz PG (2006) Recombinant expression of selectively sulfated proteins in Escherichia coli. Nat Biotechnol 24:1436–1440

12. Liu CC, Brustad E, Liu W et al (2007) Crystal structure of a biosynthetic sulfo-hirudin complexed to thrombin. J Am Chem Soc 129:10648–10649

13. Liu CC, Choe H, Farzan M et al (2009) Mutagenesis and evolution of sulfated antibodies using an expanded genetic code. Biochemistry 48:8891–8898

14. Liu CC, Mack AV, Tsao M-L et al (2008) Protein evolution with an expanded genetic code. Proc Natl Acad Sci U S A 105:17688–17693

15. Lajoie MJ, Rovner AJ, Goodman DB et al (2013) Genomically recoded organisms expand biological functions. Science 342:357–360

16. Chatterjee A, Sun SB, Furman JL et al (2013) A versatile platform for single- and multiple-unnatural amino acid mutagenesis in Escherichia coli. Biochemistry 52:1828

17. Liu CC, Cellitti SE, Geierstanger BH et al (2009) Efficient expression of tyrosine-sulfated proteins in E. coli using an expanded genetic code. Nat Protoc 4:1784–1789

18. Yu Y, Hoffhines AJ, Moore KL et al (2007) Determination of the sites of tyrosine O-sulfation in peptides and proteins. Nat Methods 4:583–588

Chapter 13

Site-Specific Protein Labeling with Tetrazine Amino Acids

Robert J. Blizzard, True E. Gibson, and Ryan A. Mehl

Abstract

Genetic code expansion is commonly used to introduce bioorthogonal reactive functional groups onto proteins for labeling. In recent years, the inverse electron demand Diels-Alder reaction between tetrazines and strained trans-cyclooctenes has increased in popularity as a bioorthogonal ligation for protein labeling due to its fast reaction rate and high in vivo stability. We provide methods for the facile synthesis of a tetrazine containing amino acid, Tet-v2.0, and the site-specific incorporation of Tet-v2.0 into proteins via genetic code expansion. Furthermore, we demonstrate that proteins containing Tet-v2.0 can be quickly and efficiently reacted with strained alkene labels at low concentrations. This chemistry has enabled the labeling of protein surfaces with fluorophores, inhibitors, or common posttranslational modifications such as glycosylation or lipidation.

Key words Tetrazine, Bioorthogonal ligations, Genetic code expansion, Trans-cyclooctene, Fluorescent labeling, Protein labeling, Inverse electron demand Diels-Alder

1 Introduction

Bioorthogonal ligations are commonly used to modify biomolecules predominantly through the addition of fluorescent labels to monitor cellular location [1, 2], attach proteins to materials [3], inhibitors to control protein activity [4], and probes to determine protein conformation [5]. These ligation reactions are defined as bioorthogonal because they occur between two functional groups that are reactive with each other but unreactive to all other functional groups found in biological molecules. Numerous biomolecules including nucleic acids [6], proteins [1, 2, 7, 8], lipids [9], and sugars [10, 11] have been studied using bioorthogonal labeling. A key step in the use of bioorthogonal reactions is the incorporation of necessary functional groups into the biomolecule(s) of choice. Since its development, genetic code expansion has proven to be a valuable strategy to introduce over a dozen different bioorthogonal functional groups onto proteins as noncanonical amino acids (ncAAs) [12].

Edward A. Lemke (ed.), *Noncanonical Amino Acids: Methods and Protocols*, Methods in Molecular Biology, vol. 1728, https://doi.org/10.1007/978-1-4939-7574-7_13, © Springer Science+Business Media, LLC 2018

Given the wide range of applications and reaction conditions used to label proteins, a variety of bioorthogonal reactions have been developed and implemented with genetic code expansion. One of the most popular reactions is the [3+2] azide-alkyne cycloaddition commonly known as click chemistry [13]. The click reaction is used for its high reaction yields and small biologically stable functional groups. Because of the slow rate of the azide-alkyne cycloaddition and its dependence on the toxic Cu^+ catalyst, the azide-alkyne reaction is less suitable for labeling in living systems [14]. While copper-free click chemistry between azides and strained alkynes overcomes the reliance on Cu^+ ions, the rate of reaction is more sluggish than copper-dependent click chemistry [14]. To overcome these challenges, the inverse electron demand Diels-Alder reaction (IEDDA) between tetrazines and strained alkenes has recently been adopted for its high reaction rate and its orthogonality in vivo [13].

The fastest reported bioorthogonal ligation reactions are IEDDA reactions such as those between tetrazines and strained alkenes [13]. As a result, ncAAs with such functional groups are ideal targets for incorporation into proteins through genetic code expansion. Thus far, these include strained alkenes, strained alkynes [15], triazines [16], and tetrazines [7, 17]. Functional group selection ultimately depends on the properties of the functional group and the desired application. In general, the reaction rate of IEDDA reactions increases as the alkene becomes more strained, and the tetrazine becomes more electron deficient. Highly reactive tetrazines and strained alkenes, such as trans-cyclooctene (TCO) and strained trans-cyclooctene (sTCO), are often toxic to or degrade in living cells by reacting with water or nucleophiles. While less reactive tetrazines and less strained alkenes such as amino-tetrazines and cyclopropenes, respectively, are stable under biological conditions, they react with more sluggish reaction rates. The use of moderately strained alkenes, such as norbornene and bicyclononyne, can achieve rate constants >1000 M^{-1} s^{-1} with highly activated tetrazines. Unfortunately, the most strained alkenes known (TCO and sTCO), which can reach reaction rates exceeding 10^6 M^{-1} s^{-1} [15], are often susceptible to degradation via isomerization to cis-cyclooctene in vivo, though some TCO variants are stable for extended periods of time [15, 18]. The electron withdrawing 1,4 pyridal substituted tetrazines are similarly unstable under biological conditions and not suitable for genetic code expansion [15, 19]. Identifying the best reaction partners to afford desired reaction rates with acceptable levels of decomposition for application conditions is therefore a delicate balancing act for IEDDA reactions. To obtain the fastest bioorthogonal reaction that will work in cells with minimal side reactions, a moderately activated tetrazine containing amino acid can be reacted with the more reactive TCO and sTCO labeling reagents. When the tetrazine containing amino acid, 4-(6-methyl-*s*-tetrazin-3-yl) phenylalanine (Tet-v2.0), is incorporated into proteins with genetic code expansion, clean in vivo reactions can be achieved with rates approaching $10^3 \, M^{-1} \, s^{-1}$ and $10^5 \, M^{-1} \, s^{-1}$ for TCO and sTCO labels respectively.

Herein, we describe a protocol for the synthesis of the tetrazine amino acid (Tet-v2.0), for its site-specific incorporation into proteins with genetic code expansion, and for its reaction when incorporated into proteins with TCO- and sTCO-functionalized labels.

2 Materials

2.1 Tet-v2.0 Synthesis

1. *N-tert*-butoxycarbonyl-4-cyano-L-phenylalanine (Chem-Impex Internationall Inc., Cat# 04112).
2. Anhydrous hydrazine.
3. Acetonitrile.
4. Nickel (II) trifluoromethanesulfonate.
5. 2 M sodium nitrite (aqueous).
6. 4 M hydrochloric acid (aqueous).
7. Anhydrous magnesium sulfate.
8. 4 M hydrochloric acid in 1,4-dioxane.
9. Deionized water.
10. Ethyl acetate.
11. 1% acetic acid in ethyl acetate.
12. Hexanes.
13. 60 Å silica gel.
14. Inert gas (Argon used here).
15. Light mineral oil.
16. Flash chromatography system or chromatography column.
17. Rotary evaporator.
18. pH strips or pH probe.
19. Desiccator with anhydrous calcium sulfate.

2.2 Expression

1. Electrocompetent DH10B cells: available as ElectroMAX DH10B cells from Thermo Fisher Scientific.
2. pBad plasmid (Invitrogen, Cat# V430-01) containing your gene of interest interrupted with an amber stop codon at your preferred location (*see* **Notes 3** and **4**).
3. pDule-D12-Tet-v2.0 plasmid (available through Addgene: plasmid# 85496) (*see* **Note 5**).

4. 18-Amino Acid Mix: Weigh out 5 g of the following reagents and dissolve in 1 L: glutamic acid sodium salt, aspartic acid, lysine HCl, arginine HCl, histidine HCl·H_2O, alanine, proline, glycine, threonine, serine, glutamine, asparagine·H_2O, valine, leucine, isoleucine, phenylalanine, tryptophan, methionine. Filter-sterilize and store at 4 °C as 50 mL aliquots (*see* **Note 6**).

5. 25× M salts: Per 1 L of solution dissolve 88.73 g sodium phosphate dibasic (625 mmol), 85.1 g potassium phosphate monobasic (625 mmol), 66.9 g ammonium chloride (1.25 mol), and 17.8 g sodium sulfate (125 mmol). Autoclave to sterilize.

6. 5000× Trace metals: Make 30 mL stock solutions of the following metal salts and filter sterilize: Calcium chloride dihydrate (8.82 g, 400 µM). Manganese chloride tetrahydrate (5.93 g, 200 µM). Zinc sulfate heptahydrate (8.62 g, 200 µM). Cobalt chloride hexahydrate (1.32 g, 40 µM). Copper(II) chloride (807 mg, 40 µM). Nickel(II) chloride (777 mg, 40 µM). Sodium selenite (1.45 g, 40 µM). Sodium molybdenate dihydrate (1.03 g, 40 µM). Boric acid (371 mg, 40 µM). Iron(III) chloride (486 mg 10 µM) (*see* **Note 7**). To make 50 mL of 5000× Trace metals mix 500 µL of all 30 mL stock solutions sans the iron(III) chloride of which 25 mL should be added. Bring the final volume to 50 mL with sterile water.

7. Non-induction Media [20]: To a clean sterile bottle add 50 mL 5% (w/v) aspartate pH 7.5, 40 mL 18-amino acid mix, 40 mL 25× M salts, 2 mL 1 mM magnesium sulfate, 200 µL 5000× trace metals, and 12.5 mL 40% (w/v) glucose. Fill to 1 L with sterile water. When making all media, antibiotics should be added as appropriate (*see* **Note 8**).

8. Auto-induction Media [20]: To a clean sterile bottle add 50 mL 5% aspartate pH 7.5, 50 mL 10% glycerol, 40 mL 18-amino acid mix, 40 mL 25× M salts, 2 mL 1 mM magnesium sulfate, 200 µL 5000× trace metals, 1.25 mL 40% (w/v) glucose, and 2.5 mL 20% (w/v) arabinose. Fill to 1 L with sterile water.

9. LB Media: Make according to the vendor's instructions and autoclave to sterilize. Alternatively, if commercial LB is not used, LB can be made by dissolving NaCl (10 g/L), tryptone (10 g/L), and yeast extract (5 g/L) in purified water with a resistivity of 18.2 MΩ cm and autoclaving to sterilize. Arabinose must be added to the media to a concentration of 0.2% (w/v) to express genes in the pBad vector. This can be performed prior to expression or after sufficient optical density is reached depending on the optimal expression conditions of your protein.

2.3 Purification

1. TALON Wash Buffer: To make a 5× stock solution of 1 L solution dissolve 87.66 g sodium chloride (1.5 mol), 10.76 g sodium phosphate monobasic hydrate (78 mmol), and 24.42 g sodium phosphate dibasic (172 mmol) in purified water with a resistivity of 18.2 MΩ cm. The final solution should be pH adjusted to pH 7 (*see* **Note 9**).

2. TALON Elution Buffer: To make a 5× stock solution of 1 L solution dissolve 87.66 g sodium chloride, 10.76 g sodium phosphate monobasic, and 24.42 g sodium phosphate dibasic in purified water with a resistivity of 18.2 MΩ cm. The final solution should be pH adjusted to pH 7.

2.4 Reaction

1. Cy5-TCO from Click Chemistry Tools.

2. TAMRA-sTCO synthesized in house (synthesis not described) (*see* **Note 1**).

3. mPEG$_{5000}$-sTCO synthesized in house (synthesis not described) (*see* **Note 2**).

3 Methods

3.1 Synthesis of Tet-v2.0 (Scheme 1) [21]

All the reactions should be performed in a fume hood.

1. Oven or flame dry a 20 mL pressure reaction vessel and stir bar. Allow them to cool to room temperature in a desiccator over anhydrous calcium sulfate.

2. Prepare a 60 °C bath using light mineral oil on a heating stir plate.

3. Weigh out *N-tert*-butoxycarbonyl-4-cyano-L-phenylalanine (150 mg, 0.52 mmol) and Nickel (II) trifluromethanesulfonate (92.2 mg, 0.26 mmol) and add them to the room temperature reaction vessel.

4. Flush the reaction vessel with inert gas (argon or nitrogen).

5. Add acetonitrile (0.270 mL, 5.2 mmol) and begin stirring the mixture at room temperature.

6. Add anhydrous hydrazine (0.811 mL, 25.83 mmol) with stirring and flush the vessel again with inert gas (*see* **Note 10**).

7. Seal the reaction vessel with a Teflon screw cap, and introduce the vessel to the 60 °C oil bath (*see* **Note 11**).

8. Allow the reaction to stir while heating for 24 h.

9. Remove the reaction from heat, and allow it to cool to room temperature.

10. Carefully open the reaction vessel and begin flushing immediately with inert gas (*see* **Note 12**).

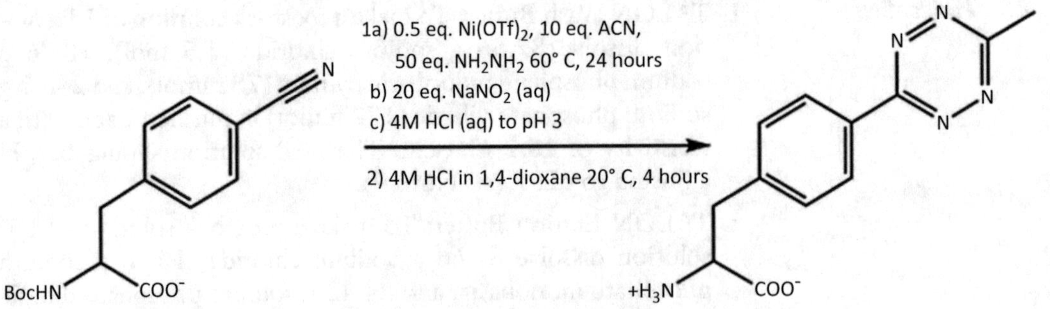

Scheme 1 Synthesis of Tet-v2.0 from Boc-protected 3-cyanophenylalanine

11. Add aqueous 2 M $NaNO_2$ (5.2 mL, 10.4 mmol) dropwise into the reaction vessel while stirring over 3 min.

12. Transfer the reaction mixture into a 500 mL separatory funnel (*see* **Note 13**).

13. Add 200 mL EtOAc and 200 mL DI H_2O to the separatory funnel. Shake the separatory funnel to mix the two phases, and allow them to separate. The desired product will extract into the aqueous phase. Drain the aqueous phase and organic phase into separate 500 mL Erlenmeyer flasks (*see* **Note 14**).

14. Repeat **step 13** using the original aqueous layer and 100 mL fresh EtOAc each time. Repeat this step until the organic phase has very little pink color in it after shaking and separation of the phases (*see* **Note 15**).

15. To a 500 mL Erlenmeyer flask containing the aqueous phase, add 4 M HCl dropwise until pH 3 is achieved. CAUTION! Adding acid to the aqueous phases will evolve toxic NO_x gases. Make sure to do this step in a well-ventilated area (*see* **Note 16**).

16. Add the acidified aqueous phase to the 500 mL separatory funnel along with 100 mL fresh EtOAc. Shake the separatory funnel to mix the two phases, and let them settle out. The product will extract into the organic phase. Drain the two layers into separate 500 mL Erlenmeyer flasks (*see* **Note 17**).

17. Repeat **step 16**, using the original acidified aqueous layer and 50 mL fresh EtOAc. Repeat this step until very little color is seen in the organic layer after shaking the separatory funnel.

18. Using the separatory funnel, extract the combined organic layers twice with 200 mL fresh DI H_2O, and once with 100 mL of saturated aqueous NaCl solution. Use the same technique as described in **steps 13** and **16** to extract the combined organic layers.

19. Dry the combined organic layers over excess anhydrous $MgSO_4$.

20. Using a Whatman #1 filter paper and a Buchner funnel, vacuum filter off the $MgSO_4$ into a round-bottomed flask. Rinse the $MgSO_4$ in the filter with 5–10 mL of fresh EtOAc.

21. Concentrate the filtrate using rotary evaporation.

22. After taking the mass of the dry product, redissolve in a minimal amount of EtOAc.

23. Add silica gel to the organic solution, and remove the EtOAc using rotary evaporation (see **Note 18**).

24. With the product coating the silica gel, transfer the mixture to a flash chromatography column and carry out flash chromatography (see **Note 19**).

25. Verify the product-containing fractions via TLC. Combine and concentrate these fractions using rotary evaporation (see **Note 20**).

26. To the flask containing the combined product fractions, add 4 M HCl in 1,4-dioxane (2.2 mL, 8.8 mmol) and a small Teflon stir bar. Purge the reaction vessel with inert gas, and stir the reaction mixture at room temperature for 4–8 h (see **Note 21**).

27. Concentrate the reaction mixture using rotary evaporation. Redissolve in EtOAc and re-concentrate. Add minimal pentane, shake, and remove the pentane using vacuum three or more times to remove trapped HCl.

Transfer the pink powder to a vial and store it at room temperature. The product can be identified by NMR. Typically, molar yields of 50–60% can be achieved (see **Note 22**).

3.2 Expression

1. The following plasmids should be co-transformed into DH10B cells: pDule-D12-Tet-v2.0 and a pBad plasmid containing your gene of interest. The pDule plasmid contains two variants (pDule1 and pDule2) that differ only by antibiotic resistance genes (Tetracycline and Spectinomycin respectively) (see **Note 23**). The pBad plasmid is ampicillin resistant and should contain your gene of interest interrupted by a TAG amber codon at the desired site of Tet-v2.0 incorporation. If you cannot express your gene of interest in the pBad plasmid, other expression vectors are acceptable, but media, antibiotics, and cell line must be varied accordingly.

2. To transform DH10B cells obtain electrocompetent cells in 80 μL aliquots and thaw on ice. Add 0.5 μL of both the pDule and pBad plasmids with at a concentration between 50 and 200 ng/μL. Mix the cells and DNA by tapping on the tube 3–4 times. Transfer the cells to a 0.2 cm electroporation cuvette and electroporate at 2500 V. Immediately add SOC media

(1 mL) and allow the cells to recover for 1 h at 37 °C. Plate the cells after recovery on LB plates containing ampicillin (100 μg/mL) and tetracycline (25 μg/mL) (*see* **Note 24**).

3. After 12–18 h, colonies should have grown large enough to pick. The plasmid containing cells can be stored at −80 °C by generating glycerol cell stocks. To do so, pick individual colonies into 5 mL cultures of LB media with ampicillin (100 μg/mL) and tetracycline (25 μg/mL) and grow 12–18 h at 37 °C. In a 1.7 mL microfuge tube mix 800 μL grown LB media and 200 μL sterile 80% glycerol in water (v/v). Store cell stocks at −80 °C.

4. Using glycerol cell stocks or colonies of DH10B cells transformed with both the pBAD plasmid containing your TAG interrupted gene of interest as well as the Tet-v2.0 pDule1 plasmid, inoculate a 5 mL culture of Non-induction Media. The Non-induction Media should contain the antibiotics ampicillin (100 μg/mL) and tetracycline (25 μg/mL). Allow the culture to grow 16–24 h (*see* **Note 25**).

5. Use 0.5 mL of the saturated 5 mL Non-induction Media culture to inoculate 50 mL cultures of Auto-induction Media or LB. Cultures should be grown in 250 mL flasks and the inoculating volume should be <1% of the final culture volume. Media should be supplemented with 1 mM of Tet-v2.0 (*see* **Notes 26–28**).

6. Grow inoculated cultures at 37 °C for 40–48 h. After growth centrifuge at 5000 rcf for 10 min to pellet cells. Decant supernatant (*see* **Notes 29** and **30**).

3.3 Purification

1. Resuspend the cells in TALON wash buffer and lyse using a Microfluidics M-110P microfluidizer or another appropriate lysis method. Clarify the cell lysate through centrifugation (21,000 rcf, 1 h). Decant the supernatant into 50 mL conical tubes (*see* **Notes 31** and **32**).

2. Prepare TALON resin by washing three times with 1 mL of water per 300 μL bed volume resin. We typically use 50 μL resin per 50 mL expression volume (*see* **Note 33**). Resin should be added to clarified lysate and incubated at 4 °C while rocking for at least 1 h to bind protein to the resin (*see* **Note 34**).

3. Incubated resin should be applied to a column and the cell extract allowed to flow through. The resin should be washed with >20 column volumes of TALON wash buffer. Protein should then be eluted in 3–10 column volumes TALON elution buffer (*see* **Note 35**).

4. Protein concentration should be measured using Bradford, BCA, or other appropriate assays.

5. Protein should be desalted into appropriate buffer using PD-10 columns (GE Healthcare, Cat# 52130800), dialysis, or other appropriate methods.

6. Protein identity and incorporation can be confirmed by SDS PAGE analysis or mass spectrometry (Fig. 1) (*see* **Note 36**).

3.4 Reaction

1. Dissolve TAMRA-sTCO and Cy5-TCO to a concentration of 2 mM in methanol. Dilute protein to a concentration of 10 μM in appropriate reaction buffer, such as Tris-buffered saline (25 mM Tris-HCl pH 7.4, 150 mM NaCl). In a 0.7 mL microfuge tube, add 20 μL of protein and 2 μL fluorescent dye (2 mM, 2000% molar equivalence), and mix by gentle pipetting. Perform a quick spin in a desktop centrifuge to ensure all the components are collected together in the bottom of the tube. Allow the reaction to proceed for 30 min prior to analysis via SDS-PAGE (Fig. 2) (*see* **Note 37**). The reaction should achieve 99% completion in less than a second for the reaction between protein and sTCO. For the reaction between TCO and protein the reaction should achieve 99% completion in less than a minute.

2. Dilute mPEG$_{5000}$-sTCO in water to a concentration of 0.5 mg/mL or 100 μM. In a 0.7 mL microfuge tube, add 20 μL protein at a concentration of 10 μM and 10 μL mPEG$_{5000}$-sTCO. Perform a quick spin in a desktop centrifuge to ensure mixing. Allow the reaction to proceed for up to 30 min prior to analysis via SDS-PAGE (Fig. 3) (*see* **Note 36**). The reaction should achieve 99% completion within 2 s.

3. To perform sub-stoichiometric labeling, dilute mPEG$_{5000}$ in water to concentrations ranging from 2 to 320 μM. In a 0.7 mL microfuge tube, add 10 μL protein at a concentration of 10 μM and 5 μL of diluted mPEG$_{5000}$. Allow the reaction to proceed for 30 min and analyze via SDS-PAGE (Fig. 4). Reaction times to achieve 99% completion vary (*see* **Note 38**).

4 Notes

1. TAMRA-sTCO was synthesized [7] from commercially available amine-labeled tetramethylrhodamine and activated sTCO acquired from the Joseph Fox lab at the University of Delaware.

2. mPEG$_{5000}$-sTCO was synthesized in a similar manner to TAMRA-sTCO [7] from commercially available amine-functionalized PEG and activated sTCO acquired from the Joseph Fox lab at the University of Delaware.

3. pBad is an arabinose dependent expression plasmid that relies on the AraBAD promoter. pBad is ampicillin resistant.

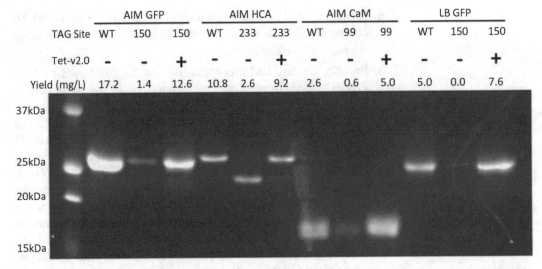

Fig. 1 SDS-PAGE gel of expressions of GFP, HCA, and CaM both wild-type (wt) and stop codon interrupted (TAG). The amount of protein loaded on each lane was not equivalent, but rather proportional to the amount of protein purified. Since the purification technique was not adjusted for the expressions in the presence and absence of Tet-v2.0, this may lead to an overrepresentation of full-length protein produced in the absence of Tet-v2.0. The HCA 233 expression without ncAA shows a significant band at a molecular weight of ~25 kDa. This is consistent with the TAG interrupted mutant which is of a similar size to the TAG suppressed mutant and could potentially adhere to the TALON resin due to its histidine rich zinc binding site. Mass spectra of the lower molecular weight band are consistent with truncation. GFP was expressed in both LB and Auto-induction Media

Your gene of interest should be cloned into this expression vector with an amber stop codon at the desired site.

4. When designing the plasmid with your gene of interest (pBad in this case) care should be taken to put a C-terminal His tag as opposed to an N-terminal His tag. With an N-terminal His tag, purification of the protein may copurify truncated protein. With a C-terminal His tag this is not likely to occur as the His tag is not expressed.

5. pDule-D12-Tet-v2.0 is a tetracycline resistant plasmid containing the *M. jannaschii* tRNA$_{CUA}$ as well as an *M. jannaschii* aaRS evolved for the incorporation of Tet-v2.0.

6. Amino acids should be dissolved in the order listed to decrease the amount of time it takes to dissolve.

7. Iron (III) chloride must be dissolved in 0.1 M hydrochloric acid and then filtered through a 0.2 μm filter to remove insoluble residues.

8. When adding antibiotics to media, we keep 1000× stocks of the antibiotics of interest. Ampicillin can be dissolved at a concentration of 100 μg/μL in water, whereas tetracycline can be stored at 25 μg/μL in *N*,*N*-dimethyl formamide (DMF).

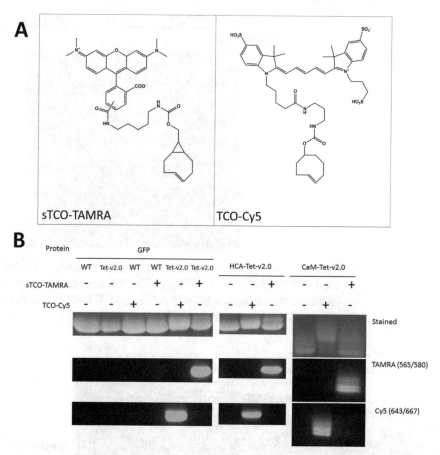

Fig. 2 Fluorescent labeling of tetrazine containing protein. (**a**) Structures of the fluorescently tagged TCO and sTCO (**b**) SDS-PAGE gel of the reaction between GFP-wt, GFP150-Tet-v2.0, HCA233-Tet-v2.0, CaM99-Tet-v2.0, and fluorescently labeled TCO/sTCO variants. Three gel images correspond to imaging of the stained gel, imaging with a rhodamine filter, and imaging with a Cy5 filter, respectively. GFP-wt is unreactive to the TCO/sTCO labels whereas the tetrazine labeled protein results in fluorescent labeling

9. TALON wash and elution buffers are useful for purifying the proteins used here, but are not compatible with every purification. Purification buffers should be adjusted as needed.

10. Fully dissolve the *N-tert*-butoxycarbonyl-4-cyano-L-phenylalanine in the acetonitrile before adding hydrazine. Direct contact between the solid reagents and liquid hydrazine can cause a vigorous and exothermic reaction. Upon addition of all four starting materials, the reaction mixture will turn translucent violet. This color will change to opaque brown in 5–30 min after heat is introduced.

11. Ensure the liquid level of the reaction mixture is below the level of the oil bath, but that the "headspace" in the reaction flask is above the bath. This will allow the hydrazine to condense in the cool headspace and to return to the reaction mixture. The reaction vessel should have significant headspace—about

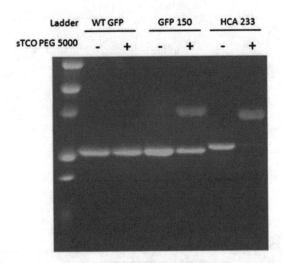

Fig. 3 SDS-PAGE gel analysis of the reaction between GFP-wt, GFP150-Tet-v2.0, and HCA233-Tet-v2.0 with mPEG$_{5000}$-sTCO. Tetrazine containing protein shifts upward in mass when reacted with mPEG$_{5000}$-sTCO in contrast to the unreacted protein without tetrazine amino acids

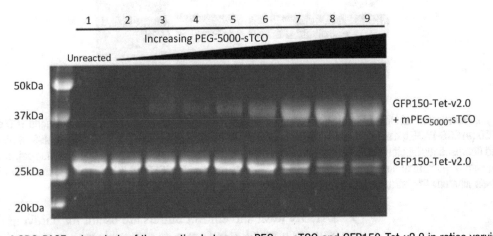

Fig. 4 SDS-PAGE gel analysis of the reaction between mPEG$_{5000}$-sTCO and GFP150-Tet-v2.0 in ratios varying from 10% molar equivalence of mPEG$_{5000}$-sTCO in lane 2 to 1600% molar equivalence of mPEG$_{5000}$-sTCO in lane 9. For comparison, the concentration of GFP150-Tet-v2.0 is held constant at 6.7 μM in all the reactions. As the ratio of mPEG$_{5000}$-sTCO to GFP150-Tet-v2.0 increases, the amount of labeled protein is seen to increase until the protein is fully labeled

10 times more headspace volume than the reaction mixture volume.

12. Opening the pressurized reaction vessel will release dissolved gases from the reaction mixture. Significant foaming will occur, so be careful and open the vessel very slowly. Immediately flushing the vessel with inert gas prevents the side reactions from atmospheric O$_2$.

13. Rinse the reaction vessel several times into the separatory funnel with DI water in order to minimize product loss during this first transfer step.

14. When shaking the separatory funnel to mix the aqueous and organic phases, it is important to vent the separatory funnel away from yourself often so pressure does not build up in the separatory funnel.

15. The pink compound being extracted into the organic phase is not your product. It is dimethyltetrazine—the main byproduct of this synthesis. It is very strongly colored and very abundant, so you will never see your organic phase separate out clear after shaking it with the aqueous phase in the separatory color. Use your best judgment in moving on to the next step, and keep in mind that this is only a crude purification step. Any residual dimethyltetrazine will be later separated out using column chromatography.

16. The goal of this step is protonation of the carboxylic acid, making the molecule neutral overall. This protonation will cause the product to migrate from the aqueous phase to the organic phase. Adding 4 M HCl too quickly during this step can cause degradation of the product. Stir the solution very vigorously during this step, and add 4 M HCl at a rate of 1 drop per second or slower. Monitor the pH of the solution using pH strips or a pH probe during the addition to ensure you do not overshoot the target pH of 3. If you drop the pH much lower than 3, the "boc" amine-protecting group will be cleaved and product can be lost.

17. Use a total ethyl acetate volume of 250 mL or less during these two steps. Minimizing solvent usage in this step will make the subsequent concentration step much quicker and easier. Note that doing several rinses with a smaller amount of ethyl acetate in each rinse is preferable to doing a few rinses with a larger volume of ethyl acetate in each rinse.

18. As a general rule, add about 5× the mass of crude product in the round-bottom flask in silica. Adding the silica to the organic solution and then re-concentrating will suspend the crude product onto the silica.

19. Dimethyltetrazine, the main byproduct, will elute at a solvent ratio of 10:90 1% HOAc in EtOAc: Hexanes. Tet 2.0 will elute at a solvent ratio of approximately 25:75 1% HOAc in EtOAC:Hexanes. The recommendation is to start at 100% hexanes for 1 min and shift to 10:90 1% HOAc in EtOAc: Hexanes over 2 min. Once at 10:90 1% HOAc in EtOAc: Hexanes keep this isocratic until dimethyltetrazine has finished coming off the column. Then run a gradient up to 30:70 1% HOAc in EtOAc: Hexanes over 5 min. If no more of the

pink tetrazine band can be seen coming off the column, you can proceed to transition up to 100:0 1% HOAc in EtOAc: Hexanes over 3 min. Several polar byproducts will elute around and above 50:50 1% HOAc in EtOAc: Hexanes. Run pure 1% HOAc in EtOAc for 1 min, and then transition instantaneously to pure hexanes, and run that isocratically for 1 min.

20. Fractions can be assessed for purity by TLC in 50:50 1% HOAc in EtOAC:Hexanes. Tet 2.0 will have an Rf of 0.3, and dimethyltetrazine will have an Rf of 0.7.

21. This step removes the "boc" amine-protecting group, and creates the hydrochloride salt of the title product. As a general rule for this step, add 1 mL of the 4 M HCl in 1,4-dioxane solution for every 50 mg of boc-protected product.

22. You will have some acetic acid, water, and ethyl acetate contaminating your product. These impurities can make the solid very difficult to remove from the round-bottomed flask, and you'll inevitably fail to quantitatively transfer the product to a vial. Dry the product on a high vac line overnight and this will aid in transfer.

23. Some cell lines have innate resistance to antibiotics and so either the pDule1 or pDule2 must be used. For example, the BL21AI cells are natively resistant to tetracycline and are incompatible with pDule1.

24. It is likely that plating the entirety of the transformed cells will create a lawn on the plate. For this reason, we recommend plating 25–100 μL of the transformed cells on a single plate and the rest on a second plate after centrifuging for ~1 min at 2000 rcf or below to pellet cells followed by resuspension in a smaller volume.

25. The purpose of the Non-induction Media is to ensure equal growth of all cultures and equal cell density prior to the inoculation of Auto-induction Media. In the interest of comparative analysis between expressions in different cultures, this step is highly recommended. It is worth noting that the use of Non-induction Media is not necessary if the intended purpose is solely protein expression. With that said, it should certainly be used if you are trying to express a new protein containing a ncAA of interest as it allows accurate comparisons with the WT protein expression and the expression of protein in the absence of amino acid. WT and cultures lacking ncAA but containing your TAG interrupted gene should be tested when expressing a given protein the first few times as they demonstrate both that your protein can be produced in the given conditions, and that no full-length protein is produced in the absence of Tet-v2.0.

26. Lower concentrations of Tet-v2.0 can be used but protein production is dependent on amino acid concentration in media. Concentrations in the range 0.2–1 mM are ideal. Less than that, and protein production will be impaired. We recommend dissolving Tet-v2.0 in DMF prior to the addition to media to ensure amino acid solubility. Tet-v2.0 is soluble in DMF up to 100 mM and *E. coli* growth is not unaffected by the presence of 1% DMF in media. When measuring Tet-v2.0 for use in media, it is important to note that the Tet-v2.0 is the HCl salt, and the mass of the HCl salt should be used.

27. Auto-induction Media contains limited glucose so as not to inhibit the AraBAD Promoter in the late stages of cell growth. Inoculation of Auto-induction Media with large amounts of Non-induction Media (>1% v/v) can result in higher than optimal glucose concentration in the Auto-induction Media.

28. The aeration of tetrazine media can affect the redox state of the tetrazine amino acid, while we haven't found this to inhibit protein production, it can affect the final reactivity of the protein of interest. Generally, Tet-v2.0 becomes reduced during log phase growth in media with insufficient aeration which can alter the color of the media from bright pink to a colorless media. The dihydro-tetrazine (reduced form) is readily reoxidized by exposure to air.

29. Maximum protein yield occurs after 24–36 h; however, the maximum reactive protein was found to be produced in the 40–48 h range.

30. Expression conditions should be varied to express your protein. If you cannot successfully express and purify the wild-type protein, you will most likely have no luck expressing your protein with Tet-v2.0 under the same conditions.

31. Residual Tet-v2.0 will be present in your cells. It is possible to remove this by washing pelleted cells with buffer prior to the reaction. His-TAG purification of protein as described above is usually sufficient to remove unincorporated Tet-v2.0. However, if you wish to perform a reaction in live cells, or cell lysate, a washing step is necessary. To wash the cells resuspend the cells in buffer and pellet (5000 rcf, 10 min). Repeat 2–4 times.

32. Cell pellets can be stored at −80 °C for extended periods of time before lysis and purification.

33. The ratio of 1 μL TALON resin per 1 mL expression volume results in protein of higher purity, though not all expressed protein can bind to the resin. Using a higher amount of resin will result in better yields but also more nonspecific binding to resin, and therefore lower purity.

34. The TALON resin contains cobalt ions as compared to the more commonly used nickel resin. Cobalt resins often result in higher protein purity but lower protein yields.

35. The wash and elution volumes listed here are recommendations. Volumes should be adjusted to reflect the desired concentration and purity of your protein of interest.

36. Calmodulin exhibits diffuse bands in an SDS-PAGE gel due to residual calcium binding in the denatured state [22].

37. Reactions between tetrazines and strained alkenes such as TCO or sTCO are bimolecular and so the amount of time it takes for the reaction to go to completion depends on the concentrations of the reagents. The time to completion listed above is dependent on the above concentrations being used a tenfold lower concentration for one reagent means that the reaction will take ten times as long. A tenfold lower concentration of both reagents means the reaction will take 100 times as long. Adjust accordingly.

38. Reaction times for sub-stoichiometric reactions tend to increase as the ratio of sTCO label to protein approaches 1:1. This is because, toward the end of the reaction, the concentration of both protein and label is lower. In contrast, reactions with a ratio closer to 1:5 have excess protein as the reaction nears completion.

References

1. Peng T, Hang HC (2016) Site-specific bioorthogonal labeling for fluorescence imaging of intracellular proteins in living cells. J Am Chem Soc 138:14423–14433. https://doi.org/10.1021/jacs.6b08733

2. Uttamapinant C, Howe JD, Lang K et al (2015) Genetic code expansion enables live-cell and super-resolution imaging of site-specifically labeled cellular proteins. J Am Chem Soc 137:4602–4605. https://doi.org/10.1021/ja512838z

3. Longo J, Yao C, Rios C et al (2014) Reversible biomechano-responsive surface based on green fluorescent protein genetically modified with unnatural amino acids. Chem Commun 51:232–235. https://doi.org/10.1039/C4CC07486F

4. Tsai Y-H, Essig S, James JR et al (2015) Selective, rapid and optically switchable regulation of protein function in live mammalian cells. Nat Chem 7:554–561. https://doi.org/10.1038/nchem.2253

5. Fleissner MR, Brustad EM, Kálai T et al (2009) Site-directed spin labeling of a genetically encoded unnatural amino acid. Proc Natl Acad Sci 106:21637–21642. https://doi.org/10.1073/pnas.0912009106

6. Someya T, Ando A, Kimoto M, Hirao I (2015) Site-specific labeling of RNA by combining genetic alphabet expansion transcription and copper-free click chemistry. Nucleic Acids Res 43:6665–6676. https://doi.org/10.1093/nar/gkv638

7. Blizzard RJ, Backus DR, Brown W et al (2015) Ideal bioorthogonal reactions using a site-specifically encoded tetrazine amino acid. J Am Chem Soc 137:10044–10047. https://doi.org/10.1021/jacs.5b03275

8. Lang K, Davis L, Torres-Kolbus J et al (2012) Genetically encoded norbornene directs site-specific cellular protein labelling via a rapid bioorthogonal reaction. Nat Chem 4:298–304. https://doi.org/10.1038/nchem.1250

9. Erdmann RS, Takakura H, Thompson AD et al (2014) Super-Resolution Imaging of the Golgi in Live Cells with a Bioorthogonal Ceramide Probe. Angew Chem Int Ed 53:10242–10246. https://doi.org/10.1002/anie.201403349

10. Agarwal P, Beahm BJ, Shieh P, Bertozzi CR (2015) Systemic fluorescence imaging of zebrafish glycans with bioorthogonal chemistry. Angew Chem Int Ed 54:11504–11510. https://doi.org/10.1002/anie.201504249

11. Machida T, Lang K, Xue L et al (2015) Site-specific glycoconjugation of protein via bioorthogonal tetrazine cycloaddition with a genetically encoded trans-cyclooctene or bicyclononyne. Bioconjug Chem 26:802–806. https://doi.org/10.1021/acs.bioconjchem.5b00101

12. Dumas A, Lercher L, Spicer CD, Davis BG (2015) Designing logical codon reassignment – expanding the chemistry in biology. Chem Sci 6:50–69. https://doi.org/10.1039/C4SC01534G

13. Mayer S, Lang K (2017) Tetrazines in inverse-electron-demand Diels–Alder cycloadditions and their use in biology. Synthesis 49:830–848. https://doi.org/10.1055/s-0036-1588682

14. Baskin JM, Prescher JA, Laughlin ST et al (2007) Copper-free click chemistry for dynamic in vivo imaging. Proc Natl Acad Sci 104:16793–16797. https://doi.org/10.1073/pnas.0707090104

15. Lang K, Davis L, Wallace S et al (2012) Genetic encoding of bicyclononynes and trans-cyclooctenes for site-specific protein labeling in vitro and in live mammalian cells via rapid fluorogenic Diels–Alder reactions. J Am Chem Soc 134:10317–10320. https://doi.org/10.1021/ja302832g

16. Kamber DN, Liang Y, Blizzard RJ et al (2015) 1,2,4-Triazines are versatile bioorthogonal reagents. J Am Chem Soc 137:8388–8391. https://doi.org/10.1021/jacs.5b05100

17. Seitchik JL, Peeler JC, Taylor MT et al (2012) Genetically encoded tetrazine amino acid directs rapid site-specific in vivo bioorthogonal ligation with trans-cyclooctenes. J Am Chem Soc 134:2898–2901. https://doi.org/10.1021/ja2109745

18. Hoffmann J-E, Plass T, Nikić I et al (2015) Highly stable trans-cyclooctene amino acids for live-cell labeling. Chem A Eur J 21:12266–12270. https://doi.org/10.1002/chem.201501647

19. Karver MR, Weissleder R, Hilderbrand SA (2011) Synthesis and evaluation of a series of 1,2,4,5-tetrazines for bioorthogonal conjugation. Bioconjug Chem 22:2263–2270. https://doi.org/10.1021/bc200295y

20. Studier FW (2005) Protein production by auto-induction in high-density shaking cultures. Protein Expr Purif 41:207–234. https://doi.org/10.1016/j.pep.2005.01.016

21. Yang J, Karver MR, Li W et al (2012) Metal-catalyzed one-pot synthesis of tetrazines directly from aliphatic nitriles and hydrazine. Angew Chem Int Ed 51:5222–5225. https://doi.org/10.1002/anie.201201117

22. Stateva SR, Salas V, Benaim G et al (2015) Characterization of phospho-(tyrosine)-mimetic calmodulin mutants. PLoS One 10:e0120798. https://doi.org/10.1371/journal.pone.0120798

Part III

Engineering in Eukaryotes

Chapter 14

Mapping of Protein Interfaces in Live Cells Using Genetically Encoded Crosslinkers

Lisa Seidel and Irene Coin

Abstract

Understanding the topology of protein-protein interactions is a matter of fundamental importance in the biomedical field. Biophysical approaches such as X-ray crystallography and nuclear magnetic resonance can investigate in detail only isolated protein complexes that are reconstituted in an artificial environment. Alternative methods are needed to investigate protein interactions in a physiological context, as well as to characterize protein complexes that elude the direct structural characterization. We describe here a general strategy to investigate protein interactions at the molecular level directly in the live mammalian cell, which is based on the genetic incorporation of photo- and chemical crosslinking noncanonical amino acids. First a photo-crosslinking amino acid is used to map putative interaction surfaces and determine which positions of a protein come into proximity of an associated partner. In a second step, the subset of residues that belong to the binding interface are substituted with a chemical crosslinker that reacts selectively with proximal cysteines strategically placed in the interaction partner. This allows determining inter-molecular spatial constraints that provide the basis for building accurate molecular models. In this chapter, we illustrate the detailed application of this experimental strategy to unravel the binding modus of the 40-mer neuropeptide hormone Urocortin1 to its class B G-protein coupled receptor, the corticotropin releasing factor receptor type 1. The approach is in principle applicable to any protein complex independent of protein type and size, employs established techniques of noncanonical amino acid mutagenesis, and is feasible in any molecular biology laboratory.

Key words Photo-crosslinking, Chemical crosslinking, Protein-protein interactions, Spatial constraints, Binding path, Proximity maps, 3D structures, GPCRs, Ligand-receptor complex, Live mammalian cell

1 Introduction

All signaling pathways in living organisms involve interactions of proteins with other proteins, peptides, and organic molecules of several kinds. Understanding the molecular mechanisms of protein interactions is a major goal of the modern biomedical science. While biophysical methods such as crystallography, nuclear magnetic resonance (NMR), and cryo-electron microscopy (cryo-EM) can provide precious high-resolution images of protein complexes

Edward A. Lemke (ed.), *Noncanonical Amino Acids: Methods and Protocols*, Methods in Molecular Biology, vol. 1728, https://doi.org/10.1007/978-1-4939-7574-7_14, © Springer Science+Business Media, LLC 2018

extracted from the cell and reconstituted in vitro, alternative methods are needed to investigate structural features of protein interactions in the cellular context. This is important not only to validate the biological significance of conformational states observed in vitro, but also to investigate the large number of proteins that elude direct structural characterization.

The genetic encoding of crosslinking amino acids is emerging as a powerful approach to investigate protein-protein interactions directly in the live cell. Using the expanded genetic code technology [1–5], synthetic crosslinkers installed on the side chain of noncanonical amino acids (ncAAs) can be incorporated into proteins in a site-specific manner directly by the ribosomal machinery. Amino acyl tRNA synthetase/tRNA (AARS/tRNA) pairs have been developed to incorporate both photo-activatable and chemical crosslinkers into proteins in the form of ncAAs, both in bacterial and eukaryotic expression systems. NcAAs for photo-crosslinking include the classic p-benzoyl-Phe (Bpa) and p-azido-Phe (Azi) crosslinkers [6–11], other azide-based ncAAs [12–15], and ncAAs featuring diazirine crosslinkers, such as AbK [16], DizPK [15, 17], and Tdf [18, 19] (Fig. 1). Photo-activatable moieties are chemically inert under physiological conditions. When activated by UV-light they convert into highly reactive species that covalently bind neighboring molecules in an unspecific fashion. Photo-crosslinking is triggered at a controlled time point and the photochemical reaction takes place rapidly at the ncAA site only, which provides direct information on the topology of the interaction [20].

Chemical crosslinkers are instead designed to react with functional groups naturally present in proteins. Chemical selectivity can be achieved by targeting the thiol group of cysteine, which is the strongest nucleophile among the proteinogenic amino acids. NcAAs for chemical crosslinking bear a chemical functionality that is on one hand stable enough to survive the presence of nucleophiles naturally occurring in the cell and on the other hand reactive

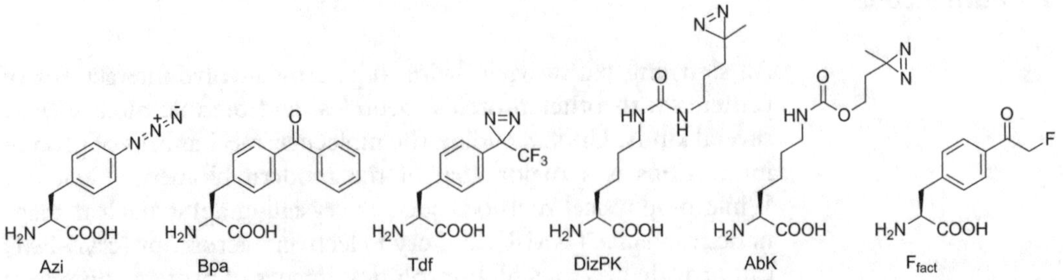

Fig. 1 Genetically encoded crosslinking amino acids. *p*-Azido-L-phenylalanine (Azi), *p*-benzoyl-L-phenylalanine (Bpa), 4′-(3-(trifluoromethyl)-3H-diazirine-3-yl)-L-phenylalanine (Tdf), ((3-(3-methyl-3H-diazirine-3-yl)propamino) carbonyl-L-lysine (DizPK), 3′-azibutyl-*N*-carbamoyl-L-lysine (AbK), and *p*-2-fluoroacetyl-L-phenylalanine (F$_{fact}$)

enough to undergo nucleophilic attack by a thiol group structurally constrained in its spatial proximity. This way, the chemical crosslinker can be incorporated intact into the protein of interest (POI) and will react intra- or inter-molecularly with cysteine residues only when the otherwise slow kinetic of the reaction is enhanced by spatial factors. NcAAs for proximity-enhanced chemical crosslinking include p-2'-fluoroacetyl-Phe (F_{fact}, [21]) and a series of haloalkane ncAAs [22]. Importantly, these amino acids form irreversible covalent bonds with cysteine, which unlike disulfide bonds are stable under reducing conditions. Crosslinking ncAAs targeting Lys and His have also been presented [23].

By combining photo- and chemical-crosslinking ncAAs we have pioneered a novel approach to map protein interaction interfaces in live mammalian cells [24, 25]. The concept is to use a photo-crosslinking ncAA as a "nonspecific proximity sensor" to map systematically entire protein surfaces and determine at a single-amino acid resolution which positions of the POI come into proximity of the interacting partner (Fig. 2a). In a second step, the subset of residues of the POI that belong to the binding interface is substituted with an ncAA enabling for chemical-crosslinking with cysteine residues strategically placed in the interaction partner. Crosslinked ncAA-Cys pairs reveal pairs of amino acids of the two proteins coming close to each other in the associated complex (Fig. 2b). This information provides inter-molecular spatial constraints to eventually build molecular models of the protein complex. The approach is in principle applicable to any complex independent of protein type and size, employs established techniques of noncanonical mutagenesis, and is feasible in any molecular biology lab. Importantly, the method provides information from intact complexes of fully posttranslationally modified proteins in the physiological environment of the live cell, which are not accessible by reductionist biophysical approaches.

We describe here an application of this experimental strategy to determine the binding mode of the 40-mer neuropeptide hormone Urocortin1 (Ucn1) to its membrane receptor, the class B G-protein coupled (GPCR) corticotropin releasing factor receptor type 1 (CRF1R). CRF1R is a key regulator of the organism's stress response and represents a pharmacological target to cure stress related and adrenal disorders [26, 27]. In general, there is nowadays a large interest in defining binding sites of orthosteric and allosteric ligands on GPCRs. As GPCRs are addressed by a large part of the currently marketed drugs, understanding how ligands interact with them to elicit a specific cellular response is a fundamental step on the way toward designing more specific and efficient drugs. However, GPCRs are difficult to crystallize and high-resolution 3D structures of ligand-receptor complexes are available only for a small number of the 800 GPCRs present in

A) Step 1: Photo-crosslinking mapping

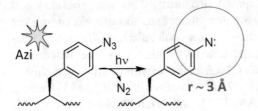

B) Step 2: Pin-point reciprocal vicinity points

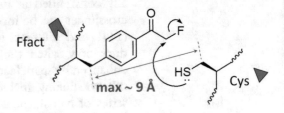

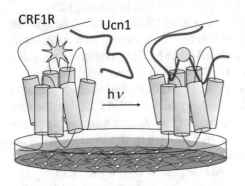

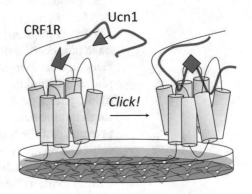

Highlight hits

Apply constraints
to MM

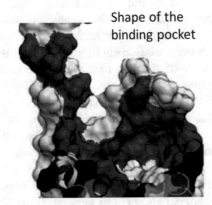
Shape of the
binding pocket

Molecular model

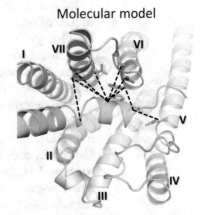

Fig. 2 Unraveling topology of protein interactions, here represented by the interaction between a GPCR, the CRF1R, and its peptide ligand Ucn1. (**a**) Step1. The putative binding pocket of the POI is mapped with the photo-crosslinking ncAA Azi (in yellow). If the Azi site lies in proximity of the bound ligand, a covalent complex is formed upon UV-irradiation. Crosslinking hits reveal the shape of the binding pocket. (**b**) Step 2. Crosslinking positions identified in step 1 are replaced by the chemical crosslinker F_{fact} (in blue). Each POI-mutant is treated with modified ligands that bear strategically placed cysteines (blue triangle). A crosslinking hit identifies a pair of residues that are located within a maximal reciprocal $C\beta$-$C\beta$ distance of ~9 Å in the ligand-receptor complex. These constraints are applied to energy-based molecular modeling (MM) experiments

the human genome. In the case of the CRF1 receptor, only structures of isolated domains have been solved and no structural information about recognition of the natural ligands is available for the 7-transmembrane (7TM) domain, where receptor activation takes place.

Crosslinking ncAAs are applied in two steps (Fig. 2). First, the photo-crosslinking ncAA Azi is systematically incorporated throughout 145 positions in the juxtamembrane domain of the receptor, which cover the putative ligand binding pocket. All mutants, grouped into sets of 12 or 24, are expressed in mammalian cells and irradiated with UV light upon addition of Ucn1 to the culture medium. To ensure that mutated receptors are still able to bind the ligand, a control reaction is performed using a photo-activatable ligand. In the case of CRF1R, a suitable control peptide is Bpa[12]-Ucn1, which has been described in the literature to behave similarly to natural Ucn1, and crosslinks efficiently to the receptor [28]. After crosslinking, cells are lysed and the solubilized membranes are analyzed by SDS-PAGE and western blot. A polyclonal antibody targeted to the ligand allows both identifying the subset of mutant receptors that covalently captured the ligand (Fig. 3a) and analyzing the control reaction (Fig. 3b), while an antibody targeted to the receptor is used to assess receptor expression (Fig. 3c). The "photo-crosslinking hits", 35 in the case of the CRF1R-Ucn1 complex, represent the positions of the receptor that come close to the bound ligand within the ligand-receptor complex, and therefore form the ligand binding pocket.

Once the binding surface has been defined using the photo-crosslinker, the second step is to reveal how the ligand is oriented and positioned within the pocket. Receptor sites corresponding to the 23 most intense photo-crosslinking hits are now replaced with the chemical-crosslinking ncAA F_{fact}, while single cysteines are introduced into five different positions of the ligand. Each ligand is combined with each F_{fact}-receptor for a total of 115 combinations. Combinations for which chemical crosslinking takes place identify inter-molecular pairs of amino acids that lay at a maximal reciprocal $C\beta$-$C\beta$ distance of about 9 Å. For the CRF1R-Ucn1 complex, nine F_{fact}-Cys hits are identified, providing nine spatial constraints for energy-based molecular modeling of the receptor-ligand complex [24].

It is clear that this approach is not alternative but rather complementary to high-resolution methods. However, it allows building the most accurate models when partial structural data or preliminary models already exist and, importantly, provides unique information from the natural system. Photo-crosslinking mapping of ligand binding sites is nowadays widely applied in the GPCR field [24, 25, 29–32].

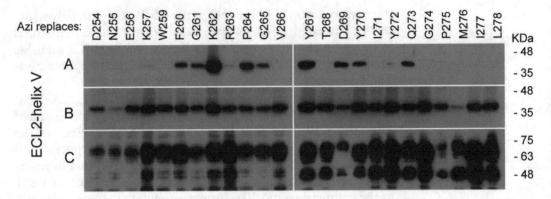

Fig. 3 Photo-crosslinking mapping of the extracellular loop 2 (ECL2) and helix V of CRF1R to identify sites of Ucn1 binding. Western blots of cell lysates resolved on SDS-PAGE. The amino acid of CRF1R replaced by Azi is indicated at the top. The complete scanning of the juxtamembrane domain of CRF1R is published in ref. 24. (a, b) Azi-CRF1R mutants are crosslinked to either wild-type Ucn1 (a), or Bpa[12]-Ucn1 (b). All the samples in these panels are deglycosylated using PNGaseF. The covalent Ucn1-Azi-CRF1R complex runs at an apparent molecular weight of ~40 kDa, and is detected using the anti-Ucn1 antibody. Free Ucn1 (4 kDa) runs at the front and is not detected. Bands detected in (a) reveal receptor positions coming into proximity of the ligand in the associated receptor-ligand complex, whereas bands in (b) assess that mutant receptors conserve the ability to bind the ligand. In panel (c) the expression levels of the Azi-CRF1R mutants are estimated using the anti-FLAG antibody against the C-terminal tag. These samples are not deglycosylated. The mature receptor runs at ~70 kDa. The band at ~50 kDa represents the high-mannose form. In this way one can assess, for instance, that M276Azi-CRF1R is correctly expressed and well targeted to the cell membrane (c), but it is not intact with respect to ligand binding (b). On the other hand, D269Azi-CRF1R is somewhat less efficiently expressed than other mutants (c), but it binds the ligand efficiently (b). In addition, position 269 of CRF1R lies close to bound Ucn1 (a). Reproduced from Coin et al. 2013 with permission from Cell

2   Materials

2.1   DNA

1. Plasmid construct encoding the orthogonal pair for incorporation of Azi : Azi-tRNA synthetase (EAziRS) and the cognate suppressor tRNA (*Bst*-Yam) (*see* **Note 1**).

2. Plasmid construct encoding an engineered *Bst*-Yam tRNA and the corresponding F_{fact}-tRNA synthetase (EKetoRS) (*see* **Note 1**).

3. Systematic library of plasmids encoding for the POI, in which the codon corresponding to each position to be investigated is replaced by the TAG codon. The POI is fused to a C-terminal FLAG-tag.

2.2   Cell Culture

1. Culture medium: Dulbecco's Modified Eagle Medium (DMEM), supplemented with 10% fetal bovine serum and 100 U/mL penicillin/streptomycin.

2. 0.5 M NaOH.

3. Azi.

4. F_{fact} (home-made).

5. Polyethylenimine "Max" (PEI max). Stock solution: 10 mg/mL PEI max, 100 μL aliquots, store at −80 °C.

6. Lactate-buffered saline (LBS): 20 mM sodium lactate pH 4.0, 150 mM NaCl.

7. HEPES dissociation buffer (HDB): 12.5 mM 4-(2-hydroxyethyl)-1-piperazineethanesulfonic acid (HEPES) pH 7.4, 140 nM NaCl, 5 mM KCl.

2.3 Crosslinking and Lysis

1. Binding buffer: 0.1% BSA, 0.01% Triton-X 100, 5 mM MgCl$_2$ in HDB.

2. Ligands for photo-crosslinking: In addition to the native ligand, a derivative containing a photo-activatable group, typically Bpa, at a permissive site is needed as a control substance (*see* **Note 2**). We used Ucn1 and Bpa^{12}-Ucn1. Stock solutions: 100 μM in DMSO, store at −20 °C.

3. Ligands for chemical crosslinking: Ligand derivatives bearing a thiol group in strategic positions, preferably cysteine for peptide ligands. We used Cys$^{6/8/10/12/14}$-Ucn1. Stock solutions: 100 μM in DMSO, store at −20 °C.

4. UV crosslinker BioBudget Technologies, BLX-365, 5 × 8 W tubes, 365 nm.

5. Protease inhibitor cocktail (Roche) supplemented with 0.5 mM EDTA.

6. Lysis buffer: 50 mM HEPES pH 7.5, 150 mM NaCl, 10% glycerol, 1% Triton X-100, 1.5 mM MgCl$_2$, 1 mM EGTA, 1 mM DTT. Store at 4 °C (short-time) or −20 °C (long-time).

2.4 SDS-PAGE and Western Blot

1. PNGase F deglycosylation kit (NEB).

2. 4× Sample buffer: 63 mM Tris-HCl pH 6.8, 2% SDS, 10% glycerol, 0.04% bromphenolblue.

3. Separating gel: 10% acrylamide/bisacrylamide (37.5/1), 375 mM Tris-HCl pH 8.8, 0.1% sodium dodecyl sulfate (SDS), 0.1% ammonium persulfate (APS), 0.075% tetramethylethylenediamine (TEMED).

4. Stacking gel: 4% acrylamide/bisacrylamide (37.5/1), 125 mM Tris-HCl pH 6.8, 0.1% SDS, 0.1% APS, 0.1% TEMED.

5. Running buffer: 25 mM Tris-HCl pH 8.8, 190 mM glycine, 0.1% SDS.

6. Immobilon®-P PVDF membrane.

7. Transfer buffer: 25 mM Tris-HCl pH 8.8, 190 mM glycine, 20% methanol.

8. Skim milk powder.

9. TBS-T: 20 mM Tris-HCl pH 7.4, 0.15 M NaCl, 0.1% Tween 20.

10. Rabbit-anti-Ucn1 antibody (home-made [33]) (*see* **Note 3**).

11. Anti-FLAG-HRP M2 antibody conjugate (Sigma).

12. Goat-anti-rabbit-HRP secondary antibody.

13. ECL reagent home-made: 25% luminol/0.1 M Tris-HCl pH 8.6:100% *p*-coumaric acid/DMSO:30% H_2O_2 (1,000:100:0.3), mix immediately before use (*see* **Note 4**).

14. Stripping buffer: 0.2 M glycine pH 2.2, 1% Tween 20, 0.1% SDS.

3 Methods

3.1 Azi-Mediated Photocrosslinking to Map the Ligand Binding Site

1. Seed 500,000 HEK 293T cells in 2 mL of culture medium per well in 6-well plates. Prepare one well for each POI mutant. Incubate the cells for 24 h at 37 °C, 95% humidity, and 5% CO_2 until they reach 70% confluency.

2. Prepare a fresh 1,000× stock solution of 0.5 M Azi in 0.5 M NaOH. Dilute Azi in medium to a final concentration of 0.5 mM (*see* **Note 5**). 1–2 h before transfection, change the cell medium with the fresh medium containing the ncAA.

3. Dilute the PEI max stock solution 1:10 in LBS. This working solution can be stored at 4 °C for 2 weeks. Add 0.5 μg of the E2Azi/*Bst*-Yam plasmid to 0.5 μg of each POI-TAG mutant plasmid and dilute the DNA in 100 μL of LBS. Add 3 μL of PEI max working solution (3 μg PEI max per μg DNA) and mix thoroughly. Incubate for 10 min at RT. Take out 400 μL of medium from the well to be transfected and use it to neutralize the LBS-buffered DNA solution. Carefully dribble the suspension onto the cells (*see* **Note 6**).

4. Harvest the cells 48 h after transfection. Remove the medium. Add 800 μL of 0.5 mM EDTA in HDB. Incubate for 10 min at RT. Detach the cells by pipetting and transfer them into a 1.5 mL reaction tube containing 200 μL of 5 mM $MgCl_2$ in HDB. Centrifuge for 3 min at $800 \times g$. Discard the supernatant.

5. Resuspend the cell pellets in 200 μL of HDB. Split the samples in half by transferring 100 μL into new reaction tubes. One half is going to be treated with the test substance, the other one with the control substance. Centrifuge for 3 min at $800 \times g$. Discard the supernatant.

6. Resuspend the cell pellets in 100 μL of binding buffer containing 100 nM ligand. Incubate for 10 min at RT (*see* **Note 7**).

7. Irradiate the open tubes with 365 nm UV light for 20 min at maximal power (*see* **Note 8**).

8. Centrifuge for 3 min at 800 × g. Discard the supernatant. Resuspend each cell pellet in 40 µL of HDB containing protease inhibitor. Flash-freeze the samples in liquid N_2. Samples can be stored at this stage at −80 °C (*see* **Note 9**).

9. Thaw the cells approx. 30 s at 37 °C. Immediately put the tubes on ice and keep them cooled from now on. Mix and centrifuge for 10 min at 2,500 × g and 4 °C. Discard the supernatant.

10. Resuspend the cell pellets in 40 µL of lysis buffer containing protease inhibitor. Lyse the cells for 30 min and mix repeatedly. Centrifuge for 10 min at 13,000 × g and 4 °C. Collect the supernatants. Cell lysates can be stored at this stage at −20 °C (*see* **Note 9**).

11. To estimate the relative expression level and the glycosylation state of the different mutants, a portion of each mutant should be analyzed using anti-FLAG antibody without performing any deglycosylation (*see* **Notes 2, 10** and **11**). Mix 2.5 µL of lysate, 3.5 µL of 4× sample buffer, 2 µL of 1 M DTT and 4 µL of water. Incubate the samples for 30 min at 37 °C before loading on SDS-PAGE gel (do not boil!).

12. For N-glycosylated membrane proteins: Perform deglycosylation using PNGase F by following the supplier's instructions (*see* **Note 10**). Use 2.5 µL of lysate in a total volume of 10 µL. After deglycosylation add 3.5 µL of 4× sample buffer before loading on SDS-PAGE gel.

13. Load the samples onto a 10% SDS-polyacrylamide gel and separate them for 20 min at 80 V followed by 60 min at 150 V.

14. Transfer the proteins onto a PVDF-membrane for 2 h at 70 V in a wet chamber.

15. Block the membrane in 5% skim milk/TBS-T for 1 h at RT.

16. Incubate the membrane with rabbit-anti-Ucn1 1:5,000 in 5% skim milk/TBS-T overnight at 4 °C. This step should not be performed when probing the membrane with anti-FLAG-HRP.

17. Wash the membrane 3 × 10 min in TBS-T. Treat the membrane either with anti-rabbit-HRP secondary antibody 1:15,000 in 5% skim milk/TBS-T or anti-FLAG-HRP 1:5,000 in TBS-T for 1 h at RT. Wash the membrane 3 × 10 min in TBS-T.

18. Develop the western blots for 2 min in ECL reagent. Take the digital picture using open aperture, no external light and 5 min exposure time. A signal ("hit") indicates that the residue replaced by Azi within the POI is located in close proximity to the bound ligand (Fig. 3; *see* **Note 12**).

3.2 F$_{fact}$-Mediated Chemical Crosslinking to Determine Spatial Constraints

1. Seed 1.7 Mio HEK293T cells in 4 mL of culture medium in a 6 cm dish. Prepare one dish for each hit mutant. Incubate the cells for 24 h at 37 °C, 95% humidity and 5% CO$_2$ until they reach 70% confluency.

2. Prepare a fresh 1,000× stock solution of 0.25 M F$_{fact}$ in water. Dilute F$_{fact}$ in medium to a final concentration of 0.25 mM. 1–2 h before transfection, change the cell medium with the fresh medium containing the ncAA.

3. Add 2 µg of the EKetoRS/*Bst*-Yam F plasmid to 1 µg of each hit plasmid of the POI-TAG mutant library and dilute the DNA in 200 µL of LBS. Add 9 µL of PEI max working solution and mix thoroughly. Incubate for 10 min at RT. Take out 800 µL of medium from the dish to be transfected and use it to neutralize the LBS-buffered DNA solution. Carefully dribble the suspension onto the cells.

4. Harvest the cells 48 h after transfection. Remove the medium. Add 1 mL of 0.5 mM EDTA in HDB. Incubate for 10 min at RT. Detach the cells by pipetting and transfer them into a 1.5 mL reaction tube containing 250 µL of 5 mM MgCl$_2$ in HDB. Centrifuge for 3 min at 800 × *g*. Discard the supernatant.

5. Resuspend the cell pellets in 600 µL of HDB. Split the samples in 6 parts by transferring 100 µL into new reaction tubes. Centrifuge for 3 min at 800 × *g*. Discard the supernatant.

6. Resuspend the cell pellets in 100 µL of binding buffer containing 100 nM ligand. Treat each mutant with each of the Cys-peptides as well as the control peptide Bpa^{12}-Ucn1. Incubate for 90 min at RT.

7. Centrifuge for 3 min at 800 × *g*. Discard the supernatant. Resuspend the cell pellets in 20 µL of HDB containing protease inhibitor. Flash-freeze the samples in liquid nitrogen and store them at −80 °C if a break is needed (*see* **Note 9**).

8. Thaw the cells approx. 30 s at 37 °C. Put the tubes on ice immediately and keep them cooled from now on. Mix and centrifuge for 10 min at 2,500 × *g* and 4 °C. Discard the supernatant.

9. Resuspend the cell pellets in 20 µL of lysis buffer containing protease inhibitor. Lyse the cells for 30 min while repeatedly mixing every few minutes. Centrifuge for 10 min at 13,000 × *g* and 4 °C. Collect the supernatants, store them at −20 °C (*see* **Note 9**).

10. Mix 2.5 µL of lysate, 3.5 µL of 4× sample buffer, 2 µL of 1 M DTT and 4 µL of water. Incubate the samples for 30 min at 37 °C (*see* **Note 13**).

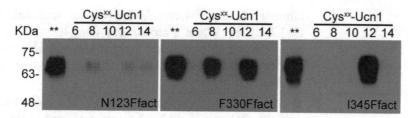

Fig. 4 F_{fact}-Cys chemical crosslinking to determine inter-molecular spatial constraints. Western blots of cell lysates resolved on SDS-PAGE and probed with an anti-Ucn1 antibody. The amino acid of CRF1R replaced by F_{fact} is indicated at the bottom of each panel. Each receptor mutant was incubated with five Ucn1 analogs bearing Cys at the position indicated in the upper string. The samples of the first lane (*) were UV-crosslinked to Bpa^{12}-Ucn1 as a positive control and reference for signal intensity. Samples are not deglycosylated and the covalent ligand-receptor complex is detected at an apparent molecular weight of ~70 kDa. The complete set of F_{fact}-CRF1R –Cys-Ucn1 combination is published in ref. 24. Reproduced from Coin et al. 2013 with permission from Cell

11. Analyze the samples via SDS-PAGE and western blot as described in **steps 13–18** in Subheading 3.1 using anti-Ucn1 antibody (Fig. 4; *see* **Note 14**). Strip the membrane 2 × 10 min in stripping buffer. Wash the membrane once in PBS for 5 min, and 2 × 5 min in TBS-T. Repeat the antibody treatment using anti-FLAG-HRP.

4   Notes

1. Both EAziRS and EKetoRS are derived from *E. coli* TyrRS [9]. Both synthetases are paired with the amber suppressor tRNA derived from the *B. stearothermophilus* tRNA^{Tyr} (*Bst*-Yam) [34]. E2Azi bears the following mutations with respect to wt TyrRS: Tyr37Leu, Asp182Ser, Phe183Met, and Leu186Ala. EKetoRS bears the following mutations: Tyr37Iso, Asp165Gly, Asp182Gly, Phe183Met, Leu186Ala. Both synthetases bear the additional mutation Asp165Arg, which improves the recognition of the suppressor tRNA [35]. Optimal plasmid constructs for the expression of ncAARS/tRNA pairs have the synthetase under control of a strong mammalian promoter (CMV or PGK), while tRNA transcription is regulated by an external Pol III promoter (H1 or U6 promoter). NcAA incorporation yields improve using tandem repeats of the tRNA expression cassette, usually 3 or 4 repeats [24].

2. It is necessary to check whether the Azi-mutant proteins are expressed and functional. The expression levels can be determined using an anti-FLAG-tag antibody, targeted to the C-terminal Flag tag in the POI. Expect the expression rate to decrease when TAG is introduced closer to the C-terminus of the POI. We used Bpa^{12}-Ucn1 as a positive control for ligand

binding, as Bpa^{12}-Ucn1 is known to crosslink efficiently CRF1R [28]. When a crosslinked complex with Bpa^{12}-Ucn1 was detected, the receptor mutant was still able to bind the peptide ligand and the mutation site was considered permissive for Azi.

3. The availability of a suitable antibody is a crucial part of the method. Polyclonal antibodies are preferred. This minimizes the risk that a crosslinking event within the epitope masks it for antibody recognition. Likewise, Cys-ligand derivatives are more likely to be still recognized by polyclonal antibodies.

4. Good results have been obtained also using SuperSignal West Pico Chemiluminescent substrate (Thermo Scientific). Shall the signal be really weak, a portion (~1/10) of a more concentrated reagent solution (femtomole sensitivity) can be added to the regular ECL reagent. We do not recommend using pure "femto" ECL reagents since they give very high background.

5. Always prepare Azi solutions fresh, as this ncAA has a short half-life in aqueous environment and the EAziRS cannot distinguish between intact and degraded ncAA. Lower concentrations down to 0.05 mM Azi in the culture medium were still sufficient to produce Azi-containing proteins.

6. We compared transfection efficiencies of lipofectamine 2000, PEI and PEI max. We tested different pH values during the formation of DNA complexes either in serum-free DMEM or LBS. Both polyethylenimines gave higher transfection rates than lipofectamine, with the best combination being PEI max in LBS as reported in the literature [36]. Once PEI-DNA complexes are formed in a serum-free environment, the neutralized transfection mixture can be applied to the cells in the presence of serum in the culture medium.

7. Adjust the incubation time of the ligand to your POI. Consider ligand kinetics, internalization processes, downstream effects and decay. The concentration of the ligand depends on its affinity. We have used a ligand concentration equal to about 50× its affinity to ensure receptor saturation.

8. Make sure to open the lid of the tubes and place them directly under the UV source, leaving as little space as possible. Higher crosslinking yields and better quality of western blots are obtained if crosslinking is performed on adherent cells. In this case, aspirate the medium, add in each well the ligand in 1 mL binding buffer, and irradiate the cells in the plate, before proceeding with harvesting and lysis.

9. If the work flow needs to be interrupted, the best time to store the harvested cells is after this flash freezing step. Sample storage at this point is safe for several weeks at −80 °C. It is also

possible to store the final lysates at −20 °C. However, in our experience the best western blots, with respect to signal background-ratio and band acuity, are obtained using fresh lysates.

10. Membrane proteins often occur in different glycosylation states. A preliminary glycosylation occurs in the ER and Golgi, whereas only fully-glycosylated (mature-glycosylated) proteins reach the cell surface. Even within the frame of a specific glycosylation state, glycosylation is not homogeneous and western blot analysis of such proteins gives usually smeared bands. Therefore, although the ligand only addresses the fully glycosylated receptor, removal of N-linked glycosylic groups by PNGaseF yields sharper and more intense bands of the ligand-receptor complex in western blot analysis using the anti-ligand antibody. On the other hand, to evaluate the expression level of the receptor using anti-FLAG, deglycosylation is not recommended. In this case, it is important to know the proportion of the mature glycosylated receptor relative to all other glycosylated forms.

11. UV irradiation can impair the quality of anti-FLAG western blots. If smearing bands are a problem, try generating the samples for anti-FLAG detection in a separate run. Repeat **steps 1–4** in Subheading 3.1 and directly proceed with **step 9** in Subheading 3.1 without UV treatment.

12. Group mutants in sets of 12 for each gel. Try to avoid loading mutants bearing the Azi-substitution in different parts of the receptor on the same gel, as signal intensities may vary significantly depending on the position of the ncAA within the protein sequence. In case background signals make it difficult to identify proper hits, use the closest hit of a series as a reference sample and run it together with the 12 samples of the following set.

13. Deglycosylation is not necessary for these experiments, as in our hands the F_{fact}-crosslinked complexes gave stronger signals in western blots than Azi-crosslinked complexes, and they did not need to be amplified. On the contrary, working with non-deglycosylated samples allows estimating differences in signal intensities more accurately.

14. Group samples in sets of 2 × 6. Load all samples taken from the same POI mutant next to each other. Bpa^{12}-Ucn1 serves as a positive control. It can happen that the F_{fact}-Cys crosslinking samples yield higher or lower intensities than the photo-crosslinked positive control. We consider positive for F_{fact}-Cys crosslinking samples that give bands of very similar or higher intensity than Bpa^{12}-Ucn1, when proofed with the anti-Ucn1 antibody.

References

1. Liu CC, Schultz PG (2010) Adding new chemistries to the genetic code. Annu Rev Biochem 79:413–444

2. Neumann H (2012) Rewiring translation - genetic code expansion and its applications. FEBS Lett 586(15):2057–2064

3. Wang Q, Parrish AR, Wang L (2009) Expanding the genetic code for biological studies. Chem Biol 16(3):323–336

4. Lang K, Chin JW (2014) Cellular incorporation of unnatural amino acids and bioorthogonal labeling of proteins. Chem Rev 114(9):4764–4806

5. Ai HW (2012) Biochemical analysis with the expanded genetic lexicon. Anal Bioanal Chem 403(8):2089–2102

6. Chin JW, Santoro SW, Martin AB, King DS, Wang L, Schultz PG (2002) Addition of p-azido-L-phenylalanine to the genetic code of Escherichia coli. J Am Chem Soc 124(31):9026–9027

7. Chin JW, Martin AB, King DS, Wang L, Schultz PG (2002) Addition of a photocrosslinking amino acid to the genetic code of Escherichiacoli. Proc Natl Acad Sci U S A 99(17):11020–11024

8. Deiters A, Cropp TA, Mukherji M, Chin JW, Anderson JC, Schultz PG (2003) Adding amino acids with novel reactivity to the genetic code of Saccharomyces cerevisiae. J Am Chem Soc 125(39):11782–11783

9. Chin JW, Cropp TA, Anderson JC, Mukherji M, Zhang Z, Schultz PG (2003) An expanded eukaryotic genetic code. Science 301(5635):964–967

10. Liu W, Brock A, Chen S, Schultz PG (2007) Genetic incorporation of unnatural amino acids into proteins in mammalian cells. Nat Methods 4(3):239–244

11. Wang WY, Takimoto JK, Louie GV, Baiga TJ, Noel JP, Lee KF, Slesinger PA, Wang L (2007) Genetically encoding unnatural amino acids for cellular and neuronal studies. Nat Neurosci 10(8):1063–1072

12. Yanagisawa T, Ishii R, Fukunaga R, Kobayashi T, Sakamoto K, Yokoyama S (2008) Multistep engineering of pyrrolysyl-tRNA synthetase to genetically encode N(epsilon)-(o-azidobenzyloxycarbonyl) lysine for site-specific protein modification. Chem Biol 15(11):1187–1197

13. Nguyen DP, Lusic H, Neumann H, Kapadnis PB, Deiters A, Chin JW (2009) Genetic encoding and labeling of aliphatic azides and alkynes in recombinant proteins via a pyrrolysyl-tRNA synthetase/tRNA(CUA) pair and click chemistry. J Am Chem Soc 131(25):8720–8721

14. Hao Z, Song Y, Lin S, Yang M, Liang Y, Wang J, Chen PR (2011) A readily synthesized cyclic pyrrolysine analogue for site-specific protein "click" labeling. Chem Commun (Camb) 47(15):4502–4504

15. Lin S, Zhang Z, Xu H, Li L, Chen S, Li J, Hao Z, Chen PR (2011) Site-specific incorporation of photo-cross-linker and bioorthogonal amino acids into enteric bacterial pathogens. J Am Chem Soc 133(50):20581–20587

16. Ai HW, Shen W, Sagi A, Chen PR, Schultz PG (2011) Probing protein-protein interactions with a genetically encoded photo-crosslinking amino acid. Chembiochem 12(12): 1854–1857

17. Zhang M, Lin S, Song X, Liu J, Fu Y, Ge X, Fu X, Chang Z, Chen PR (2011) A genetically incorporated crosslinker reveals chaperone cooperation in acid resistance. Nat Chem Biol 7(10):671–677

18. Hino N, Oyama M, Sato A, Mukai T, Iraha F, Hayashi A, Kozuka-Hata H, Yamamoto T, Yokoyama S, Sakamoto K (2011) Genetic incorporation of a photo-crosslinkable amino acid reveals novel protein complexes with GRB2 in mammalian cells. J Mol Biol 406(2):343–353

19. Tippmann EM, Liu W, Summerer D, Mack AV, Schultz PG (2007) A genetically encoded diazirine photocrosslinker in Escherichia coli. Chembiochem 8(18):2210–2214

20. Tanaka Y, Bond MR, Kohler JJ (2008) Photocrosslinkers illuminate interactions in living cells. Mol Biosyst 4(6):473–480

21. Xiang Z, Ren H, Hu YS, Coin I, Wei J, Cang H, Wang L (2013) Adding an unnatural covalent bond to proteins through proximity-enhanced bioreactivity. Nat Methods 10(9):885–888

22. Xiang Z, Lacey VK, Ren H, Xu J, Burban DJ, Jennings PA, Wang L (2014) Proximity-Enabled Protein Crosslinking through Genetically Encoding Haloalkane Unnatural Amino Acids. Angew Chem Int Ed Engl 53(8):2190–2193

23. Chen XH, Xiang Z, Hu YS, Lacey VK, Cang H, Wang L (2014) Genetically encoding an electrophilic amino acid for protein stapling and covalent binding to native receptors. ACS Chem Biol 9(9):1956–1961

24. Coin I, Katritch V, Sun T, Xiang Z, Siu FY, Beyermann M, Stevens RC, Wang L (2013) Genetically encoded chemical probes in cells

reveal the binding path of urocortin-I to CRF class B GPCR. Cell 155(6):1258–1269

25. Coin I, Perrin MH, Vale WW, Wang L (2011) Photo-cross-linkers incorporated into G-protein-coupled receptors in mammalian cells: a ligand comparison. Angew Chem Int Ed Engl 50:8077–8081

26. Binder EB, Nemeroff CB (2010) The CRF system, stress, depression and anxiety-insights from human genetic studies. Mol Psychiatry 15(6):574–588

27. Turcu AF, Spencer-Segal JL, Farber RH, Luo R, Grigoriadis DE, Ramm CA, Madrigal D, Muth T, O'Brien CF, Auchus RJ (2016) Single-dose study of a corticotropin-releasing factor receptor-1 antagonist in women with 21-hydroxylase deficiency. J Clin Endocrinol Metab 101:1174

28. Kraetke O, Holeran B, Berger H, Escher E, Bienert M, Beyermann M (2005) Photoaffinity cross-linking of the corticotropin-releasing factor receptor type 1 with photoreactive urocortin analogues. Biochemistry 44(47):15569–15577

29. Huang LY, Umanah G, Hauser M, Son C, Arshava B, Naider F, Becker JM (2008) Unnatural amino acid replacement in a yeast G protein-coupled receptor in its native environment. Biochemistry 47(20):5638–5648

30. Grunbeck A, Huber T, Sachdev P, Sakmar TP (2011) Mapping the ligand-binding site on a G protein-coupled receptor (GPCR) using genetically encoded photocrosslinkers. Biochemistry 50(17):3411–3413

31. Ray-Saha S, Huber T, Sakmar TP (2014) Antibody epitopes on g protein-coupled receptors mapped with genetically encoded photo-activatable cross-linkers. Biochemistry 53(8): 1302–1310

32. Valentin-Hansen L, Park M, Huber T, Grunbeck A, Naganathan S, Schwartz TW, Sakmar TP (2014) Mapping substance P binding sites on the neurokinin-1 receptor using genetic incorporation of a photoreactive amino acid. J Biol Chem 289(26):18045–18054

33. Bittencourt JC, Vaughan J, Arias C, Rissman RA, Vale WW, Sawchenko PE (1999) Urocortin expression in rat brain: evidence against a pervasive relationship of urocortin-containing projections with targets bearing type 2 CRF receptors. J Comp Neurol 415(3):285–312

34. Sakamoto K, Hayashi A, Sakamoto A, Kiga D, Nakayama H, Soma A, Kobayashi T, Kitabatake M, Takio K, Saito K, Shirouzu M, Hirao I, Yokoyama S (2002) Site-specific incorporation of an unnatural amino acid into proteins in mammalian cells. Nucleic Acids Res 30(21):4692–4699

35. Takimoto JK, Adams KL, Xiang Z, Wang L (2009) Improving orthogonal tRNA-synthetase recognition for efficient unnatural amino acid incorporation and application in mammalian cells. Mol Biosyst 5(9):931–934

36. Fukumoto Y, Obata Y, Ishibashi K, Tamura N, Kikuchi I, Aoyama K, Hattori Y, Tsuda K, Nakayama Y, Yamaguchi N (2010) Cost-effective gene transfection by DNA compaction at pH 4.0 using acidified, long shelf-life polyethylenimine. Cytotechnology 62(1): 73–82

Chapter 15

Generation of Stable Amber Suppression Cell Lines

Simon J. Elsässer

Abstract

Noncanonical amino acid mutagenesis via amber suppression provides the means to tailor proteins inside living cells. A wide range of noncanonical amino acids have been incorporated using the *Methanococcus* pyrrolysyl-tRNA synthetase/tRNACUA (PylRS/PylT) in mammalian cell systems in proof of principle experiments, for (1) minimal genetically encoded fluorescence or affinity tagging, (2) photo-control of enzymes, (3) genetically encoded posttranslational protein modifications. We have developed a general and efficient method to genomically integrate the PylRS/PylT amber suppression machinery using PiggyBac-mediated transposition. A general protocol for the generation of stable amber suppression cell lines is described here. Using the modular plasmid system, homogenous and highly efficient amber suppression in a wide range of cell lines can be achieved.

Key words Genetic code expansion, Mammalian cells, Unnatural amio acids, Noncanonical amino acids, tRNA, Aminoacyl-tRNA-Synthetase, PiggyBac transposase, Site-specific protein labeling

1 Introduction

A successful method for noncanonical amino acid (ncAA) mutagenesis via stable amber suppression in mammalian cells entails expression of PylT and PylRS genes at high levels from one or more integrated copies in the genome. To accomplish this, we have established a versatile vector system for PiggyBac-mediated transgenesis in mammalian cells [1]. PiggyBac is a short inverted terminal repeats (ITR) transposon isolated from the moth *Trichoplusia ni* [2]. It integrates and excises inserts of up to 200 kb with high precision at genomic TTAA nucleotide sequences. A single protein, termed PiggyBac Transposase (PBase), mediates cut-and-paste transposition of any insert cassette flanked by ITRs, making the system easily adaptable for genetic engineering. PiggyBac transposase is highly active in mammalian cells, achieving high integration efficiency of one or more targeting cassettes (1–30% in mouse embryonic stem cells) [3, 4].

Edward A. Lemke (ed.), *Noncanonical Amino Acids: Methods and Protocols*, Methods in Molecular Biology, vol. 1728, https://doi.org/10.1007/978-1-4939-7574-7_15, © Springer Science+Business Media, LLC 2018

pPiggyBac 4xPylT/PylRS

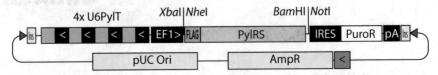

pPiggyBac 4xPylT/mCherry-TAG-EGFP

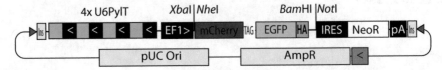

Fig. 1 Vectors used for PiggyBac transposase-mediated integration of amber suppression machinery. Two variants with complementary selection markers are used to integrate both Pyrrolysyol-tRNA syntethase (PylRS) gene and the gene of interest (here the double-fluorescent reporter mCherry-TAG-EGFP-HA). An N-terminal FLAG tag is added to PylRS. The double fluorescent reporter carries a C-terminal HA tag for detecting full-length product. The vector backbone contains inverted repeats for recognition by PiggyBac transposase (pink triangles), insulator elements ("Ins," gray box), a tandem cassette of four PylT genes driven by a U6 Pol III promoter, strong EF1-α Pol II promoter, and IRES driving a Puro or Neo resistance gene

To allow flexible combination of PylRS variants and proteins of interest (POI), we have designed targeting cassettes that encode for PylRS and POI, and complementary antibiotic selection markers on two independent vectors (Fig. 1). Each vector contains a Pol II transcript driven by an EF1-α promoter and a tandem repeat of four PylT genes driven by U6 Pol III promoters, arranged head-to-head (Fig. 1). A Puromycin or Neomycin resistance marker is expressed from an IRES in the Pol II transcript. The expression cassette is flanked by universal chicken β-globin insulators and PiggyBac ITRs. PiggyBac transposase is supplied from a third non-integrating plasmid (referred to as pPBase here). The integration process entails introducing all three plasmids into the target cells using a suitable transfection or electroporation method. PiggyBac transposase, transiently expressed from pPBase, mediates reversible integration of both targeting cassettes at genomic TTAA nucleotide sequences. Although a high copy number of donor plasmid in the transfected cell skews the reversible process toward integration, the genomic copies remain unstable as long as PiggyBac transposase remains active. Thus, care must be taken to apply a sufficiently long drug selection period until the transposition activity has ceased. Each targeting cassette may be integrated in one or several copies at distinct locations. This random process is tunable by varying plasmid ratios in the transfection and stringency of selection conditions. Of note, the repeated process of integration and excision of individual copies does not introduce mutations in the genomic TTAA targeting sites because of the high precision of cutting and pasting by the transposase enzyme. Once integrated, we have observed long-term stable expression and amber suppression efficiency, facilitated by both the silencing-resistant EF1-α promoter and flanking insulator sequence.

Here, the process of cloning the gene for the POI into a PiggyBac targeting vector, transfecting and selecting cells is described, and an optimized protocol for mESC is provided.

2   Materials

1. Chemically competent *E. coli* Stbl3 or NEB® Stable cells (*see* **Note 1**).

2. Standard cloning enzymes; e.g., *Nhe*I, *Not*I, DNA polymerase, T4 DNA Ligase.

3. Standard media and plates for *E. coli* transformation.

4. Agarose gel electrophoresis equipment.

5. Mini and endotoxin-free Maxi plasmid preparation kits.

6. EF1_up primer: AACTGGGAAAGTGATGTCG.

7. IRES_down primer: AGACCCCTAGGAATGCTCGT.

8. 10 mg/mL Puromycin stock solution, store at −20 °C.

9. 50 mg/mL G418 Sulfate stock solution, store at +4 °C.

10. 1 mg/mL Polyethyleneimine (PEI) transfection solution (*see* **Note 2**).

11. Mouse embryonic stem cell transfection reagent TransIT-X2® (Mirus).

12. Cell culture medium (formulation depending on requirements of cell line).

13. ES cell culture medium: 500 mL KO DMEM, 6 mL of 100× Ala-Gln, 6 mL of 100 μM 2-mercaptoethanol stock solution (35 μL of 2-mercaptoethanol in 50 mL PBS, sterile filtered, stored at 4 °C), 6 mL of 100× nonessential amino acids, 60 μL 10 Mio U/mL LIF (Leukaemia inhibitory factor), 95 mL ES grade FBS.

14. OPTI-MEM I Reduced Serum Medium (ThermoFisher Scientific).

15. Super PiggyBac Transposase Expression Vector pPBase (PB200PA, System Bioscience Inc).

16. PiggyBac targeting vector pPB 4×PylT/PylRS with Puromycin resistance marker (*see* **Note 3**).

17. PiggyBac targeting vector for amber suppression reporter or gene of interest with Neomycin resistance marker, and positive control, e.g., pPB 4×PylT/mCherry-TAG-EGFP (*see* **Note 3**).

3 Methods

3.1 Cloning the Gene of Interest into PiggyBac Targeting Vector

Two vectors are needed to accomplish stable integration of both the Pyrrolysyl-tRNA-Synthetase gene and the gene of interest containing an amber codon in the desired position. The PiggyBac targeting cassettes embedded in these vectors differ in the antibiotic resistance gene (Fig. 1). Importantly, the use of two independent vectors allows stepwise generation of cell lines and adjustment of expression levels of the PylRS and gene of interest by varying antibiotic concentrations. New PylRS variants and genes of interest can be easily cloned into the targeting vectors by restriction cloning, replacing the existing insert.

1. Design external primers to amplify your gene of interest (GOI) with overhangs matching the restriction site in the PiggyBac targeting vector (e.g., *Nhe*I and *Not*I). Include a Kozak sequence and the Met start codon (e.g., CCACC<u>ATG</u>). To simplify detection of the amber suppressed protein, include a N- or C-terminal peptide tag (e.g., HA) (*see* **Note 4**).

2. Design internal primers for an overlap extension PCR reaction to replace a sense codon with an amber codon at the desired position of the ncAA (*see* **Note 5**).

3. Perform two PCR reactions using primer pairs to amplify the two fragments of the GOI 3′ and 5′ of the introduced TAG codon. As a control, amplify full-length GOI without amber codon using external primers only.

4. Purify amplified PCR products using agarose gel electrophoresis.

5. To produce full-length GOI insert with amber codon, perform a second PCR reaction using a mix of 3′ and 5′ fragment of the GOI as template and external primers only.

6. Digest pPB 4×PylT/mCherry-TAG-EGFP vector (e.g., using *Nhe*I, *Not*I), and purify backbone (8 kb) using agarose gel electrophoresis.

7. Use restriction cloning and ligation (or homology-based cloning, e.g., Gibson Assembly®, In-Fusion®) to insert PCR products into the digested vector.

8. Transform ligation products into Stbl3 or NEB Stable cells and plate on LB/Amp agar plates.

9. Pick colonies, mini-prep and validate GOI by restriction digest and Sanger sequencing (using EF1_up and IRES_down sequencing primers).

10. Maxi-prep validated clones for transfection into cell lines. Plasmid should be eluted from column using sterile Elution Buffer or ddH₂O, under clean or sterile conditions.

3.2 Transfection of Adherent Cell Lines

PiggyBac vectors for integration and pPBase vector are introduced into the cells using lipofection. The amount of the three vectors has been optimized to 1 μg pPB 4×PylT/PylRS, 1 μg pPB 4×PylT/GOI, 0.5 μg pPBase (2:2:1 ratio). In addition to your GOI, perform a control transfection with 2:2:1 pPB 4×PylT/PylRS:pPB 4×PylT/mCherry-TAG-EGFP:pPBase to allow easy assessment of amber suppression efficiency by fluorescence or western blot readout.

1. For each transfection, seed 2.5×10^5 cells in a 6-well plate 24 h before transfection in medium without antibiotics. Grow in an incubator at 37 °C with 5% CO_2 atmosphere.

2. Prepare plasmid mix, 2.5 μg total DNA, in 250 μL OPTI-MEM.

3. Vortex gently to mix.

4. Add 7.5 μL 1 mg/mL PEI.

5. Vortex gently to mix.

6. Incubate for 20–30 min to form transfection complex.

7. Add transfection mix dropwise to entire area of the well.

8. Incubate for 12–24 h before changing medium to fresh medium (may contain Pen/Strep).

9. Split the cells and start selection 48 h after transfection.

3.3 Optimized Protocol for Transfection of Mouse Embryonic Stem Cells

*Mouse embryonic stem cells (ESC) and other primary cells are difficult-to-transfect and PEI can be toxic to those cells. Thus, an optimized protocol for transfection of mouse ESC is described here. 150,000 cells are transfected with 1 μg plasmid and 3 μL TransIT-X2® (Mirus, see **Note 6**) in suspension, yielding an excellent transfection efficiency of 50–70% and low toxicity. The amount of vectors is 0.4 μg 4×PylT/PylRS, 0.4 μg 4×PylT/GOI, 0.2 μg pPBase (2:2:1 ratio).*

1. Add 2 mL ES Medium to the wells of a gelatin-coated 12-well plate and let it equilibrate in an incubator at 37 °C with 5% CO_2 atmosphere.

2. Prepare plasmid mix, 1 μg total DNA, in 100 μL OPTI-MEM.

3. Vortex gently to mix.

4. Add 3 μL TransIT-X2®.

5. Vortex gently to mix.

6. Incubate for 15–30 min before **step 10**.

7. Trypsinize the cells, resuspend in ESC medium, count and dilute to 0.6×10^6 cells/mL in ESC medium.

8. Take 12-well plate out of incubator.

9. Add 250 µL cell suspension to each 100 µL transfection mix and for each transfection continue to **steps 11** and **12** within 1 min.

10. Mix well by pipetting up and down three times.

11. Transfer entire volume to a well of the preequilibrated 12-well plate, distributing cells evenly.

12. Incubate the cells in an incubator at 37 °C with 5% CO_2 atmosphere. Exchange medium after 24 h (may contain Pen/Strep) and start selection 48 h after transfection.

3.4 Selection of Polyclonal Pool

1. Split the cells 48 h after transfection 1:6 into 6 wells of the original size. Make a series of Puromycin and G418 concentrations to find the highest feasible selection condition (*see* **Note 7**). For example:

Puro G418	0.5 µg/mL 500 µg/mL	1 µg/mL 1000 µg/mL	1 µg/mL 1500 µg/mL
Puro G418	1 µg/mL 2000 µg/mL	2 µg/mL 2000 µg/mL	5 µg/mL 2000 µg/mL

2. Maintain the cells under selection for at least 7 days to ensure stable integration, replacing medium every 2–4 days (*see* **Note 8**).

3. Collect surviving cells from wells with highest drug concentration and expand (*see* **Note 9**).

4. Single clones can be derived from polyclonal pool by single cell sorting or serial dilution.

3.5 Selection of Individual Clones

1. Split the cells 48 h after transfection, split each well into 3 × 10 cm dishes. Add three different concentrations of antibiotic selection markers.

2. Maintain the cells under selection for at least 7 days to ensure stable integration, changing medium every 2–3 days.

3. Collect single colonies at the highest antibiotic concentration and expand.

3.6 Functional Test of Amber Suppression

1. Expand polyclonal pool after selection or monoclonal cell line from a single colony and make frozen stocks as soon as sufficient cells are available (*see* **Note 10**).

2. Seed 2.5×10^5 cells per well in several wells of a 6-well plate and let it attach.

3. After 24 h replace medium with fresh medium, supplementing one well with ncAA (typical range 0.2–1 mM for most ncAA). Additional wells can be used to test a range of concentrations for optimization (e.g., 0.1, 0.25, 0.5, 1 mM).

4. Harvest cells 24–48 h after addition of ncAA for downstream analysis such as western blotting, immunofluorescent staining, or FACS analysis.

4   Notes

1. Recombination-deficient *E. coli* strains are used to maintain the 4× U6PylT tandem repeats.

2. Note that PEI length and branching may vary, and using the specific product from Polyscience (Polyethylenimine 25 kDa linear, cat# 23966-2) is important for high transfection efficiency. To prepare PEI stock solution:

   - Heat 50 mL ddH$_2$O to ~80 °C (e.g., by briefly boiling in a microwave).

   - Dissolve 50 mg PEI powder.

   - Let it cool to room temperature.

   - Neutralize to pH 7.0, filter sterilize (0.22 μm), aliquot and store at −20 °C; a working stock (e.g., 1 mL aliquot) can be kept at 4 °C for extended periods.

3. Vectors can be requested from the Centre for Chemical & Synthetic Biology (CCSB) at MRC Laboratory of Molecular Biology.

4. A C-terminal tag will detect full-length protein only. An N-terminal tag allows detection of both the full-length protein and any truncated product terminated at the amber codon. Either can be useful for troubleshooting purposes: An N-terminal tag can be used to determine the readthrough efficiency at the amber codon by comparing the ratio of full-length to truncated product. If the truncated product is rapidly degraded, no N-terminal tag will be detected in the absence of ncAA [1]. A C-terminal can be used to identify secondary translation initiation events after the amber codon that lead to production of a protein fragment lacking the N-terminus [5].

5. Placement of the amber codon requires careful consideration: The position may be predetermined if a specific functional residue is intended to be replaced with a noncanonical amino acid (e.g., for genetically encoding a site-specific posttranslational modification [1]). In other cases, where the placement is more flexible, it is beneficial to test three or more positions to find one that is efficiently suppressed. Several factors may be helpful to consider: First, the position of the amber codon within the open reading frame may crucially affect the production of desired full-length protein product and unwanted byproducts. Placing the amber codon immediately after or close to the ATG start codon will favor secondary initiation, producing an unwanted N-terminally truncated protein product [5]. Placing the amber codon close to the 3′ end of the

open reading frame may lead to the production of a stable truncation product in the absence of noncanonical amino acid. Thus as a general rule, inserting noncanonical amino acids in or before the first folded domain of the POI, but not within the first ten residues will be most useful. Second, the local context of the amber codon may affect efficiency. Thus moving the amber codon even by one position may have an impact. In our hands, no particular consensus has been found for upstream or downstream bases or motifs that enhance amber suppression efficiency.

6. In a limited survey of transfection reagents, we determined the following transfection efficiency ranking for mouse embryonic stem cells: TransIT-X2® > JetPRIME® > TransIT-2020® > TransIT-LT1® > PEI.

7. For a new cell line it is highly recommended to perform "kill curves" with Puromycin and G418 prior to this selection experiment and/or to include a control transfection without pPBase plasmid to establish antibiotic concentrations that effectively kill cells without stable integrants.

8. Cells at lower antibiotic concentrations may expand rapidly and need to be split once or twice within the 7-day selection period. At higher antibiotic concentrations, the cells may grow considerably slower but eventually recover. Amber suppression efficiency crucially depends on high expression levels of PylT and GOI, thus it is important to check growth in the wells with higher antibiotic concentrations.

9. If the cells initially survived only at low antibiotic concentrations, aliquots of this pool can be subsequently exposed to higher concentrations to select for higher expressing cells within the population.

10. Some cell lines such as HEK293 cells have a highly unstable chromosome composition and may lose transgenes spontaneously. Periodic reselection with the original antibiotic concentrations ensures the maintenance of the integrated amber suppression machinery.

Acknowledgment

I thank Jason Chin, Russell Ernst, Roberto Zanchi, and members of the Elsässer lab for helpful discussions.

References

1. Elsässer SJ, Ernst RJ, Walker OS, Chin JW (2016) Genetic code expansion in stable cell lines enables encoded chromatin modification. Nat Methods 13:158–164. https://doi.org/10.1038/nmeth.3701

2. Fraser MJ, Ciszczon T, Elick T, Bauser C (1996) Precise excision of TTAA-specific lepidopteran transposons piggyBac (IFP2) and tagalong (TFP3) from the baculovirus genome in cell lines from two species of Lepidoptera. Insect Mol Biol 5:141–151

3. Wang W, Lin C, Lu D et al (2008) Chromosomal transposition of PiggyBac in mouse embryonic stem cells. Proc Natl Acad Sci U S A 105: 9290–9295. https://doi.org/10.1073/pnas.0801017105

4. Lu X, Huang W (2014) PiggyBac mediated multiplex gene transfer in mouse embryonic stem cell. PLoS One 9:e115072–e115017. https://doi.org/10.1371/journal.pone.0115072

5. Kalstrup T, Blunck R (2015) Reinitiation at non-canonical start codons leads to leak expression when incorporating unnatural amino acids. Sci Rep 5:1–9. https://doi.org/10.1038/srep11866

Chapter 16

Trapping Chromatin Interacting Proteins with Genetically Encoded, UV-Activatable Crosslinkers In Vivo

Christian Hoffmann, Heinz Neumann, and Petra Neumann-Staubitz

Abstract

The installation of unnatural amino acids into proteins of living cells is an enabling technology that facilitates an enormous number of applications. UV-activatable crosslinker amino acids allow the formation of a covalent bond between interaction partners in living cells with nearly perfect spatial and temporal control. Here, we describe how this method can be employed to map chromatin interactions and to follow these interactions across the cell cycle in synchronized yeast populations. This method thereby provides unprecedented insights into the molecular events controlling chromatin reorganization in mitosis. As similar tools are available for other organisms, it should be possible to derive similar strategies for these and for other synchronizable processes.

Key words Genetic code expansion, Noncanonical amino acids, UV-crosslinking, Chromatin, Cell cycle, Mitosis, Histones

1 Introduction

Identification of veritable chromatin-interacting proteins is hampered by the highly charged nature of DNA which attracts unspecific binding of proteins. These proteins can only be extracted by harsh chemical treatment, making it extremely difficult to identify true chromosomal proteins. Many studies have tried to address this problem by various strategies leading to the identification of an unlikely high number of possible components that require validation [1].

The use of genetically encoded, UV-activatable crosslinker amino acids may contribute significantly to resolve this issue. These unnatural amino acids are introduced site-specifically into proteins of living cells using evolved, orthogonal amber suppressor tRNA/aminoacyl-tRNA synthetase pairs [2]. The formation of a covalent bond between the interaction partners upon irradiation (while present in an intact cell) allows the subsequent use of harsh, denaturing conditions to remove unspecific interaction partners.

Edward A. Lemke (ed.), *Noncanonical Amino Acids: Methods and Protocols*, Methods in Molecular Biology, vol. 1728, https://doi.org/10.1007/978-1-4939-7574-7_16, © Springer Science+Business Media, LLC 2018

Activation of the crosslinker with UV-light provides perfect temporal control over the crosslinking reaction, thereby facilitating time-resolved experiments with high precision.

Crosslinker amino acids of various types can be encoded to study interactions of proteins with other proteins, nucleic acids, or lipids (Fig. 1). The PylS/PylT pair of *Methanosarcina barkeri* or *mazei* can be used to encode diazirine-derivatives of lysine in a wide range of host systems [3–5]. Phenylalanine-based crosslinker amino acids, e.g., pBPA or pAzF, are encoded by the *E. coli* TyrRS/tRNA$_{CUA}$ pair in yeast, metazoan cell lines, and animals [6], whereas in bacteria the TyrRS/tRNA$_{CUA}$ pair from *Methanococcus jannaschii* is used.

Trapping a structurally well-characterized interaction in vivo by installing crosslinker amino acid along the interaction surface is relatively straightforward. More challenging is the identification of novel interactors or the systematic analysis of an interaction surface. For example, we have recently mapped the yeast histone chaperone complex FACT for contacts to core histones in living yeast by a pBPA-crosslinking scan [7]. This revealed the C-terminus of the Pob3 subunit as the dominant histone H2A/B interaction site. The crosslinks identified in such scans can be used to model the structure of the complex, if the crosslinking site on the target protein is determined. This has been achieved, for example, for urocortin-1 bound to its GPCR [8].

By scanning the α2-helix of H2A with pBPA, we have been able to identify an interaction with the H4 tail of another nucleosome in *S. cerevisiae*. Combining this crosslinking approach with cell cycle synchronization revealed that this interaction contributes a condensin-independent driving force of mitotic chromosome condensation [9]. Time-resolved in vivo crosslinking with genetically encoded UV-crosslinkers is ideally suited to address

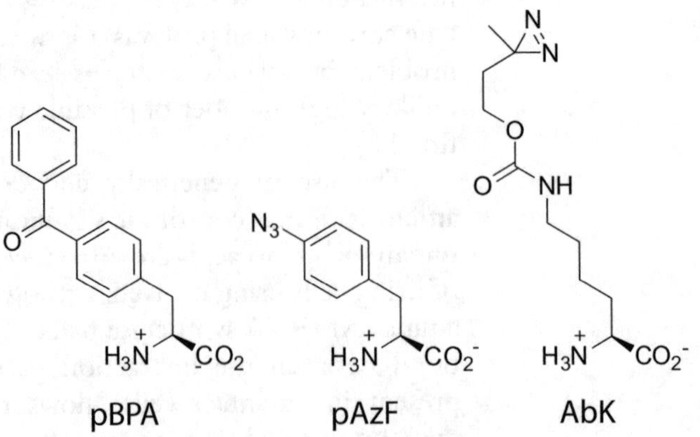

Fig. 1 Available genetically encoded UV-crosslinker amino acids (pBPA: *p*-benzoyl-phenylalanine, pAzF: *p*-azido-phenylalanine, AbK: 3′-azibutyl-*N*-carbamoyl-lysine)

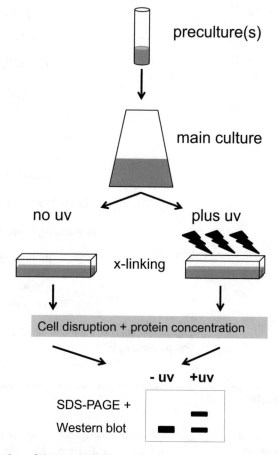

Fig. 2 Overview of the general UV-crosslinking procedure

some of the open questions in chromosome structure, composition, and condensation. Most critically it has the potential to reveal the network of interactions within mitotic chromosomes and to identify the sequence of events involved in the mechanism of condensation.

Based on the number of samples and the abundance of the crosslink partner, different strategies are available to identify a potential interaction partner. Therefore, we present a detailed description of the crosslinking procedure (Subheading 3.1, Fig. 2), followed by strategies to enrich crosslink products to facilitate identification (Subheadings 3.2 and 3.3, Fig. 3). Subheading 3.4 describes the combination of the crosslinking method with cell cycle synchronization of yeast cultures. The mapping of interaction interfaces requires the screening of a large set of amber mutants. In Subheading 3.5 we describe how crosslinking experiments are adapted and scaled to a 48-well block format.

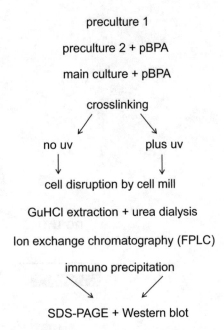

preculture 1

preculture 2 + pBPA

main culture + pBPA

crosslinking

no uv   plus uv

cell disruption by cell mill

GuHCl extraction + urea dialysis

Ion exchange chromatography (FPLC)

immuno precipitation

SDS-PAGE + Western blot

Fig. 3 Flow chart of the major steps of protocol 2 "Enrichment of histone cross-link partners by IP"

2   Materials

2.1   Basic Method: Trapping Histone-Protein Interactions by UV-Crosslinking in Yeast

1. Synthetic complete drop-out (SC) medium lacking the components for auxotrophic selection of plasmids (*see* **Note 1**) of yeasts strain containing 2% (w/v) glucose.

2. Yeast strain containing plasmids to express *Ec*BPARS/tRNA$_{CUA}$ and a histone amber mutant of choice. For more details *see* [9].

3. 100 mM p-benzoyl-L-phenylalanine (pBPA): solve in ddH$_2$O (dd, double deionized) containing 0.12 M NaOH by vigorously shaking (should be prepared ahead of the experiment and can be stored at −20 °C) (*see* **Note 2**).

4. Phosphate-buffered saline (PBS) pH 7.4 (10× stock: 1.37 M NaCl, 27 mM KCl, 81 mM Na$_2$HPO$_4$, 18 mM KH$_2$PO$_4$ pH 7.4).

5. Protease inhibitors: Protease inhibitor cocktail (PIC, 1000× stock): dissolve 90 mg Pefablock, 0.36 mg Leupeptin, 44 mg *o*-Phenanthrolin, 1.7 mg Pepstatin A in 5 mL DMSO; 200 mM PMSF in DMSO (100× stock solution).

6. UV-lamp such as VILBER Lourmat VL-208.BL, with 2 × 8 W 365 nm tubes.

7. Stainless steel tray with flat dents (approx. Ø = 15 mm, approx. 5 mm deep) fitting 200 μL volume.

8. Square ice bucket, which fits the UV-lamp and corresponding cover to shield the UV light.

9. Roedel-mix: 2 M NaOH, 7.5% (v/v) β-mercapthoethanol.

10. 50% (w/v) TCA (trichloroacetic acid).

11. Acetone.

12. 4× SDS loading buffer: 200 mM Tris/HCl pH 6.8, 8% (w/v) SDS, 40% glycerol (v/v), 0.04% Bromophenol blue (w/v), 0.2% (w/v) DTT (or 4% (v/v) β-mercaptoethanol).

13. Protein gel such as NuPAGE™ Novex™ 4–12% Bis-Tris protein gel (ThermoFisherScientific).

14. Protein standard such as PageRuler™ prestained protein ladder, 10–180 kDa (ThermoFisherScientific).

15. SDS-PAGE running buffer such as NuPAGE® MOPS (ThermoFisherScientific:"1× SDS running buffer: 50 mM MOPS, 50 mM Trisbase, 0.1% (w/v) SDS, 1 mM Na$_2$-EDTA, pH 7.7; prepare a 20× stock". Without pH adjustments).

16. Coomassie stain such as Instant Blue™ by Sigma-Aldrich.

17. Blotting paper 0.35 mm thick (e.g., Sartorius blotting papers BF2).

18. Transfer Membrane such as Immobilon®-P 0.45 μm by EMD Millipore (PVDF) (see **Notes 3 and 4**).

19. Transfer buffer such as NuPAGE® Transfer buffer (ThermoFisherScientific: 1× buffer: 25 mM Bicine, 25 mM Bis-Tris (free base), 1 mM Na$_2$-EDTA, pH 7.2; prepare a 20× stock. Without pH adjustments), with 10% (v/v) methanol final.

20. Ponceau S solution: 0.1% (w/v) Ponceau S (w/v) in 1% (w/v) acetic acid.

21. Tris buffered saline (TBS): 50 mM Tris-HCl pH 7.5, 150 mM NaCl.

22. Blocking solution depending on the primary antibody either in 5% (w/v) skimmed milk or 3% BSA in TBS.

23. Primary antibody.

24. TBS-Tween: 50 mM Tris-HCl pH 7.5, 150 mM NaCl, 0.05% (v/v) Tween 20.

25. Secondary antibody, e.g., as horseradish peroxidase (HRP) conjugate.

26. Substrate for the HRP such as Amersham™ ECL Select™ Western blotting detection reagent (GE Healthcare Life Sciences).

27. Chemoluminescence imager (e.g., Celvin® S by Biostep) or photographic film and developer reagents.

2.2 Enrichment of Crosslink Products by Nuclei Isolation (in Addition to Items Listed in Subheading 2.1)

1. Reducing buffer: 10 mM Tris-HCl pH 7.5, 30 mM DTT.

2. Lyticase buffer: YPD medium [2% peptone, 1% yeast extract, 2% glucose] containing 1 M sorbitol, 1 mM DTT, 1× PIC, 10 mM sodium butyrate, and 1 mM PMSF.

3. Lyticase: 100 kU/mL stock solution in ddH$_2$O; Sigma-Aldrich L2524.

4. Nuclei isolation buffer (NIB): 10 mM Tris-HCl, pH 7.5, 18% Ficoll-400, 20 mM KCl, 1 mM Na$_2$-EDTA, 0.15 mM spermine, 0.5 mM spermidine, 1× PIC, 10 mM sodium butyrate, and 1 mM PMSF.

5. YPD-medium containing 1 M sorbitol.

6. 0.4 N H$_2$SO$_4$.

2.3 Enrichment of Histone Crosslink Partners by IP (in Addition to Items 1–8 and 12–27 Listed in Subheading 2.1)

1. Liquid nitrogen.

2. Lysis buffer: PBS pH 7.4 containing 2× PIC and 2 mM PMSF.

3. Extraction buffer: 7 M guanidinium hydrochloride, 20 mM Tris-HCl, 10 mM DTT, pH 7.5.

4. Dialysis buffer: 7 M urea, 100 mM NaCl, 1 mM Na$_2$-EDTA, 10 mM Tris-HCl, 5 mM β-mercapthoethanol, pH 8.

5. Buffer A: 7 M urea, 1 mM Na$_2$-EDTA, 10 mM Tris-HCl pH 8, 1 mM DTT.

6. Elution buffer B: 7 M urea, 1 M NaCl, 1 mM Na$_2$-EDTA, 10 mM Tris-HCl pH 8, 1 mM DTT.

7. 1.2× RIPA buffer: 12 mM Tris-HCl, 1.2 mM Na$_2$-EDTA pH 8, 1.2% Triton X-100 (v/v), 0.12% Na-deoxycholate (w/v), 0.12% (w/v) SDS, 168 mM NaCl.

8. Affinity beads: E.g., HA-beads such as Pierce™ HA-Epitope tag antibody agarose conjugate (2-2.2.14) (ThermoFisherScientific).

9. 1× RIPA buffer: 1× volume ddH$_2$O, 5× volume 1.2× RIPA buffer.

10. High salt TBS: 50 mM Tris-HCl pH 7.5, 300 mM NaCl.

11. Coomassie stain such as Instant Blue™ by Sigma-Aldrich.

2.4 Crosslinking in Blocked/Synchronized Cells (in Addition to Items Listed in Subheading 2.1)

1. Yeast mating pheromone α-factor (Sigma-Aldrich, T6901), 1.6 mg/mL in 0.1 N HCl.

2. Nocodazole (Sigma-Aldrich, M1404), 1.5 mg/mL in DMSO.

3. Materials for FACS analysis: *see* [10].

2.5 Mapping, Libraries, 48-Well Crosslinking (in Addition to Items 1–8 and 12–27 Listed in Subheading 2.1)

1. Synthetic drop-out (SC) agar petri dishes lacking the components for plasmid selection (e.g., specific amino acids) and containing 2% (w/v) glucose.

2. 48-well culture blocks: Riplate® SW 48, PP, 5 mL (43001 - 0062, heat sterilize before use).

3. PCR plate, 96-well (Thermo Scientific, AB0600L).

4. Breathable film for sealing culture blocks (AeraSeal™ film; Sigma-Aldrich).

5. Centrifuge with adapters for 48 deep-well blocks such as Eppendorf 5810R with MTP Flex buckets.

3 Method

3.1 Basic Method: Trapping Histone-Protein Interactions by UV-Crosslinking

3.1.1 Culturing, Crosslinking, and Preparing Samples for SDS-PAGE

1. Culture the cells in 5 mL of the appropriate SC-medium supplemented with 1 mM pBPA ON (overnight) at 30°C and 200 rpm in a test tube.

2. Dilute the ON culture 1:10 in 20 mL fresh SC-medium containing 1 mM pBPA and let the cells continue to grow for 8 h.

3. Harvest the cells for 4 min at 3500 rpm ($1350 \times g$), room temperature (RT).

4. Resuspend the cells in 200 μL PBS.

5. Fill the sample in a dent of the metal tray and crosslink for 15 min at 4°C.

6. Remove the sample from the dent and harvest the cells for 4 min at 3500 rpm ($1350 \times g$) and RT. Remove the supernatant (*see* **Note 5**).

7. Resuspend the cells in 1 mL ddH$_2$O containing 2 mM PMSF and 2× PIC.

8. Add 160 μL of Roedel-mix, mix carefully, and incubate the sample on ice for 10 min.

9. Add 300 μL 50% TCA and incubate for 10 min on ice.

10. Spin the sample at 4°C for 5 min full speed and discard the supernatant.

11. Wash the pellet two times with 0.5–1 mL acetone and remove the supernatant completely.

12. Air-dry the pellet to remove acetone (*see* **Note 6**).

13. Dissolve the pellet in 75 μL ddH$_2$O and mix with 25 μL 4× SDS loading buffer.

14. Boil the sample for 10 min at 95°C.

15. Spin the sample for 1 min at full speed. The supernatant can be loaded on an SDS-PAGE.

3.1.2 *SDS-PAGE and Western Blot*

1. Load up to 20 μL of the sample and 7 μL of the protein standard on an SDS-PAGE gel and let it run until the Bromophenol blue band elutes from the bottom of the gel (*see* **Note 7**).

2. For Western blot analysis prepare the transfer membrane for blotting, e.g., wet the PVDF membrane briefly in methanol.

3. Assemble gel and membrane for wet blotting as indicated by the manufacturer of your blotting module and blot for 1 h at 250 mA constant at 4°C.

4. After transfer, stain the membrane with Ponceau S solution and destain with water to verify successful transfer.

5. Cover the membrane in blocking solution for 30 min at RT using a rocking shaker.

6. Add the primary antibody to the blocking solution in the desired dilution and incubate >2 h at RT or ON at 4°C using a rocking shaker.

7. Thoroughly wash the membrane four times with TBS-Tween for a total of approx. 15 min on the rocking shaker.

8. Cover the membrane in blocking solution containing the secondary antibody (typically in a 1:10,000 dilution) and incubate for 1 h at RT with shaking.

9. Wash the membrane again thoroughly on the rocking shaker with TBS-Tween for approx. 15 min.

10. Last wash in TBS without Tween 20 (*see* **Note 8**).

11. Prepare the substrate for the HRP detection following the instruction of the manufacturer and detect the antibody-coated protein bands, e.g., with a chemoluminescence imager.

To identify the potential interaction partner, look for a band within the crosslink pattern of the combined molecular weight of histone and protein interaction partner. Ideally, verify the suspected crosslink partner with an antibody against it by a second SDS-PAGE followed by Western blot analysis for the target protein.

If the antibody of the desired protein cannot detect its target protein in a cell extract, the desired crosslink can be enriched using method 2 "Enrichment of crosslinks products by nuclei isolation" or 3 "Enrichment of histone crosslink partners by IP." The nuclei prep protocol is useful to handle several samples in parallel, whereas the more time-consuming IP protocol is highly efficient in enriching crosslinks.

If the crosslink pattern is too complex or the antibody against the desired crosslink partner is not specific enough to detect its target protein in a cell extract, the histone interaction partner can be genetically tagged. Then method 1 can be repeated using two yeast strains: (a) A "wild-type" strain: Yeast strain containing plasmids to express EcBPARS/tRNA$_{CUA}$ and a histone amber mutant of choice and (b) A genomically tagged (e.g., GFP or myc on the interaction candidate) yeast strain otherwise of the identical genetic background as the "wild-type" strain. The patterns of both crosslinking experiments should differ by a shifted band in the tagged sample.

3.2 Enrichment of Crosslink Products by Nuclei Isolation

We used this method to enrich an H2A-H4 crosslink product previously [9]. It is possible to process multiple samples in parallel by this method (compared to the IP protocol).

1. Grow yeast cells harboring plasmids to encode EcBPARS/tRNA$_{CUA}$ and a histone amber mutant in 100 mL SC medium with 2% glucose but without uracil and leucine (depending on the plasmid selection markers) at 30°C to an OD$_{600}$ of 1.

2. Harvest cells and resuspend pellets in 10 mL Reducing buffer. Incubate for 10 min at 30 °C.

3. Pellet cells (5000 × g, 2 min) and resuspend in lyticase buffer containing 100 U/mL lyticase. Incubate for 20–30 min at 30°C in a water bath. The cells should have become spheroplasts by this time (see **Note 9**).

4. Collect spheroplasts by centrifugation (3000 × g, 2 min) (see **Note 10**). Wash once with YPD + 1 M sorbitol and once with 1 M sorbitol, both ice-cold. All the following steps are done on ice or at 4°C.

5. Resuspend spheroplasts in 5 mL nuclei isolation buffer and disrupt in a prechilled glass homogenizer (size S) with 20 even strokes.

6. Remove large cellular debris by centrifugation (3000 × g, 2 min). Collect supernatant and discard pellet.

7. Centrifuge the supernatant at 16,500 × g for 10 min. Remove the supernatant.

8. Extract the histones (and their crosslinks) from the pellet by thorough mixing in 0.4 N H$_2$SO$_4$ for 30 min (see **Note 11**).

9. Continue acid extraction overnight at 4°C by placing the samples on a rotating wheel.

10. Centrifuge at 16,500 × g for 10 min and collect the supernatant in a new tube.

11. Precipitate histones by adding 15% TCA (final concentration) and set on ice for 2 h.

12. Centrifuge at 16,500 × g for 10 min, wash twice with acetone, and dissolve pellet after air drying in 1× SDS loading buffer (see **Note 6**).

13. For SDS-PAGE and Western Blot analyses please see Subheading 3.1.2 SDS-PAGE and Western blot.

3.3 Enrichment of Histone Crosslink Partners by IP

This method efficiently enriches all crosslink products emanating from a particular site on a histone. This will subsequently help to identify crosslink products using antibodies directed against an expected crosslink partner by Western blot (Fig. 3). The following protocol describes the preparation of a crosslinked sample containing a C-terminally HA-tagged histone. The same sample without UV-irradiation is prepared in parallel and used as a negative control.

3.3.1 Culturing of Yeast Cells

1. Start a 5 mL overnight (ON) culture in a test tube with the appropriate medium and incubate at 30°C and 200 rpm.

2. Add the 5 mL ON-culture to 100 mL (in 2 L flask) of appropriate minimal medium supplemented with 1 mM pBPA and incubate at 30°C and 200 rpm ON.

3. The following morning add 900 mL minimal medium supplemented with 1 mM pBPA to the ON-culture and continue incubating for 8 h.

3.3.2 Crosslinking

1. Harvest the yeast cells for 10 min at 4800 rpm and 4°C.

2. Resuspend the pellet in 20 mL PBS.

3. Fill the sample in the instrument tray placed on ice and crosslink samples three times for 5 min, 4°C with gentle manual panning in between.

4. Harvest the yeast cells for 4 min at 3500 rpm and RT (see Note 5).

3.3.3 Cell Breakage and Preparing the Cell Lysate for Dialysis

1. Resuspend the cell pellet in 15 mL lysis buffer (see Note 12).

2. Flash freeze the cells by dripping the cell extract slowly into liquid nitrogen.

3. Collect the droplets with the strainer in a cooled 50 mL Falcon tube (see Note 13).

4. Disrupt the cells using the cell mill (e.g., Retsch ZM 200) (see Note 14). It is important to quickly scrape the frozen cell powder with a liquid N_2-cooled spatula into a beaker.

5. Let the cell lysate thaw at RT (see Note 15).

6. Transfer the lysate to a high-speed centrifuge tube, add Triton X-100 to a final concentration of 1% (v/v), and extract for 5 min at RT on a tube roller.

7. Spin at 18,000 rpm ($40,000 \times g$) for 15 min at 4 °C and discard the supernatant.

8. Wash the pellet with 30 mL lysis buffer.

9. Resuspend the pellet in 20 mL extraction buffer and incubate for 1 h at 37°C with vigorous shaking.

10. Spin at 18,000 rpm ($40,000 \times g$) for 15 min at 4°C and keep the supernatant.

11. Fill the supernatant into a dialysis membrane tube (MWCO 6–8 kDa, e.g., Spectra/Por®) and dialyse two times against 1 L dialysis buffer at RT. If a precipitate after dialysis is visible, spin for 3 min at 3500 rpm ($2500 \times g$) and RT and use the supernatant for ion exchange purification.

3.3.4 Ion Exchange Purification by FPLC

1. For ion exchange purification use a 5 mL GE HiTrap SP FF column and prepare the FPLC and the column sequentially by each step until a baseline is reached: Wash with 20% ethanol. Wash with ddH$_2$O. Wash with 100% elution buffer B.

2. Prior to loading the sample equilibrate the column as follows [11]:

   H2A or H2B: 10% elution buffer B.

   H3 or H4: 20% elution buffer B.

3. Load the sample onto the column (e.g., via sample pump, super loop or syringe) with a flow speed of 2 mL/min and collect the flow through in 10 mL fractions.

4. Wash the column under equilibration conditions for the desired histone—no fractionation is required.

5. Elute histones and crosslink products with 100% elution buffer B; choose a fraction size of 0.5 mL and a flow speed to 1 mL/min. Pool peak fractions (*see* **Note 16**).

6. If the volume of the pooled fractions exceeds 8 mL, measure the protein concentrations in the sample (e.g., Bradford test) and reduce the volume to at least 8 mL or >1 mg/mL protein concentration via concentrator MWCO 10 kDa (*see* **Note 17**).

3.3.5 Immuno-precipitation

1. Add fivefold volume of 1.2× RIPA buffer to the sample.

2. Prepare HA-beads: Wash 150 µL slurry of the HA-agarose beads with 1 mL 1× RIPA in a 1.5 mL Eppendorf tube, spin for 1 min at 2000 rpm ($400 \times g$), and remove the supernatant.

3. Transfer the beads completely to the RIPA-diluted sample and incubate on a tube roller at 4°C for 3 h.

4. Spin the HA-beads for 1 min at 3500 rpm ($1200 \times g$) and remove the supernatant. Transfer the beads to a 1.5 mL Eppendorf tube.

5. Wash the HA-beads (for 5 min each time) once with 1 mL 1× RIPA buffer, once with 1 mL high salt TBS and once 1 mL 1× RIPA buffer using an overhead shaker.

6. Remove the remaining supernatant completely with a Hamilton syringe.

7. Instantly add 100 µL 1× SDS-loading buffer and boil the sample at 95°C for 10 min.

8. Spin the sample at 13,300 rpm ($17,400 \times g$) for 2 min. The supernatant of the sample is ready to be loaded on an SDS-PAGE gel (*see* **Note 5**).

3.3.6 SDS-PAGE, Coomassie Stain, and Western Blot

See Subheading 3.1.2 "SDS-PAGE and Western blot."

3.4 Crosslinking in Synchronized Cell Populations

3.4.1 Block and Release with α-Factor

1. Grow yeast (e.g., BY4741 transformed with plasmids to encode *Ec*BPARS/tRNA$_{CUA}$ and a histone with an amber mutation) in 200 mL SC medium lacking appropriate components for plasmid maintenance (e.g., uracil and leucine) supplemented with 1 mM pBPA overnight at 30°C.

2. Dilute the cells into 200 mL fresh YPD medium with pBPA to an OD_{600} of 0.2 and shake for 2 h at 30°C.

3. Add α-factor to a final concentration of 5 μg/mL. As a control, harvest 1 mL cells for FACS analysis (spin 1 min at $1000 \times g$, resuspend pellet in 1.5 mL ddH$_2$O and slowly add 3.5 mL 95% ethanol).

4. Shake the cells at 30°C, 200 rpm until >95% of cells display "shmoo" morphology (monitored by light microscopy), usually 1.5–2 h. Take another sample for FACS analysis.

5. Release the cells from block by washing two times with 400 mL YPD (pellet cells by centrifugation (2 min, 4800 rpm ($4600 \times g$)) and resuspend in YPD).

6. Resuspend the cells in pre-warmed YPD at $OD_{600} = 0.4$. Shake at 30°C, 200 rpm.

7. Take samples of 12 OD_{600} units every 15 min, treat with UV light as described in Subheading 3.1.1, **steps 4–12** and analyze by SDS-PAGE and Western blot (Subheading 3.1.2).

3.4.2 Block and Release with Nocodazole

1. Grow yeast (e.g., BY4741 transformed with plasmids to encode *Ec*BPARS/tRNA$_{CUA}$ and a histone with an amber mutation) in 200 mL SC medium lacking uracil and leucine supplemented with 1 mM pBPA overnight at 30°C.

2. Dilute the cells into 500 mL fresh YPD medium with pBPA to an OD_{600} of 0.4 and shake for 2 h at 30°C, 200 rpm.

3. Adjust OD_{600} again to 0.4 and add nocodazole to a final concentration of 15 μg/mL (from 100× stock in DMSO). Harvest 1 mL cells for FACS analysis (spin 1 min at $1000 \times g$, resuspend pellet in 1.5 mL ddH$_2$O and slowly add 3.5 mL 95% ethanol).

4. Continue shaking at 30°C for 1.5 h and harvest another sample for FACS analysis.

5. Release the cells from block by washing four times with 500 mL YPD + 1% DMSO (*see* **Note 18**).

6. Resuspend the cells in 500 mL pre-warmed YPD at $OD_{600} = 0.4$ and continue shaking at 30°C, 200 rpm.

7. Take samples of 12 OD_{600} units every 15 min, treat with UV light as described in Subheading 3.1.1, **steps 4–12** and analyze by SDS-PAGE and Western blot (Subheading 3.1.2).

Of course, cells can also be crosslinked while arrested by α-factor or nocodazole to compare crosslinking efficiencies at different cell cycle stages. FACS analyses are important to monitor successful synchronization.

3.5 Mapping, Libraries, 48-Well Crosslinking

This protocol was previously used by our lab to survey interaction sites of the FACT complex with histones in vivo. It required streamlining of mutagenesis, transformation, growth, crosslinking, and protein analysis procedures to allow the efficient screening of more than 200 amber mutants of the FACT complex [7].

3.5.1 Creating Amber Mutant Library of Your Gene of Interest

This section describes an adaptation of the standard Quikchange protocol for 96-well format.

1. Cloning of the library master plasmid encoding your protein of interest fused to an epitope tag for Western blot detection. Check protein expression by Western blot prior to library creation.

2. Design primers for Quikchange mutagenesis of the library master plasmid (*see* **Note 19**) and order primers in 96-well blocks (forward and reverse block): Primers are designed in an overlapping fashion covering the region of interest (10 bp upstream and at least 17 bp downstream of the desired TAG substitution are complementary to the plasmid template).

3. Set up Quikchange reactions according to a standard protocol in 96-well PCR plate with Phusion or TurboPfu polymerase in a total volume of 20 μL. After DpnI digest (5 U/rxn) for 1 h at 37°C, transfer 5 μL to a fresh 96-well PCR-plate, and add 50 μL of chemically competent bacteria for transformation. Perform heat shock according to a standard protocol in a thermocycler. Transfer transformed bacteria to a 48-well block and perform recovery according to standard procedures (*see* **Note 20**).

4. Plate pre-culture on an agar petri dish with LB-medium containing the appropriate antibiotics for the selection of transformants and incubate overnight at 37°C.

5. Perform plasmid preparation of transformants and test for successful TAG substitution by sequencing.

3.5.2 Culturing of Yeast Cells

1. Transform parent yeast strain harboring the plasmid encoding EcBPARS/tRNA$_{CUA}$ individually with the library plasmids (*see* **Note 21**).

2. Inoculate overnight pre-cultures for crosslinking in 48-well blocks containing 3.5 mL SC medium with 1 mM pBPA. Seal the block with a breathable film and incubate at 30°C overnight with shaking at 220 rpm.

3.5.3 Crosslinking

1. Determine the optical density of the pre-cultures and inoculate two fresh 48-well blocks containing 3.5 mL SC medium supplemented with 1 mM pBPA at an OD_{600} of 0.15–0.2. One block is subjected later to UV radiation for crosslinking, while the other block serves as non-treated control.

2. Grow cells to late log-phase at 30°C with rotational shaking at 220 rpm (approximately OD_{600} of 1.2) and harvest expression cultures by centrifugation ($805 \times g$ for 2 min at 4 °C).

3. Remove the supernatant by swiftly inverting the block over a collection vessel and place the blocks the right way up on ice. For crosslinking, subject one entire 48-well block to UV radiation for 15 min on ice while keeping the control block dark on ice.

4. Store samples either at −20°C or subject them immediately to Western blot analysis (see below).

3.5.4 SDS-PAGE and Western Blot Analysis

1. For crude sample preparation, resuspend the pellets in 75 μL ddH_2O and transfer them to a fresh 96-well PCR plate. Add 25 μL 4× SDS loading buffer, seal the plate with adhesive PCR film, and boil the samples for 10 min at 95°C in a thermocycler (*see* **Note 22**). Spin the PCR plate for 10 min at full speed.

2. Load 5–10 μL of the supernatant on a SDS-PAGE gel and perform electrophoresis.

3. Follow the procedure as described in 3.1.2 SDS-PAGE and Western blot.

4 Notes

1. We usually use plasmids providing uracil (pRS426 containing the gene to encode the histone amber mutant) and leucine (pESC-BPARS encoding the pBPA-specific variant of EcYRS and the cognate amber suppressor tRNA) auxotrophy.

2. Optional: solution can be sterile filtered through a 0.2 μm syringe filter.

3. We have made very good experience with the specific membrane, which gave significantly lower levels of unspecific signals.

4. If fluorescent secondary antibodies (such as anti-mouse/anti-rabbit-IgG-Cy3/Cy5) conjugates are used, use the Transfer membrane: Amersham Hybond™ LFP 0.2 PVDF (GE Healthcare Life Sciences, 10600022) for low background signals.

5. This would be a good time to interrupt. Samples can be stored at −20°C.

6. Sample may be heated carefully but make sure not to overheat them as this may affect resolution in SDS-PAGE.

7. Do not run the gel too fast (above 100 V) as this affects resolution, which is critical for the detection of individual crosslink bands in the Western blot.

8. The presence of detergent during the detection by chemoluminescence may increase background.

9. This can be checked by microscopy: After adding 1% SDS to a sample, which will lyse the spheroplasts, "ghosting" should be observed.

10. If the cells were efficiently spheroplasted, the supernatant should be turbid.

11. The acid extraction of histones from the nuclei may not be efficient for some crosslink products. In this case it may be preferable to solubilize nuclei directly in 1× SDS loading buffer.

12. We do not recommend using a smaller volume of buffer because this will result in a significant loss of material during the cell disruption step if the Retsch ZM 200 cell grinder is used. Other means of cell lysis may require adjusting the volume of lysis buffer.

13. Optional: droplets can be stored at −80°C.

14. Cool the grinder in excess liquid nitrogen, assemble and start the mill at maximum speed. Add the frozen droplets and stop the machine after 5 s. Make sure to wear appropriate protective gear.

15. In urgent cases: defrost cell powder very carefully in a microwave.

16. Optional: (pooled) peak fractions can be stored at 4°C.

17. High protein concentrations improve immunoprecipitation efficiency, but avoid concentrating your sample too much as this may cause precipitation.

18. It is important to work fast here because the cells will resume cycling once the nocodazole is removed. Centrifuge the cells briefly at room temperature (for 2 min at $5000 \times g$) when washing the cells. Avoid unnecessary interruptions. It is not required to work sterile at this point.

19. Identify sites suitable for pBPA incorporation in your protein with the help of structural or functional data. Select surface exposed residues, non-conserved sites are preferable. Substitution of every 10th residue for a protein interaction library is manageable.

20. If the plasmid conveys ampicillin resistance, which is usually the case for yeast shuttle vectors, this step can be omitted and the cells can be plated directly after the heat shock.

21. Transformation of yeast cells can be done in 48-well blocks, however, it is advisable to plate transformations on individual agar plates after transformation to avoid potential problems with contaminations. Therefore, prepare competent cells from a yeast strain harboring the tRNA/synthetase plasmid. Add the cells to individual library plasmids (1–2 μg each) and perform standard heat shock procedure using a water bath. Spin blocks and remove the supernatant. Resuspend the cells in 200 μL ddH$_2$O and plate on appropriate drop-out agar plates.

22. The quality of samples obtained by boiling cells in 1× Sample buffer is clearly inferior to the one obtained by the Roedel procedure. It may therefore be necessary to resort to individual handling at this stage as described in Subheading 3.1.1 **steps 7–15** if the resolution of crosslink products in SDS-PAGE is insufficient.

Acknowledgments

The authors are grateful for financial support by the Cluster of Excellence and DFG Research Center Nanoscale Microscopy and Molecular Physiology of the Brain (CNMPB) and the German Research Foundation (DFG) [NE1589/5-1 to H.N.].

References

1. Ohta S, Wood L, Bukowski-Wills JC, Rappsilber J, Earnshaw WC (2011) Building mitotic chromosomes. Curr Opin Cell Biol 23(1):114–121

2. Neumann H (2012) Rewiring translation - genetic code expansion and its applications. FEBS Lett 586(15):2057–2064

3. Ai HW, Shen W, Sagi A, Chen PR, Schultz PG (2011) Probing protein-protein interactions with a genetically encoded photo-crosslinking amino acid. Chembiochem 12(12):1854–1857

4. Chou CJ, Uprety R, Davis L, Chin JW, Deiters A (2011) Genetically encoding an aliphatic diazirine for protein photocrosslinking. Chem Sci 2(3):480–483

5. Zhang M, Lin S, Song X, Liu J, Fu Y, Ge X, Fu X, Chang Z, Chen PR (2011) A genetically incorporated crosslinker reveals chaperone cooperation in acid resistance. Nat Chem Biol 7(10):671–677

6. Chin JW, Cropp TA, Anderson JC, Mukherji M, Zhang Z, Schultz PG (2003) An expanded eukaryotic genetic code. Science 301(5635):964–967

7. Hoffmann C, Neumann H (2015) In vivo mapping of FACT-histone interactions identifies a role of pob3 C-terminus in H2A-H2B binding. ACS Chem Biol 10(12):2753–2763

8. Coin I, Katritch V, Sun T, Xiang Z, Siu FY, Beyermann M, Stevens RC, Wang L (2013) Genetically encoded chemical probes in cells reveal the binding path of urocortin-I to CRF class B GPCR. Cell 155(6):1258–1269

9. Wilkins BJ, Rall NA, Ostwal Y, Kruitwagen T, Hiragami-Hamada K, Winkler M, Barral Y, Fischle W, Neumann H (2014) A cascade of histone modifications induces chromatin condensation in mitosis. Science 343(6166):77–80

10. Haase SB (2004) Cell cycle analysis of budding yeast using SYTOX green. Curr Protoc Cytom. Chapter 7:Unit 7.23

11. Luger K, Rechsteiner TJ, Richmond TJ (1999) Preparation of nucleosome core particle from recombinant histones. Methods Enzymol 304:3–19

Chapter 17

Genetically Encoding Unnatural Amino Acids in Neurons In Vitro and in the Embryonic Mouse Brain for Optical Control of Neuronal Proteins

Ji-Yong Kang, Daichi Kawaguchi, and Lei Wang

Abstract

Deciphering neuronal networks governing specific brain functions is a longstanding mission in neuroscience, yet global manipulation of protein functions pharmacologically or genetically lacks sufficient specificity to reveal a neuronal protein's function in a particular neuron or a circuitry. Photostimulation presents a great venue for researchers to control neuronal proteins with high temporal and spatial resolution. Recently, an approach to optically control the function of a neuronal protein directly in neurons has been demonstrated using genetically encoded light-sensitive Unnatural amino acids (Uaas). Here, we describe procedures for genetically incorporating Uaas into target neuronal proteins in neurons in vitro and in embryonic mouse brain. As an example, a photocaged Uaa was incorporated into an inwardly rectifying potassium channel Kir2.1 to render Kir2.1 photo-activatable. This method has the potential to be generally applied to many neuronal proteins to achieve optical regulation of different processes in brains. Uaas with other properties can be similarly incorporated into neuronal proteins in neurons for various applications.

Key words Unnatural amino acid, Expansion of the genetic code, Amber suppression, Ion channel, Optogenetics, Light-activation, Optical control, Photocage, Neuronal activity, Brain

1 Introduction

Optical control of neuronal activity has helped answer many challenging questions in neuroscience that are not easily addressable with conventional methods. Photocaged agonists have been utilized to stimulate a local brain network [1, 2]. Ion channel gating was controlled with light by attaching photoisomerizable azobenze groups to the target protein [3, 4]. Light-sensitive opsin channels and pumps are widely adopted to modulate neuronal excitability in different neuronal circuitries [5–7]. These different approaches are ingenious ways to exploit the high specificity of photostimulation to manipulate neuronal function. However, new methods to photo-regulate various endogenous proteins in their native

Edward A. Lemke (ed.), *Noncanonical Amino Acids: Methods and Protocols*, Methods in Molecular Biology, vol. 1728, https://doi.org/10.1007/978-1-4939-7574-7_17, © Springer Science+Business Media, LLC 2018

environment remain desired, so that photo-control specificity can be achieved at the molecular level for general proteins.

Genetically encoding Unnatural amino acids (Uaas) is a uniquely attractive tool to engineer light-responsiveness into general proteins. An orthogonal tRNA/synthetase pair incorporates a Uaa in response to the amber stop codon during endogenous translation, generating a novel protein with the intervening Uaa at the selected site [8–11]. This method imposes no restrictions on target protein type or cellular location, nor on Uaa modification site. Such great flexibility makes it available to target a wide variety of endogenous proteins. Manipulation at the translational level intrinsically gives it genetic specificity. Using this approach, we initially incorporated Uaas with different physical properties into ion channels in primary neuron culture and in neural stem cells followed by differentiation into neurons [12, 13]. Recently, we succeeded in incorporating photoresponsive Uaas into an ion channel in vivo in the mouse brain, which paves the way for photo-control of specific neuronal proteins in mammalian brain in vivo [14].

In the study by Kang et al. [14], we incorporated a photoresponsive Uaa, 4,5-dimethoxy-2-nitrobenzyl-cysteine (Cmn), into an inwardly rectifying potassium channel Kir2.1 (Fig. 1). The side chains of Cmn incorporated in the Kir2.1 pore sterically hinder the passage of potassium ions, shutting off the channel function of Kir2.1. A brief illumination of UV-light instantly and specifically cleaves off the Cmn side chains, turning on normal Kir2.1 function. These *photo*-inducible *i*nwardly *r*ectifying potassium (PIRK) channels were expressed in rat hippocampal primary neurons as well as in the embryonic mouse neocortex in vivo, demonstrating the optical control of Kir2.1 function in different research settings.

The current protocol entails an easy-to-follow procedure to incorporate Uaas in neurons in vitro and in vivo to achieve optical control over neuronal proteins' activity. With the availability of numerous photoreactive Uaas, a wide application of the current technology would be sought for. We expect this protocol to be a clear guide for future studies indulging in photo control over neuronal proteins and optogenetic biological studies in general.

2   Materials

2.1   Culture Rat Hippocampal Primary Neurons

1. Sprague Dawley rat pups (postnatal day 1–4). One pup produces 10–12 coverslips (*see* **Note 1**).

2. Glass coverslips (circles, 12 mm in diameter).

3. 24-well tissue culture plates.

4. 100 mM Borate buffer: weigh 1.24 g boric acid and 1.90 g sodium tetraborate in ~350 mL of deionized distilled H_2O

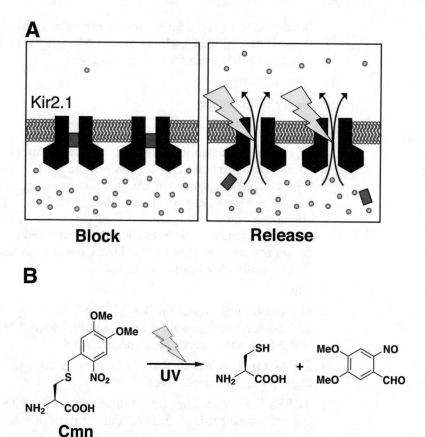

Fig. 1 Genetically encoding a photocaged Uaa to generate photo-inducible channels in neurons. (**a**) Left: Incorporation of the Uaa Cmn (in red) in the pore of Kir2.1 channels renders the channel nonconducting; Right: UV light exposure irreversibly removes the photocaging group, dimethoxynitrobenzyl, to allow permeation through the Kir2.1 channel, restoring outward K⁺ current. (**b**) Photolysis of Cmn into Cys

(ddH$_2$O), and adjust pH to 8.5 with dilute HCl. Bring the volume up to 400 mL.

5. 0.5 mg/mL poly-D-lysine: dissolve 20 mg poly-D-lysine in 40 mL of 100 mM borate buffer. With a serological pipette, mix the solution thoroughly by pipetting up and down (*see* **Note 2**).

6. Isoflurane.

7. Dissection surgery tools: scissors, forceps, blades, and spatulas.

8. Saline: 10 mM HEPES and 20 mM D-glucose added in Hank's balanced salt solution.

9. 2.5% Trypsin.

10. Glass pipettes.

11. Growth media: in 46 mL Minimum Essential Medium (MEM), add 425 μL of 2.5 M sterile glucose solution, 2.5 mL Fetal

Bovine Serum (FBS), 50 μL serum extender, 1 mL B27, and 0.5 mL of 200 mM L-glutamine. Ideally, the Growth media should be made fresh for each experiment.

12. Hemocytometer.

13. 40 μm Nylon mesh.

2.2 Calcium-Phosphate (Ca-P) Transfection

1. A set of recombinant DNA to incorporate Uaa into Kir2.1 (Fig. 2): (1) recombinant Kir2.1 gene with an amber stop codon TAG introduced at the desired site for Uaa incorporation, Cys169 in the current study. A fluorescent protein, mCitrine in the study, can be fused at the C-terminus of Kir2.1 to visualize successful amber suppression [14]. (2) tRNA/synthetase gene to decode the introduced amber stop codon and incorporate a specific Uaa, Cmn in this study, to the TAG-specified site (*see* **Notes 3** and **4**).

2. Cmn.

3. Stock solutions: make up 0.5 M BES (*N*,*N*-bis[2-hydroxyethyl]-2-aminoethanesulfonic acid) buffer, 150 mM Na_2HPO_4, and 2.8 M NaCl, sterile filter and store at 4 °C.

4. 2.5 M $CaCl_2$ solution: make it fresh with ddH$_2$O for every experiment, and sterile filter.

5. 2× BES-buffered Saline (BBS) buffer: add 1 mL of 0.5 M BES, 0.1 mL of 150 mL Na_2HPO_4, and 1 mL of 2.8 M NaCl to 7 mL ddH$_2$O. Adjust pH to 7.00 with dilute NaOH, bring the volume up to 10 mL, and sterile filter (*see* **Note 5**). Make fresh 2× BBS buffer for every experiment.

6. Transfection media: in 50 mL MEM, add 425 μL of 2.5 M sterile glucose solution. Ideally, the Transfection media should be made fresh for each experiment.

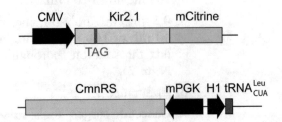

Fig. 2 Plasmids for incorporation of Cmn into Kir2.1 in neuron culture. The top plasmid encodes Kir2.1 gene containing a TAG amber codon at Cys169 site for Uaa incorporation driven by the cytomegalovirus (CMV) promoter. Fluorescent protein mCitrine was fused to the C-terminus for easy visualization of Uaa incorporation and Kir2.1 expression. The bottom plasmid encode CmnRS (the synthetase specific for Cmn) driven by the mouse phosphoglycerate kinase-1 (mPGK) promoter and the orthogonal amber suppressing leucyl tRNA driven by the H1 promoter

7. Washing buffer: mix 500 mL ddH$_2$O with 135 mM NaCl, 20 mM HEPES, 4 mM KCl, 2 mM CaCl$_2$, 1 mM MgCl$_2$, 1 mM Na$_2$HPO$_4$, and 10 mM glucose at pH 7.4 (adjusted with dilute NaOH), and sterile filter to store at 4 °C.

2.3 Whole Cell Recording with Light Activation from Primary Neurons

1. Extracellular solution: mix 1 L ddH$_2$O with 150 mM NaCl, 3 mM KCl, 5 mM MgCl$_2$, 0.5 mM CaCl$_2$, 5 mM glucose, and 10 mM HEPES, at pH 7.4 (adjusted with dilute NaOH), and sterile filter.

2. Intracellular solution: mix 20 mL ddH$_2$O with 135 mM potassium gluconate, 10 mM NaCl, 2 mM MgCl$_2$, 10 mM HEPES, 1 mM EGTA, 2.56 mM K$_2$ATP, and 0.3 mM Li$_2$GTP at pH 7.4 (adjusted with dilute KOH), and sterile filter. Aliquot 300–400 μL per tube, and freeze for long-term storage.

3. Electrophysiology rig for whole-cell patch clamping: microscope fitted with 4× and 20× objectives, differential interference contrast (DIC), and mCitrine filter (excitation: 495/10 nm, emission: 525/25 nm), manipulator, patch-clamp amplifier, digitizer, and data acquisition and analysis software.

4. Light-emitting diode (LED) with emission of 385 nm, cable light guide installed (*see* **Note 6**).

5. Light power meter.

6. Glass patch pipettes.

7. 0.5 mM BaCl$_2$ Extracellular solution: Add 25 μL of 1 M BaCl$_2$ in 50 mL Extracellular solution.

2.4 In Utero Electroporation and Uaa Microinjection

1. A set of recombinant DNA to incorporate Uaa into Kir2.1 (Fig. 3) (*see* **Note 7**).

2. Cmn-Ala (*see* **Note 8**).

3. A timed pregnant mouse at the embryonic day 14.5 (E14.5).

4. Dissection surgery tools: scissors, forceps, and ring forceps.

5. Glass capillaries.

6. Electroporator.

7. Sodium pentobarbital or isoflurane.

8. Fast Green.

9. Dulbecco's Phosphate-Buffered Saline (PBS).

2.5 Whole Cell Recording with Light Activation from Acute Brain Slices

1. Artificial cerebrospinal fluid (ACSF) solution: mix 1 L ddH$_2$O with 119 mM NaCl, 2.5 mM KCl, 1.3 mM MgCl$_2$, 2.5 mM CaCl$_2$, 1 mM NaH$_2$PO$_4$, 26.2 mM NaHCO$_3$, and 11 mM glucose. Bubble ACSF with 5% CO$_2$: 95% O$_2$ gas while stirring. Subsequently, adjust pH to 7.3 with dilute HCl. Continue bubbling the solution for the rest of the experimental day. ACSF solution is made fresh every day.

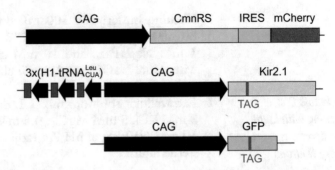

Fig. 3 Plasmids for incorporation of Cmn into Kir2.1 in the mouse neocortex in vivo. The top plasmid encodes CmnRS driven by a strong chicken beta-actin (CAG) promoter and red fluorescent protein mCherry via internal ribosomal entry site (IRES). mCherry fluorescence serves to verify the expression of CmnRS. The middle plasmid encodes Kir2.1 gene containing the amber stop codon TAG at Cys169 site, together with three copies of the orthogonal tRNA expression cassette. The bottom plasmid encodes GFP gene with an amber stop codon introduced at the permissive Tyr182 site. Green fluorescence of GFP indicates Cmn incorporation at TAG sites

2. Dissection surgery tools: scissors, forceps, blades, and spatulas.

3. Microwave.

4. 4% Low melting point agarose solution: weigh 4 g low melting point agarose into a flask and add 100 mL ddH$_2$O. Dissolve the agarose by heating up the flask in a microwave. Cool down the solution below 50 °C, and use it before it solidifies. Do not re-heat up agarose blocks.

5. Vibratome equipment.

6. Tissue adhesive.

7. Incubation ACSF: Mix 250 mL ACSF with 3 mM *myo*-inositol, 0.4 mM ascorbic acid, and 2 mM sodium pyruvate.

8. Water bath for tissue culture.

9. Glass pipettes.

10. Intracellular solution: mix 20 mL ddH$_2$O with 130 mM potassium gluconate, 4 mM MgCl$_2$, 5 mM HEPES, 1.1 mM EGTA, 3.4 mM Na$_2$ATP, 10 mM sodium creatine phosphate, and 0.1 mM Na$_3$GTP at pH 7.3 (adjusted with dilute KOH), and sterile filter. Aliquot 300–400 μL per tube, and freeze for long-term storage.

11. Slice electrophysiology rig: Electrophysiology rig for whole-cell patch clamping equipped with perfusion pump, perfusion chamber, water immersion objectives, a temperature controller, GFP filter (excitation: 480/30 nm, emission: 535/40 nm) and mCherry filter (excitation: 580/20 nm emission: 675/130 nm).

12. 0.5 mM BaCl$_2$ ACSF: Add 25 μL of 1 M BaCl$_2$ in 50 mL ACSF.

3   Methods

3.1   Culture Rat Hippocampal Primary Neurons

1. One day before the experiment, place glass coverslips in 24-well plates, one coverslip in each well. Prepare around 10–12 coverslips per pup.

2. Add ~500 μL of 70% EtOH in each well and incubate for 15–20 min to sterilize the coverslips.

3. Rinse each well with ddH$_2$O 3–5 times.

4. Add ~250 μL of 0.5 mg/mL poly-D-lysine in each well, seal the culture plate, and incubate overnight at room temperature.

5. On the day of experiment, rinse each well with ddH$_2$O 3–5 times.

6. Add ~500 μL of ddH$_2$O in each well and incubate during the dissection.

7. Warm up Saline and Growth media in a water bath at 37 °C.

8. Bring 1–4 postnatal day-old rat pups to the laboratory.

9. Anesthetize the pups in a chamber containing isoflurane (2–4%), and wait until they lose consciousness and fail to respond to tactile stimuli and to pinching of the paws.

10. Decapitate pups and dissect out neonatal brains.

11. Using the standard technique [15], dissect hippocampi from the brains in warm Saline, and collect them in a conical tube filled with 4.5 mL warm Saline (see **Note 9**).

12. Add 500 μL of 2.5% trypsin to the hippocampi (0.25% final trypsin concentration), and incubate for 10 min in a water bath at 37 °C.

13. Thoroughly rinse the tissue with Saline (3 × 10 mL) and triturate with a glass pipette.

14. Reconstitute the dissociated neurons in 1 mL of warm Growth media, and filter through a 40 μm nylon mesh.

15. Count the number of neurons with a hemocytometer.

16. In the meantime, remove ddH$_2$O from coverslips and dry them for a short period.

17. Plate neurons onto coverslips in 24-well plates at 1.0–1.5 × 10^5 cells/well density with 500 μL of Growth media.

18. Incubate the neuron culture at 35 °C in a 5% CO$_2$: 95% air humidified incubator for 2–3 weeks. For the best result, avoid disturbing the culture as much as possible during incubation (see **Note 10**).

3.2   Ca-P Transfection

1. Perform Ca-P transfection as early as 4 days in vitro (DIV) after the neuron culture preparation (see **Note 11**).

2. On the day of experiment, obtain high-quality plasmid DNAs using miniprep or maxiprep commercial kits according to the manufacturer's protocol (*see* **Note 12**).

3. Warm up Transfection media, Washing buffer, and Growth media in a water bath at 37 °C.

4. Replace the culture Growth media with 500 µL of Transfection media in each well. * **Do not discard the original Growth media** (*see* **Note 13**).

5. Make Master Mix Transfection solution (33 µL per each coverslip × number of coverslips). For each coverslip, add 1.65 µL of 2.5 M $CaCl_2$, 0.7 µg of DNA, 16.5 µL of 2× BBS buffer, and add ddH_2O to bring the volume up to 33 µL.

6. Inside the tissue culture hood, prepare Master Mix Transfection solution by first combining 2.5 M $CaCl_2$ and ddH_2O in a tube while slowly agitating.

7. Continue agitating and slowly add DNA into the solution.

8. Add 2× BBS buffer drop-wise while agitating.

9. **Immediately** add 30 µL of Transfection solution to each coverslip of neuron culture inside the culture hood.

10. Rock the culture dish a few times to mix and incubate at 35 °C in a 5% CO_2: 95% air humidified incubator for 45 min–1 h. After neurons are incubated with the Transfection solution, very fine DNA/Ca-P co-precipitates would form a layer covering neurons.

11. Replace the Transfection media with 500 µL of Washing buffer and incubate in the incubator for 15–20 min. The precipitates would disappear after the Wash step.

12. Replace Washing buffer with 500 µL of fresh Growth media.

13. Again, replace the Growth media with the saved original Growth media (*see* **Note 13**).

14. Add the Uaa Cmn (pre-mixed in 50 µL of warm Growth media) to the culture to reach 1 mM final concentration.

15. Incubate the transfected culture in the incubator for 12–48 h before assays.

3.3 Whole Cell Recording with Light Activation from Primary Neurons

1. Prepare Extracellular/Intracellular solutions.

2. Set up the electrophysiology rig.

3. Install an LED by the microscope at the rig to deliver light to the focal point from 1 cm away at a 45 ° angle.

4. Check the power of LED with a light power meter (*see* **Note 14**).

5. Pull patch pipettes from glass electrodes using a commercial micropipette puller to have 3–6 MΩ pipette resistance. Follow

the manufacturer's instruction to set up the micropipette puller (*see* **Note 15**).

6. Take out a coverslip from the neuron culture in the incubator and rinse it by placing it in a 35 mm culture dish filled with fresh Extracellular solution.

7. Put a dab of vacuum grease on the bottom of a new 35 mm culture dish and fill with fresh Extracellular solution (*see* **Note 16**).

8. Move the coverslip to the prepared dish and hold it down onto the grease.

9. Place the coverslip/dish on the electrophysiology microscope platform.

10. Using standard patch clamping techniques, patch a neuron showing mCitrine fluorescence.

11. Record neuronal activity using current clamp (I-clamp) method. First, adjust the resting potential to around −72 mV by injecting a small current. Then, inject a step current (10–200 pA) to induce continuous firing (5–15 Hz) of action potentials.

12. Manually, or using the data acquisition software, flash a pulse (a single pulse of 100 ms–1 s duration) of LED light to the neuron while recording, and see if action potentials are affected.

13. Perfuse 0.5 mM $BaCl_2$ Extracellular solution to the bath and verify if action potentials are recovered (*see* **Note 17**).

3.4 In Utero Electroporation and Uaa Microinjection (Fig. 4)

1. Plasmid DNA preparation: Purify DNA with an endotoxin-free Maxiprep kit, then perform phenol-chloroform extraction followed by ethanol precipitation to prepare highly purified plasmid DNA (condensed to 2–5 μg/μL).

2. Anesthetize a timed pregnant mouse at E14.5 with pentobarbital (50 μg/gram body weight, intraperitoneal injection) or isoflurane (2–4% inhalation).

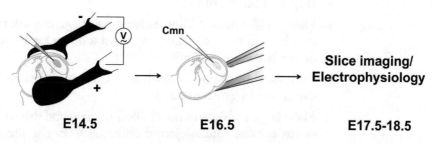

Fig. 4 Experimental procedure for Cmn incorporation in Kir2.1 in the mouse neocortex in vivo. Plasmids shown in Fig. 3 are injected into the mouse neocortex (E14.5) and electroporated in utero. Two days later, Cmn-Ala dipeptide is injected in the ventricle in the electroporated side or both sides of cerebral hemispheres. On E17.5 to E18.5, slice imaging and electrophysiological assay can be performed

3. Make a small incision at the abdominal midline, then pull out the uterine horns with forceps, ring forceps, and fingertips onto a 37 °C pre-warmed PBS-moistened cotton gauze placed around the incision (*see* **Note 18**).

4. Inject ~1 μL of plasmid DNA solution, mixed with 0.1% Fast green for visualization, into the lateral ventricle of each littermate with a glass capillary inserted through the uterine wall.

5. Electroporate the embryos with an electroporator using a tweezer electrode (33–35 V, 50 ms duration, 950 ms interval, 4–8 pulses) (*see* **Note 19**).

6. Return the uterine horns to the abdominal cavity gently with forceps, ring forceps, and finger tips. Suture the muscle wall, then the skin with surgical suture to allow the embryos to continue development in utero.

7. After 2 days, make a small incision at the abdominal midline again at E16.5, and then pull out the electroporated embryos gently with forceps, ring forceps, and fingertips onto a 37 °C pre-warmed PBS-moistened cotton gauze placed around the incision (*see* **Note 18**).

8. Inject 2–5 μL Cmn-Ala (500 mM) to the electroporated side or both sides of the lateral ventricle using a glass capillary inserted through the uterine wall.

9. Return the uterine horns again to the abdominal cavity gently with forceps, ring forceps, and fingertips. Suture the muscle wall, and then the skin with surgical suture to allow the embryos to continue development in utero.

3.5 Whole Cell Recording with Light Activation from Acute Brain Slices

1. From 1 L of ACSF solution, take 200 mL and fast-freeze at −80 °C for 20–30 min. Prechilled ACSF will form slush and be used at **step 13**. Bubble the rest 800 mL of ACSF with 5% CO_2: 95% O_2 gas at room temperature.

2. Sterilize dissection surgery tools to harvest brains from the embryos.

3. Prepare a bucket of ice.

4. Make ~100 mL of 4% low melting point agarose solution in a flask by heating in a microwave. Cool it down for about 5 min before it starts to solidify.

5. Euthanize the mouse 12–24 h after Cmn-Ala injection by CO_2 overdose [14](*see* **Note 20**).

6. Make large incisions at the abdominal area, and dissect out the electroporated/microinjected embryos from the uterus with fine scissors and forceps.

7. Harvest brains from the embryos with fine scissors and forceps.

8. Place each brain on a 10 cm culture dish placed (upside down) on the ice bucket.

9. Divide two hemispheres using a sharp blade, and place each hemisphere on the dish with the midsagittal plane touching the bottom of the dish.

10. Quickly pour the agarose solution over brains (around 500 μL per hemisphere) (*see* **Note 21**).

11. Using a blade, square cut the agarose around a brain to make a brain-embedded agarose block.

12. Using tissue adhesive, glue down the midsagittal plane surface of the agarose block on a mount for vibratome.

13. Place the mount in a vibratome chamber and fill with pre-chilled ACSF.

14. Cut 200 μm sagittal brain slices, and incubate in the Incubation ACSF at 33 °C for 42 min while bubbling with 5% CO_2: 95% O_2 gas [14](*see* **Note 22**).

15. Turn off the heat from the water bath, and continue to incubate the slices at room temp while bubbling. Start whole-cell patch recording at least 15 min after turning off the heater.

16. Prepare Intracellular solution.

17. Set up the slice electrophysiology rig, and superfuse the recording chamber with ACSF at 2 mL/min rate with a perfusion pump. Adjust the chamber temperature to around 33 °C.

18. Install an LED by the microscope at the rig to deliver light to the focal point from 1 cm away at a 45 ° angle.

19. Check the power of LED with a light power meter (*see* **Note 14**).

20. Pull patch pipettes from glass electrodes using a commercial micropipette puller to have 3–6 MΩ pipette resistance. Follow the manufacturer's instruction to set up the micropipette puller (*see* **Note 15**).

21. Carefully pick up a brain slice with a glass pipette and place in the perfusion chamber. Hold down the slice with a harp.

22. Patch a neuron showing mCherry and GFP fluorescence from the neocortical region.

23. Record PIRK activity using voltage clamp (V-clamp) method. First, hold the membrane potential at −60 mV. Then, record currents at the fixed negative membrane potential (−100 mV) or voltage ramps (−100 mV to +40 mV). Specifically, monitor Kir2.1 specific inward currents at −100 mV.

24. Manually, or using the data acquisition software, flash a pulse (a single pulse of 100 ms–1 s duration) of LED light to the neuron while recording, and see if PIRK proteins are activated.

Once PIRK is activated, inward currents at −100 mV would increase significantly.

25. Perfuse 0.5 mM BaCl$_2$ ACSF to the bath and verify if PIRK gets inactivated again (*see* **Note 17**).

---

4   Notes

1. Rat embryos are also routinely used to prepare neuron culture [16].

2. In our hands, Poly-D-Lysine showed much better results than Poly-L-Lysine in terms of the quality of neuron culture. Neurons and some cell types were reported to release proteases which break down Poly-L-Lysine, but not Poly-D-Lysine [17].

3. To make a protein light-inducible, the foremost step is to determine the most suitable amino acid site in the target protein sequence for Uaa incorporation. Structural and functional understanding of the target protein is crucial. Make UAG amber mutation at the target site for Uaa incorporation. Moreover, one can select a few candidate sites and test them in a mammalian cell line such as Human Embryonic Kidney (HEK) cells, to gauge the ease and efficiency of amber suppression as well as light activation at each candidate site [14].

4. There are a number of photoresponsive Uaas with orthogonal tRNA/synthetase available, including Uaas that can be photo-switched reversibly [18–21], so researchers may choose an optimal Uaa for individual purpose of experiments. The orthogonal tRNA for a specific Uaa is transcribed under a special type-3 polymerase III promoter, such as the H1 promoter or U6 promoter, together with a 3′-flanking sequence as described previously [10, 12]. Uaa incorporation efficiency is optimized by varying the copy number of tRNA expression cassette [14, 22].

5. Preparation of 2× BBS buffer is a critical step. Calibrate the optimal pH for 2× BBS buffer between pH 6.90–7.15, for every new DNA construct or prep. The buffer pH could be adjusted with precision to the 0.01 digits if it helps to achieve consistent results.

6. Cmn absorbs long and medium wave UV lights (280–400 nm). To avoid exposure to cytotoxic shortwave UV as well as unnecessary heat from far-red light, an LED with a single-wavelength emission is most suitable for Cmn activation.

7. For in vivo expression, use a strong promoter such as CAG (chicken beta-actin promoter with CMV enhancer). Also, three copies of the tRNA expression cassette increased Cmn incorporation efficiency, as tested previously [14].

8. In order to increase the bioavailability of Cmn, use the dipeptide Cmn-Ala (Cmn-alanine) to deliver Cmn in vivo [23]. Oligopeptide transporter PEPT2 is highly expressed in rodent brains, which helps transport dipeptides into neurons [24]. Inside neurons, the dipeptides are hydrolyzed by cellular peptidases to yield single amino acid, Cmn. Sometimes, it helps to inject Cmn-Ala in the ventricle on both the electroporated and the opposite hemisphere, to increase Cmn bioavailability through diffusion.

9. Removing meningeal tissue as much as possible without damaging the hippocampus is key to achieve healthy neuron culture with minimum microglia population.

10. Successful Uaa incorporation in neurons depends strongly on the preparation of healthy neuron culture. Every step in the protocol should be performed in a precise and meticulous manner. After culture preparation, it is best to maintain the culture in the incubator without disturbance. Opening and closing the incubator door, as well as moving the culture dish in and out of the incubator, can add up to undermining the quality of culture. Lower incubator temperature (35 °C instead of 37 °C) helps reduce excitotoxicity.

11. Younger neurons are transfected at higher efficiency, i.e., Ca-P transfection efficiency goes down as the neuron culture gets older. At the same time, Ca-P transfection of young neurons might not be suitable for some experiments, as they are still immature. Generally, 7 DIV is the earliest time to perform Ca-P transfection to record action potentials.

12. Around 1.9 of the 260/280 ratio is optimal for the purified DNA. When necessary, perform an agarose gel electrophoresis to check purity of the prepped DNA [25]. Supercoiled plasmid DNA with high purity is best for neuronal transfection.

13. It is important to save the original Growth media during Ca-P transfection, and add back to the culture to restore the original condition. Maintaining the neuron culture in fresh media after transfection would interfere with neurons' recovery, since fresh media lack key molecules for cell proliferation such as growth factors released from neurons.

14. Regularly check the LED performance using an optical power meter.

15. To test pipette resistance, first fill a pipette with Intracellular solution and position it on the electrode holder. Dip the pipette in a 35 mm culture dish filled with Extracellular solution, placed on the microscope platform. Immerse a ground electrode into the dish to complete a circuit. Turn on the amplifier/digitizer and start a data acquisition software. Monitor pipette resistance with a membrane test protocol.

16. **Steps 7–9** in Subheading 3.3 can be minimized if a coverslip-compatible chamber is equipped at the microscope.

17. BaCl$_2$ is a Kir2.1-specific inhibitor, and its addition would re-block PIRK after photo-activation. This step verifies the specificity of light-control.

18. Because this procedure includes two major surgeries, it is required to handle the exposed uterus/embryos extremely gently to reduce stress and damage. Keep the embryos warm and moist during the surgery with 37 °C pre-warmed PBS. After each surgery, look after the mouse throughout its recovery.

19. This technique could be applied not only to neocortical neurons but also to neurons in other brain regions such as striatum, diencephalon, and cerebellum, when electroporation/injection site is adjusted for each region.

20. Depending on the target protein's half-life, one can wait longer or shorter after Uaa microinjection before harvesting embryos.

21. Agarose embedding helps stabilize soft embryonic brains for vibratome cutting. Although low melting point agarose minimizes temperature shock to the brain tissue, caution is required when pouring melted agarose over cold brains. The effect of temperature shock to the tissue from agarose embedding was not reported in the previous study [14].

22. Sagittal brain slices work best for recording/imaging neocortical Uaa incorporation. However, coronal or horizontal sections might be more suitable for other brain regions.

Acknowledgment

L.W. acknowledges support from NIH (R01GM118384, RF1MH114079).

References

1. Callaway EM, Katz LC (1993) Photostimulation using caged glutamate reveals functional circuitry in living brain slices. Proc Natl Acad Sci U S A 90:7661–7665

2. Yoshimura Y, Callaway EM (2005) Fine-scale specificity of cortical networks depends on inhibitory cell type and connectivity. Nat Neurosci 8:1552–1559

3. Banghart M, Borges K, Isacoff E, Trauner D, Kramer RH (2004) Light-activated ion channels for remote control of neuronal firing. Nat Neurosci 7:1381–1386

4. Volgraf M, Gorostiza P, Numano R, Kramer RH, Isacoff EY, Trauner D (2006) Allosteric control of an ionotropic glutamate receptor with an optical switch. Nat Chem Biol 2:47–52

5. Boyden ES, Zhang F, Bamberg E, Nagel G, Deisseroth K (2005) Millisecond-timescale, genetically targeted optical control of neural activity. Nat Neurosci 8:1263–1268

6. Fenno L, Yizhar O, Deisseroth K (2011) The development and application of optogenetics. Annu Rev Neurosci 34:389–412

7. Adamantidis A, Arber S, Bains JS, Bamberg E, Bonci A, Buzsaki G, Cardin JA, Costa RM, Dan Y, Goda Y, Graybiel AM, Hausser M, Hegemann P, Huguenard JR, Insel TR, Janak PH, Johnston D, Josselyn SA, Koch C, Kreitzer AC, Luscher C, Malenka RC, Miesenbock G, Nagel G, Roska B, Schnitzer MJ, Shenoy KV, Soltesz I, Sternson SM, Tsien RW, Tsien RY, Turrigiano GG, Tye KM, Wilson RI (2015) Optogenetics: 10 years after ChR2 in neurons--views from the community. Nat Neurosci 18:1202–1212

8. Wang L, Brock A, Herberich B, Schultz PG (2001) Expanding the genetic code of Escherichia coli. Science 292:498–500

9. Wang L, Schultz PG (2005) Expanding the genetic code. Angew Chem Int Ed 44:34–66

10. Wang L (2017) Engineering the genetic code in cells and animals: biological considerations and impacts. Acc Chem Res 50:2767–2775

11. Liu CC, Schultz PG (2010) Adding new chemistries to the genetic code. Annu Rev Biochem 79:413–444

12. Wang W, Takimoto JK, Louie GV, Baiga TJ, Noel JP, Lee KF, Slesinger PA, Wang L (2007) Genetically encoding unnatural amino acids for cellular and neuronal studies. Nat Neurosci 10:1063–1072

13. Shen B, Xiang Z, Miller B, Louie G, Wang W, Noel JP, Gage FH, Wang L (2011) Genetically encoding unnatural amino acids in neural stem cells and optically reporting voltage-sensitive domain changes in differentiated neurons. Stem Cells 29:1231–1240

14. Kang JY, Kawaguchi D, Coin I, Xiang Z, O'Leary DD, Slesinger PA, Wang L (2013) In vivo expression of a light-activatable potassium channel using unnatural amino acids. Neuron 80:358–370

15. Beaudoin GM III, Lee SH, Singh D, Yuan Y, Ng YG, Reichardt LF, Arikkath J (2012) Culturing pyramidal neurons from the early postnatal mouse hippocampus and cortex. Nat Protoc 7:1741–1754

16. Kaech S, Banker G (2006) Culturing hippocampal neurons. Nat Protoc 1:2406–2415

17. Banker G, Goslin K (1998) Culturing Nerve Cells. MIT Press, Cambridge, MA

18. Beene DL, Dougherty DA, Lester HA (2003) Unnatural amino acid mutagenesis in mapping ion channel function. Curr Opin Neurobiol 13:264–270

19. Lemke EA, Summerer D, Geierstanger BH, Brittain SM, Schultz PG (2007) Control of protein phosphorylation with a genetically encoded photocaged amino acid. Nat Chem Biol 3:769–772

20. Hoppmann C, Lacey VK, Louie GV, Wei J, Noel JP, Wang L (2014) Genetically encoding photoswitchable click amino acids in Escherichia coli and mammalian cells. Angew Chem Int Ed Engl 53:3932–3936

21. Hoppmann C, Maslennikov I, Choe S, Wang L (2015) In situ formation of an azo bridge on proteins controllable by visible light. J Am Chem Soc 137:11218–11221

22. Coin I, Katritch V, Sun T, Xiang Z, Siu FY, Beyermann M, Stevens RC, Wang L (2013) Genetically encoded chemical probes in cells reveal the binding path of urocortin-I to CRF class B GPCR. Cell 155:1258–1269

23. Parrish AR, She X, Xiang Z, Coin I, Shen Z, Briggs SP, Dillin A, Wang L (2012) Expanding the genetic code of Caenorhabditis elegans using bacterial aminoacyl-tRNA synthetase/tRNA pairs. ACS Chem Biol 7:1292–1302

24. Lu H, Klaassen C (2006) Tissue distribution and thyroid hormone regulation of Pept1 and Pept2 mRNA in rodents. Peptides 27:850–857

25. Green MR, Sambrook J (2012) Molecular cloning: a laboratory manual, 4th edn. Cold Spring Harbor Laboratory Press, Cold Spring Harbor, NY

Chapter 18

Genetic Code Expansion- and Click Chemistry-Based Site-Specific Protein Labeling for Intracellular DNA-PAINT Imaging

Ivana Nikić-Spiegel

Abstract

Super-resolution microscopy allows imaging of cellular structures at nanometer resolution. This comes with a demand for small labels which can be attached directly to the structures of interest. In the context of protein labeling, one way to achieve this is by using genetic code expansion (GCE) and click chemistry. With GCE, small labeling handles in the form of noncanonical amino acids (ncAAs) are site-specifically introduced into a target protein. In a subsequent step, these amino acids can be directly labeled with small organic dyes by click chemistry reactions. Click chemistry labeling can also be combined with other methods, such as DNA-PAINT in which a "clickable" oligonucleotide is first attached to the ncAA-bearing target protein and then labeled with complementary fluorescent oligonucleotides. This protocol will cover both aspects: I describe (1) how to encode ncAAs and perform intracellular click chemistry-based labeling with an improved GCE system for eukaryotic cells and (2) how to combine click chemistry-based labeling with DNA-PAINT super-resolution imaging. As an example, I show click-PAINT imaging of vimentin and low-abundance nuclear protein, nucleoporin 153.

Key words Noncanonical amino acid, Genetic code expansion, Amber suppression, DNA-PAINT, Click chemistry, Super-resolution microscopy, Intracellular protein labeling, Click-PAINT

1 Introduction

Super-resolution microscopy (SRM) techniques have introduced a nanometer resolution to light microscopy [1–7]. Thanks to rapid development of instrumentation and software, it is now possible to reach molecular level of detail with optical microscopy in fixed and living cells [8–11]. However, to fully utilize the potential of SRM and expand its scope of biological applications, further improvements are needed [12–15]. One of them is a demand for optimized labeling techniques [16–19]. To take advantage of improved "super" resolution, labels should be as small as possible. As the resolution is approaching sub-10 nm range, any label that is similar or larger than it will affect how faithfully the label represents the structure of inter-

Edward A. Lemke (ed.), *Noncanonical Amino Acids: Methods and Protocols*, Methods in Molecular Biology, vol. 1728, https://doi.org/10.1007/978-1-4939-7574-7_18, © Springer Science+Business Media, LLC 2018

est. Furthermore, the label should be attached as close as possible to the target structure in a minimally invasive way. In addition, the optimal label needs to be compatible with high density labeling which is necessary for optimal resolution. Due to their small size and better photophysical properties certain organic dyes are thus preferred over autofluorescent proteins (AFP) for certain applications. However, their direct attachment to target proteins remains challenging. One rapidly evolving method to achieve this is to combine genetic code expansion (GCE) and click chemistry [20–22].

GCE modifies host cells to allow site-specific introduction of specially designed noncanonical amino acids (ncAA) in target proteins. This can be achieved by adding an ncAA-specific orthogonal transfer RNA/aminoacyl tRNA synthetase (tRNA/RS) pair to the host cell. To be orthogonal, the introduced tRNA/RS pair must not cross-react with any of the host tRNA/RS pairs and the introduced tRNA needs to recognize a unique codon. One of the most widely used and versatile pairs in this regard is the archaeal amber (TAG) stop codon suppressor pair for pyrrolysine (Pyl) from the *Methanosarcinaceae* family—*M. mazei* and *M. barkeri* [23, 24]. The PylRS/tRNA^{Pyl} pair incorporates ncAAs at an amber codon site which can be introduced at any position in a target protein gene by site-directed mutagenesis. PylRS and its AF mutant with a modified (enlarged) amino acid binding pocket [25] can be used to charge tRNA^{Pyl} with a plethora of ncAAs with various biorthogonal moieties in single cells and whole organisms [22, 26]. Very useful for fluorescent labeling are ncAAs carrying strained alkynes and alkenes which can be used to attach dyes directly to proteins by click reactions [27–30]. Click reactions with tetrazines and azides, strain-promoted inverse electron-demand Diels-Alder cycloaddition (SPIEDAC), and strain-promoted alkyne-azide cycloaddition (SPAAC), respectively, are not toxic and are fully biorthogonal. Due to their ultrafast speed, SPIEDAC reactions are preferred candidates for fast labeling of living eukaryotic cells and they were used for protein labeling for SRM [31]. That was a first report on the use of click-chemistry for protein labeling for SRM, but it was limited to the cell membrane. In the meantime, intracellular labeling of vimentin and actin cytoskeletal proteins was also achieved [32] but more advanced SRM applications are hampered by the limitations of the method. One of the limitations is low yield which was recently significantly enhanced by targeting *M. mazei* PylRS^{AF}/tRNA pair to the cytoplasm [33]. In that study the authors found out that PylRS contains a putative nuclear localization signal (NLS) which trapped it in the nucleus, away from the rest of translation machinery. Targeting the PylRS to the cytoplasm by adding a nuclear export signal (NES) to it did not only increase the yield, but also helped to reduce nonspecific nuclear/nucleolar labeling [33]. Another problem, especially important for microscopy is background signal coming from the nonspecific dye labeling, which is the problem inherent to all fluorescent dyes-based labeling techniques [34]. As recently reported, one way to overcome this is to combine improved GCE systems for site-specific click labeling with DNA-PAINT [33]. DNA-PAINT recently emerged as a powerful technique for SRM [35]. It is a variation of the PAINT (point accumulation for imaging in nanoscale topography) technique [36] based on the continuous exchange of fluorescent oligonucleotides on a target structure. A small docking strand is attached to the target protein while complementary imaging strands conjugated with a dye transiently bind to it (Fig. 1a). Interactions of imaging and docking strands are recorded as fluorescent blinking spots and used for image reconstruction similar to other single molecule localization techniques. However, in contrast to dye

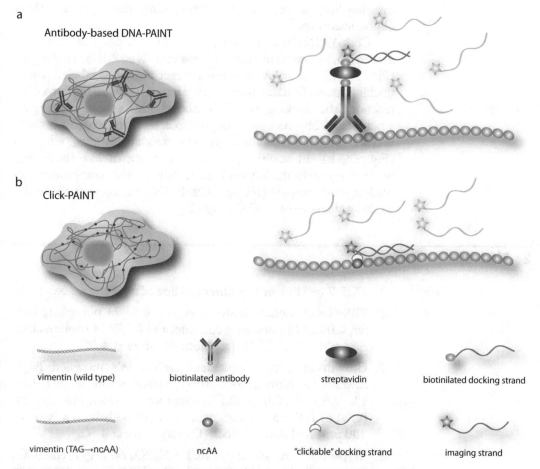

Fig. 1 Comparison of antibody-based DNA-PAINT and click-PAINT: (**a**) immunocytochemistry with antibody-docking strand conjugates is used to label target protein. In published approach [37], biotinilated primary antibody was conjugated to biotinilated docking strand by streptavidin bridge; (**b**) genetic code expansion is used to introduce ncAA in target protein. Clickable docking strand is then attached to ncAA by click chemistry. In both panels **a** and **b** imaging strands continuously and transiently bind docking strands

photoswitching-based methods, which are limited by the dye properties, provided that the pool of imaging strands is large enough, DNA-PAINT can in principle be bleaching-free. Furthermore, conjugating dyes to DNA oligonucleotides makes most of them more hydrophilic and thus biocompatible. The initial DNA-PAINT was limited to DNA imaging [35]. It was later expanded to proteins for which an antibody-based immunocytochemistry type of labeling was used to attach docking strands to target structures [37] (Fig. 1a). However, antibody-based labeling can be a problem for SRM due to a relatively large antibody size (10–15 nm). Nikić et al. recently developed an approach to overcome this issue by incorporating a docking strand directly to the protein of interest by the click-PAINT method [33] (Fig. 1b). This approach combines the best properties of its parent techniques: GCE and click chemistry bring site-specific labeling precision, while background dye labeling is reduced by conjugating the dyes to DNA oligonucleotides.

Click-PAINT is a two-step process: (1) a target protein is engineered to incorporate ncAA by using a NESPylRS/tRNA pair (Fig. 2a); (2) an ncAA-containing target protein is labeled with a docking strand functionalized with tetrazine (Fig. 2b). After a click reaction, the docking strand is labeled with a pool of imaging strands (Fig. 2b). As a proof of principle, this protocol shows click-PAINT imaging of abundant cytoskeletal protein, vimentin (Fig. 3a, b). In addition, it shows imaging of low-abundance nuclear protein, nucleoporin 153, one of the components of nuclear pore complex (NPC). Click-PAINT has enough resolution to show characteristic NPC rings (Fig. 3c).

2 Materials

1. COS-7 cell line or any other cell line of interest (*see* **Note 1**).

2. PBS (1× phosphate-buffered saline): 0.01 M phosphate buffer, 0.0027 M potassium chloride, and 0.137 M sodium chloride, pH 7.4, at 25 °C (*see* **Note 2**). Store at 4 °C.

3. Growth medium: supplement Dulbeccos's modified Eagle medium (*see* **Note 3**) with 10% (v/v) fetal bovine serum (FBS), 1% (v/v) of 100× penicillin-streptomycin stock (Gibco), 1% (v/v) of 100 mM sodium pyruvate stock (Gibco), 1% (v/v) of 200 mM L-glutamine stock (Gibco). Store at 4 °C.

4. Trypsin-EDTA: 400 mg/l KCl, KH_2PO_4 60 mg/l, $NaHCO_3$ 350 mg/l, NaCl 8000 mg/l, Na_2HPO^4-$7H_2O$ 90 mg/l, D-glucose (dextrose) 1000 mg/l, EDTA 4Na $2H_2O$ 10 mg/l, Trypsin 500 mg/l (*see* **Note 4**). Trypsin-EDTA is stored at −20 °C, but to avoid freezing-thawing I keep the bottle which is used on a daily basis at 4 °C.

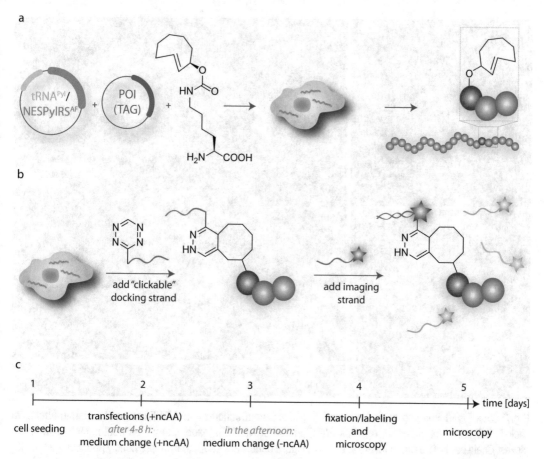

Fig. 2 Outline of the click-PAINT procedure: (**a**) host cell is transfected with plasmids for NESPylRS^AF/tRNA^Pyl and amber (TAG) mutant of protein of interest (POI). NcAA is added to the medium after transfections; (**b**) cells are fixed and clickable docking strand is added to the cells. After click labeling imaging strand is added to the cells; (**c**) timeline of the labeling/imaging procedure. Panels **a** and **b** are adapted from previously published work [33]

5. Hemocytometer.

6. Cell culture incubator at 37 °C with a humidified 5% CO_2 atmosphere.

7. Cell culture hood.

8. Water bath at 37 °C.

9. Sterile plastic ware (Falcon and microcentrifuge tubes, 100 mm cell culture dishes, serological pipettes).

10. Lab-Tek™ II Chambered Coverglass with 4-wells (Nunc™, *see* **Note 5**).

11. JetPrime transfection reagent (Polyplus, *see* **Note 6**).

12. A set of recombinant DNA for amber suppression—eukaryotic expression vectors with *M. mazei* NESPylRS^AF/tRNA^Pyl is

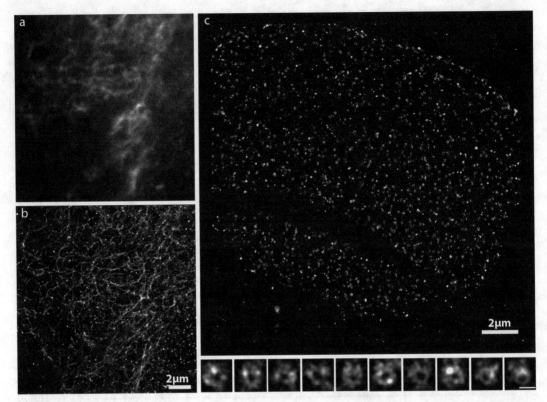

Fig. 3 Click-PAINT imaging of vimentin and Nup153: (**a**) diffraction-limited TIRF image of vimentin fibers; (**b**) click-PAINT image of vimentin; (**c**) click-PAINT image of Nup153. Insets show individual nuclear pore complexes. Scale bar is 100 nm in insets in panel **c**. Figure panels are adapted from previously published work [33]

available from Dr. Edward Lemke, EMBL, Heidelberg (*see* **Note 7**).

13. Eukaryotic expression vector for target protein amber mutant. In this manuscript I used pVimentin[N116TAG]-PSmOrange and pEGFP[N149TAG]-Nup153 vectors from previously published work [33]. Or alternatively, the expression vector of wild-type target protein and site-directed mutagenesis kit.

14. EndoFree plasmid Maxi kit (e.g., from Qiagen) (*see* **Note 8**).

15. NcAA stock: 100 mM trans-Cyclooct-2-ene-L-Lysine axial isomer (TCO*A; available from Sirius Fine Chemicals), 15% (vol/vol) DMSO, 0.2 M NaOH (*see* **Note 9**). Store at −20 °C.

16. Sterile 1 M HEPES. Store at 4 °C.

17. Vortex.

18. Docking strand (5′-ttatacatcta-3′) functionalized with 1,2,4,5-tetrazine at 5′ end (available from Biomers.net, *see* **Note 10**). Prepare 500 μM stock in distilled H_2O. Store aliquots at −20 °C.

19. Imaging strand complementary to the docking strand (5′-cta-gatgtat-3′) and functionalized with Atto 655 at the 3′ end (available from Biomers.net). Prepare 250 µM stock in distilled H$_2$O and make small aliquots (*see* **Note 11**). Store at −20 °C.

20. 2% PFA: 2% paraformaldehyde in PBS (*see* **Note 12**).

21. 0.1% Triton in PBS.

22. Imaging buffer: 1×PBS, 500 mM NaCl, pH 8.

23. GATTA-PAINT nanoruler (available from GATTAquant, optional, *see* **Note 13**).

24. TIRF-microscope (e.g., Leica GSDIM) with appropriate laser and filter cube for the imaging strand (*see* **Note 14**) and software for data analysis (*see* **Note 15**).

3 Methods

Carry out all the procedures at room temperature unless specified otherwise. Perform cell seeding and transfections under aseptic conditions in a cell culture hood. Please note that the steps after cell fixation do not require aseptic conditions.

3.1 Mutagenesis

1. If an amber mutant of the target protein is not available, perform a site-directed mutagenesis of its wild-type version to mutate the desired position to the TAG (amber) codon (*see* **Note 16**). This can be done by one of the commercial kits, e.g., a QuickChange site-directed mutagenesis kit (from Agilent). In this manuscript, vimentin-PSmOrange protein fusion carrying N116TAG mutation in Vimentin and GFP-Nup153 fusion carrying N149TAG mutation in GFP were used.

2. Verify mutagenesis results by sequencing. If correct, prepare high-quality DNA prep of a desired amber mutant (e.g., by Endofree Maxi Prep kit).

3.2 Cell Seeding

1. Warm up PBS and growth medium in a water bath at 37 °C.

2. Warm up trypsin-EDTA to room temperature.

3. Take cells out of a cell culture incubator.

4. Aspirate off growth medium.

5. Rinse with 5–10 ml of PBS and aspirate off PBS.

6. Add 2 ml of trypsin-EDTA to one 100 mm cell culture plate.

7. Put the plate back to the cell culture incubator for 3–5 min (*see* **Note 17**).

8. Check if the cells are detached from the plate surface. If yes, inactivate trypsin-EDTA with 8 ml of growth medium.

9. Pipette up and down.

10. Transfer the cell suspension to a 15 ml conical bottom tube.

11. Count the number of cells with a hemocytometer.

12. Calculate the number of cells necessary for seeding. For one well of Lab-Tek I seed 55,000 cells in a 500 μl growth medium (*see* **Note 18**). Multiply this number with the number of wells that you want to seed. For example, if you want to seed four wells in total, you will need 220,000 cells in 2 ml of growth medium. Take cell suspension in a volume corresponding to the desired number of cells and add the growth medium up to a final volume. Seed 500 μl of this mixture per one well of the Lab-Tek.

13. Incubate the cells in the cell culture incubator overnight.

3.3 Transfections

1. On the following day (15–20 h after the seeding), proceed with transfections (*see* Fig. 2c for a time course).

2. Prepare the transfection mix according to the JetPrime instructions. For one well of the Lab-Tek, add 0.3 μg of plasmid coding for NESPylRS^{AF}/tRNA^{Pyl} and 0.3 μg of plasmid coding for vimentin amber mutant to 50 μl of JetPrime buffer in a microcentrifuge tube. For Nup153 use 0.5 μg of NESPylRS^{AF}/tRNA^{Pyl} plasmid and 0.5 μg of Nup153 amber mutant (*see* **Note 19**). Please note that you can prepare a master mix by multiplying this amount with the number of wells that you want to transfect.

3. Vortex the tube for 10 s at maximum speed and then briefly spin it down at $2680 \times g$ by using a microcentrifuge.

4. Add JetPrime reagent to the tube. I use 1:2 DNA to JetPrime ratio (w/v), which is also recommended by the manufacturer. For 1 μg of the total DNA, add 2 μl of JetPrime reagent to the tube.

5. Vortex the tube for 10 s at maximum speed and then briefly spin it down at $2680 \times g$ using a microcentrifuge.

6. Incubate the transfection mix for 10 min.

7. After the incubation time is over, take the Lab-Tek with cells out of the cell culture incubator.

8. Add 50 μl of the transfection mix dropwise to the cells in one well.

9. Gently rock the Lab-Tek back and forth and from side to side.

10. Prepare an ncAA working solution. For one well of the Lab-Tek, mix 1.25 μl of ncAA stock and 3.75 μl of 1 M HEPES in a tube (*see* **Note 20**). Vortex the tube at maximum speed and

then spin down briefly at $2680 \times g$ using a microcentrifuge. Add 5 µl of the ncAA working solution per well (Fig. 2c).

11. Gently rock the Lab-Tek back and forth and from side to side.

12. Incubate at 37 °C in a cell culture incubator for 4–8 h.

13. After 4–8 h aspirate off the growth medium (*see* **Note 21**).

14. Add fresh pre-warmed (37 °C) growth medium to the Lab-Tek (500 µl per well).

15. Again prepare the ncAA working solution as in **step 10** of this section. Add 5 µl of the ncAA working solution per well and incubate the cells for an additional 24 h in the cell culture incubator (*see* **Note 22**, Fig. 2c).

16. After 24 h (e.g., the following afternoon) aspirate off the growth medium and add a fresh pre-warmed medium to the Lab-Tek. Incubate the cells in fresh medium without ncAA overnight in the cell culture incubator (*see* **Note 23**, Fig. 2c).

3.4 Click Chemistry-Based Labeling

1. Take the cells out of the cell culture incubator.

2. Aspirate off the medium.

3. Rinse with PBS 2–3 times.

4. Add 500 µl of 2% PFA per well and incubate for 10 min.

5. Aspirate off PFA.

6. Rinse with PBS 2–3 times and aspirate off the PBS.

7. Add 500 µl of 0.1% Triton per well and incubate for 15 min. During incubation prepare a 15 µM docking strand working solution (add 9 µl of docking strand stock to 300 µl of PBS; *see* **Note 24**).

8. Aspirate off 0.1% Triton.

9. Rinse with PBS 2–3 times and aspirate off PBS.

10. Add 300 µl of docking strand working solution per well.

11. Incubate for 10 min at 37 °C (*see* **Note 25**).

12. Aspirate off docking strand and rinse with PBS 2–3 times.

13. Proceed with imaging or put the cells in the fridge (*see* **Note 26**).

3.5 Imaging and Data Processing

1. Prepare an 800 pM imaging strand working solution. Take the 250 µM stock and dilute it 1:1000 in dH_2O to make an intermediate 250 nM stock solution (*see* **Note 27**). Add 0.96 µl of the intermediate 250 nM stock solution to 300 µl of imaging buffer.

2. Add 300 µl of the imaging strand working solution per well (*see* **Note 28**).

3. Take the cells to the microscope.

4. Use mOrange and GFP excitation/emission filters to identify transfected vimentin and Nup153 cells, respectively (*see* **Note 29**). For optimal results, search for bright cells showing high expression levels and characteristic expression pattern of your target proteins (for vimentin you expect to see cytoplasmic cytoskeletal network and for Nup153 you expect to see dots in the nucleus).

5. Change to the laser of the imaging strand dye. Select the appropriate filter set. Set exposure to 100 ms (*see* **Note 30**) and switch on the live mode. Adjust the TIRF illumination angle (*see* **Note 31**).

6. Check the density of blinking spots. If too high or too low, you will need to go to **step 1** in this section and adjust the concentration of imaging strand (*see* **Note 32**). Acquire 10,000–50,000 frames. The optimal number of frames depends on the sample quality. With longer imaging you will usually get better signal-to-noise ratio.

7. Do further image processing in appropriate software (*see* **Note 15**). If necessary, after acquisition and before processing, export the movies as tiff files in ImageJ.

8. To identify the spots, apply a threshold based on the maximum likelihood ratio.

9. To localize the spots, perform fitting with a symmetrical 2D Gaussian function.

10. Consolidate identical emitters into a single intensity-weighed localization (*see* **Note 33**).

11. Reconstruct a final super-resolution from binning all the detected events and convolving the resulting image with a Gaussian width according to the resolution determined by the Fourier ring correlation [38]-2σ criterion for Nup153 and 0.143 criterion for vimentin.

4   Notes

1. Click-PAINT was established with COS-7 cells but in principle any other cell line can be used. Important to keep in mind is the efficiency of the amber codon suppression which differs between different cell lines. With this system I also get good suppression efficiency in HEK293T, HeLa, and U2OSf cells.

2. To prepare PBS, I use phosphate-buffered saline tablets from Sigma.

3. I use DMEM with high glucose (4.5 g/l) and phenol red (15 mg/l) from Gibco but other manufactures offer a similar formulation.

4. I use 0.05% trypsin-EDTA from Gibco but similar products are offered by other manufacturers.

5. I used glass-bottom chambers from Nunc. Other products, including a simple coverglass without a chamber, can be used but keep in mind appropriate glass thickness. In most cases coverglass thicknesses of #1 or #1.5 will be suitable.

6. I use JetPrime which gives us very good amber suppression efficiency in COS-7, HEK293T, U2OSf cell lines. But in principle any other transfection reagent can be used. For cells that are difficult to transfect, alternative methods such as viral transduction, electroporation, etc. should be considered.

7. Expression vectors for PylRS^{AF}/tRNA^{Pyl} are available upon request from the Lemke lab (EMBL, Heidelberg). There are two versions. The conventional PylRS^{AF} system that was also used in the first click chemistry-based protein labeling SRM study [31] and the newer version NESPylRS^{AF} [33] which contains N-terminal nuclear export sequence (NES). As published recently and discussed in the introduction, it was noticed that the conventional PylRS^{AF} contains a putative nuclear localization sequence (NLS) and that as a result of it, PylRS^{AF}/tRNA^{Pyl} get trapped in the nucleus. Addition of NES to PylRS^{AF} releases an NESPylRS^{AF}/tRNA^{Pyl} pair in the cytoplasm where protein translation takes place. This reduced some of the nonspecific nuclear labeling which is particularly problematic for imaging of nuclear proteins, such as nucleoporins. Furthermore, with NESPylRS^{AF}/tRNA^{Pyl} an increase in amber suppression efficiency was noticed which is important for any study relying on GCE. Thus, I would recommend its use and I used it in this protocol. Please note that if you have other already existing expression vectors for PylRS^{AF}/tRNA^{Pyl} you can also consider adding NES directly to its N-terminus by insertional mutagenesis (e.g., overlap PCR mutagenesis). This will most likely increase the yield, however, after mutagenesis you need to make sure that the expression vector is still functional.

8. Endotoxins in purified plasmid DNA can decrease transfection efficiency. That is why the DNA needs to be of the highest quality for amber suppression experiments. Check levels of endotoxins (use endotoxin-free purification kits), use high concentrated DNA preparations, etc.

9. In this protocol I used TCO*A, an axial isomer of an ncAA containing strained alkene (trans-cyclooct-2-ene), but another "clickable" ncAAs, e.g., BCN (Bicyclo [6.1.0] nonyne-Lysine, available from SiChem) or SCO (Cyclooctyne-Lysine, available from SiChem), can be used too. For an overview of commercially available ncAAs and their reaction rates, please see

previously published work [39]. Even though it has been reported that decaging (beta elimination of the bound group and release of a lysine) can be a problem for trans-cyclooct-2-ene-L-Lysine (TCO*), it was shown that due to its stability TCO* (especially its axial isomer) is a better choice for labeling. 10–30% of a dye loss due to potential uncaging of TCO* is probably compensated for by its higher labeling and incorporation efficiency [40].

10. I have used an oligonucleotide sequence based on published work [37], but this can be adapted to your needs. Different oligonucleotides will have different binding on-off rates, so please keep that in mind during imaging.

11. Imaging strand needs to be complementary to the docking strand. I have used Atto655 but any other fluorophore can be used. You just need to order an appropriately modified imaging strand. After preparing the stock of imaging strand, make smaller aliquots. I noticed some problems with imaging strand stability (also *see* **Note 27**).

12. PFA is a toxic reagent. Wear protective gear while handling. Avoid contact with skin and eyes. Avoid inhalation, especially during preparation. I usually make a larger amount (1–2 l) and freeze small aliquots at −20 °C.

13. This is an optional control. Control DNA-PAINT nanoruler probes are commercially available (from GATTAquant) and can be used to test your microscope setup and imaging settings. In addition, custom-made DNA-PAINT nanorulers can be made to match your docking/imaging strand sequence (also from GATTAquant).

14. I used a commercial Leica GSDIM microscope but any other TIRF microscope with appropriate lasers, cameras, and filter cubes can be used. Generally speaking, any localization-based microscope that would be suitable for PAINT would work for click-PAINT too.

15. I used Localizer package for IgorPro but any other software for localization-based microscopy, such as Leica's GSDIM tools and various ImageJ plugins, can be used. Please also note that developers of original DNA-PAINT offer DNA-PAINT specific software [37, 41, 42].

16. Please note that different amber positions can lead to different expression levels. This is related to the local sequence context of the introduced amber codon [43] but this process is poorly understood in eukaryotic cells. In addition, in eukaryotic cells mRNAs containing premature amber codon can be subjected to degradation by nonsense-mediated decay [44], a process that is also poorly understood. That is why it is important to pick and test several random positions. Furthermore, for the

labeling step, it is important that the introduced amber position is at the surface and accessible for the labeling. If the structure of the protein is not known, it helps again to pick and test several random positions. It also helps to have an expression plasmid with C-terminal AFP fusion while optimizing this step [39]. In such a case, AFP will only be expressed if the Amber codon is successfully suppressed with ncAA. Last but not the least, please note that the stop codon used for the termination of the target protein sequence must not be amber (TAG). If it is, it needs to be mutated to TAA or TGA.

17. Depending on the cell line, incubation time needs to be adjusted (e.g., HEK293T cells require only 1 min). Furthermore, be careful not to leave the cells too long with trypsin-EDTA as it can be toxic.

18. The number of cells depends on the cell line, seeding surface as well as transfection reagent. Transfection reagent used in this protocol requires a confluency of 60–80% on the day of transfection.

19. For a large range of low-abundance proteins I usually use a total of 1 µg of DNA in a 1:1 ratio. For vimentin, I noticed very high expression levels and less DNA was enough (total of 0.6 µg). Please note that this step might need to be optimized for each protein.

20. Addition of HEPES is used to keep the pH stable upon addition of ncAA stock which contains NaOH. Depending on the stock solution composition, this step might not be necessary. Also, please note that for more wells, a master mix can be made, but it should always be done fresh before adding it to the cells.

21. It is advised by JetPrime instructions to change the medium. It might not be necessary and should be determined for each protein. If you notice cytotoxicity you should replace the medium.

22. Depending on the target protein this can be longer or shorter (e.g., for insulin receptor from our previous study [31] even with only 4–8 h of incubation with ncAA we got high expression levels).

23. This step helps to reduce the background as well. It may not be necessary for every protein or with more hydrophilic ncAA.

24. The concentration of the docking strand can be adapted. For example, I also saw vimentin labeling with 5 µM docking strand solution but for Nup153 because of low number of proteins in the NPC (only 32 copies of Nup153 [45] have recently been counted) a higher concentration of docking strand (15 µM) was required.

25. Please note that at this stage you should not put fixed cells to your sterile cell culture incubator. Instead you should use another incubator or warm room. Alternatively, labeling can be performed at RT. Duration of labeling can also be optimized.

26. At this step, the cells can be stored in the fridge for up to 3 days.

27. Use fresh 250 µM aliquot to prepare an intermediate 250 nM stock. Use only fresh 250 nM stock (keep it in the fridge for only few days) to make an 800 pM imaging strand solution prior to the imaging experiment.

28. For my application 800 pM worked well, but this might need to be adapted to meet the needs of your experiments. If you see too little or too many binding events, consider changing the concentration of imaging strand (also *see* **Note 32**).

29. Please note that with pVimentin-PSmOrange construct, I sometimes noticed a photoconversion from the orange to the red channel, so please keep that in mind and do not expose it to high levels of blue/green light. If necessary, shortly pre-bleach in the red channel before acquiring.

30. The acquisition rate is 10 Hz, as described in one of the original papers [37].

31. TIRF or HILO can be used and it needs to be adapted to the needs of your experiment. Illumination angle needs to be adapted in a way to maximize surface illumination and suppress transmission illumination.

32. As mentioned above (**Note 28**), you might need to optimize the density of binding events by changing the imaging strand concentration. In case of low labeling efficiency, you might need to adjust docking strand concentration as well (*see* **Note 24**).

33. In click-PAINT, sporadic long-lasting associations of docking and imaging strands were observed which led to repetitive localization in sequential frames. In order to correct this, identical emitters (falling within one standard deviation of the spot fit) were consolidated into a single intensity-weighed localization.

Acknowledgment

I would like to thank Ivana Milošević for her valuable help with the manuscript proofreading. I would like to thank all the members of the Lemke group for their help and support. I would also like to thank EMBL's Advanced Light Microscopy Facility, as well as my postdoctoral funding (EMBO Long-Term and Marie Curie IEF fellowships). I am currently supported by the Emmy-Noether programme of the Deutsche Forschungsgemeinschaft (DFG) and the Werner Reichardt Centre for Integrative Neuroscience (CIN) at the Eberhard Karls University of Tuebingen. The CIN is an Excellence Cluster funded by the DFG within the framework of the Excellence Initiative (EXC 307).

References

1. Betzig E, Patterson GH, Sougrat R, Lindwasser OW, Olenych S, Bonifacino JS, Davidson MW, Lippincott-Schwartz J, Hess HF (2006) Imaging intracellular fluorescent proteins at nanometer resolution. Science 313(5793):1642–1645. https://doi.org/10.1126/science.1127344

2. Hell SW, Wichmann J (1994) Breaking the diffraction resolution limit by stimulated emission: stimulated-emission-depletion fluorescence microscopy. Opt Lett 19(11):780–782

3. Hell SW (2007) Far-field optical nanoscopy. Science 316(5828):1153–1158. https://doi.org/10.1126/science.1137395

4. Rust MJ, Bates M, Zhuang X (2006) Sub-diffraction-limit imaging by stochastic optical reconstruction microscopy (STORM). Nat Methods 3(10):793–795. https://doi.org/10.1038/nmeth929

5. Hess ST, Girirajan TP, Mason MD (2006) Ultra-high resolution imaging by fluorescence photoactivation localization microscopy. Biophys J 91(11):4258–4272. https://doi.org/10.1529/biophysj.106.091116

6. Heilemann M, van de Linde S, Schuttpelz M, Kasper R, Seefeldt B, Mukherjee A, Tinnefeld P, Sauer M (2008) Subdiffraction-resolution fluorescence imaging with conventional fluorescent probes. Angew Chem 47(33):6172–6176. https://doi.org/10.1002/anie.200802376

7. Gustafsson MG (2005) Nonlinear structured-illumination microscopy: widefield fluorescence imaging with theoretically unlimited resolution. Proc Natl Acad Sci U S A 102(37):13081–13086. https://doi.org/10.1073/pnas.0406877102

8. Huang B, Babcock H, Zhuang X (2010) Breaking the diffraction barrier: super-resolution imaging of cells. Cell 143(7):1047–1058. https://doi.org/10.1016/j.cell.2010.12.002

9. Sahl SJ, Moerner WE (2013) Super-resolution fluorescence imaging with single molecules. Curr Opin Struct Biol 23(5):778–787. https://doi.org/10.1016/j.sbi.2013.07.010

10. Liu Z, Lavis LD, Betzig E (2015) Imaging live-cell dynamics and structure at the single-molecule level. Mol Cell 58(4):644–659. https://doi.org/10.1016/j.molcel.2015.02.033

11. Sydor AM, Czymmek KJ, Puchner EM, Mennella V (2015) Super-resolution microscopy: from single molecules to supramolecular assemblies. Trends Cell Biol 25(12):730–748. https://doi.org/10.1016/j.tcb.2015.10.004

12. Lakadamyali M (2014) Super-resolution microscopy: going live and going fast. Chemphyschem 15(4):630–636. https://doi.org/10.1002/cphc.201300720

13. Schermelleh L, Heintzmann R, Leonhardt H (2010) A guide to super-resolution fluorescence microscopy. J Cell Biol 190(2):165–175. https://doi.org/10.1083/jcb.201002018

14. Dempsey GT (2013) A user's guide to localization-based super-resolution fluorescence imaging. Methods Cell Biol 114:561–592. https://doi.org/10.1016/B978-0-12-407761-4.00024-5

15. Deschout H, Cella Zanacchi F, Mlodzianoski M, Diaspro A, Bewersdorf J, Hess ST, Braeckmans K (2014) Precisely and accurately localizing single emitters in fluorescence microscopy. Nat Methods 11(3):253–266. https://doi.org/10.1038/nmeth.2843

16. Bachmann M, Fiederling F, Bastmeyer M (2016) Practical limitations of superresolution imaging due to conventional sample preparation revealed by a direct comparison of CLSM, SIM and dSTORM. J Microsc 262(3):306–315. https://doi.org/10.1111/jmi.12365

17. Fernandez-Suarez M, Ting AY (2008) Fluorescent probes for super-resolution imaging in living cells. Nat Rev Mol Cell Biol 9(12):929–943. https://doi.org/10.1038/nrm2531

18. van de Linde S, Heilemann M, Sauer M (2012) Live-cell super-resolution imaging with synthetic fluorophores. Annu Rev Phys Chem 63:519–540. https://doi.org/10.1146/annurev-physchem-032811-112012

19. van de Linde S, Aufmkolk S, Franke C, Holm T, Klein T, Loschberger A, Proppert S, Wolter S, Sauer M (2013) Investigating cellular structures at the nanoscale with organic fluorophores. Chem Biol 20(1):8–18. https://doi.org/10.1016/j.chembiol.2012.11.004

20. Lang K, Chin JW (2014) Cellular incorporation of unnatural amino acids and bioorthogonal labeling of proteins. Chem Rev 114(9):4764–4806. https://doi.org/10.1021/cr400355w

21. Liu CC, Schultz PG (2010) Adding new chemistries to the genetic code. Annu Rev Biochem 79:413–444. https://doi.org/10.1146/annurev.biochem.052308.105824

22. Lemke EA (2014) The exploding genetic code. Chembiochem 15(12):1691–1694. https://doi.org/10.1002/cbic.201402362

23. Fekner T, Chan MK (2011) The pyrrolysine translational machinery as a genetic-code expansion tool. Curr Opin Chem Biol 15(3):387–391. https://doi.org/10.1016/j.cbpa.2011.03.007

24. Wan W, Tharp JM, Liu WR (2014) Pyrrolysyl-tRNA synthetase: an ordinary enzyme but an outstanding genetic code expansion tool. Biochim Biophys Acta 1844(6):1059–1070. https://doi.org/10.1016/j.bbapap.2014.03.002

25. Yanagisawa T, Ishii R, Fukunaga R, Kobayashi T, Sakamoto K, Yokoyama S (2008) Multistep engineering of pyrrolysyl-tRNA synthetase to genetically encode N(epsilon)-(o-azidobenzyloxycarbonyl) lysine for site-specific protein modification. Chem Biol 15(11):1187–1197. https://doi.org/10.1016/j.chembiol.2008.10.004

26. Chin JW (2014) Expanding and reprogramming the genetic code of cells and animals. Annu Rev Biochem 83:379–408. https://doi.org/10.1146/annurev-biochem-060713-035737

27. Kaya E, Vrabel M, Deiml C, Prill S, Fluxa VS, Carell T (2012) A genetically encoded norbornene amino acid for the mild and selective modification of proteins in a copper-free click reaction. Angew Chem 51(18):4466–4469. https://doi.org/10.1002/anie.201109252

28. Lang K, Davis L, Torres-Kolbus J, Chou C, Deiters A, Chin JW (2012) Genetically encoded norbornene directs site-specific cellular protein labelling via a rapid bioorthogonal reaction. Nat Chem 4(4):298–304. https://doi.org/10.1038/nchem.1250

29. Plass T, Milles S, Koehler C, Schultz C, Lemke EA (2011) Genetically encoded copper-free click chemistry. Angew Chem 50(17):3878–3881. https://doi.org/10.1002/anie.201008178

30. Plass T, Milles S, Koehler C, Szymanski J, Mueller R, Wiessler M, Schultz C, Lemke EA (2012) Amino acids for Diels-Alder reactions in living cells. Angew Chem 51(17):4166–4170. https://doi.org/10.1002/anie.201108231

31. Nikic I, Plass T, Schraidt O, Szymanski J, Briggs JA, Schultz C, Lemke EA (2014) Minimal tags for rapid dual-color live-cell labeling and super-resolution microscopy. Angew Chem 53(8):2245–2249. https://doi.org/10.1002/anie.201309847

32. Uttamapinant C, Howe JD, Lang K, Beranek V, Davis L, Mahesh M, Barry NP, Chin JW (2015) Genetic code expansion enables live-cell and super-resolution imaging of site-specifically labeled cellular proteins. J Am Chem Soc 137(14):4602–4605. https://doi.org/10.1021/ja512838z

33. Nikic I, Estrada Girona G, Kang JH, Paci G, Mikhaleva S, Koehler C, Shymanska NV, Ventura Santos C, Spitz D, Lemke EA (2016) Debugging eukaryotic genetic code expansion for site-specific click-PAINT super-resolution microscopy. Angew Chem 55(52):16172–16176. https://doi.org/10.1002/anie.201608284

34. Hughes LD, Rawle RJ, Boxer SG (2014) Choose your label wisely: water-soluble fluorophores often interact with lipid bilayers. PLoS One 9(2):e87649. https://doi.org/10.1371/journal.pone.0087649

35. Jungmann R, Steinhauer C, Scheible M, Kuzyk A, Tinnefeld P, Simmel FC (2010) Single-molecule kinetics and super-resolution microscopy by fluorescence imaging of transient binding on DNA origami. Nano Lett 10(11):4756–4761. https://doi.org/10.1021/nl103427w

36. Sharonov A, Hochstrasser RM (2006) Widefield subdiffraction imaging by accumulated binding of diffusing probes. Proc Natl Acad Sci U S A 103(50):18911–18916. https://doi.org/10.1073/pnas.0609643104

37. Jungmann R, Avendano MS, Woehrstein JB, Dai M, Shih WM, Yin P (2014) Multiplexed 3D cellular super-resolution imaging with DNA-PAINT and exchange-PAINT. Nat Methods 11(3):313–318. https://doi.org/10.1038/nmeth.2835

38. Banterle N, Bui KH, Lemke EA, Beck M (2013) Fourier ring correlation as a resolution criterion for super-resolution microscopy. J Struct Biol 183(3):363–367. https://doi.org/10.1016/j.jsb.2013.05.004

39. Nikic I, Kang JH, Girona GE, Aramburu IV, Lemke EA (2015) Labeling proteins on live mammalian cells using click chemistry. Nat Protoc 10(5):780–791. https://doi.org/10.1038/nprot.2015.045

40. Hoffmann JE, Plass T, Nikic I, Aramburu IV, Koehler C, Gillandt H, Lemke EA, Schultz C (2015) Highly stable trans-Cyclooctene amino acids for live-cell labeling. Chemistry 21(35):12266–12270. https://doi.org/10.1002/chem.201501647

41. Dai M, Jungmann R, Yin P (2016) Optical imaging of individual biomolecules in densely packed clusters. Nat Nanotechnol 11:798. https://doi.org/10.1038/nnano.2016.95

42. Dai M (2017) DNA-PAINT super-resolution imaging for nucleic acid nanostructures. Methods Mol Biol 1500:185–202. https://doi.org/10.1007/978-1-4939-6454-3_13

43. Pott M, Schmidt MJ, Summerer D (2014) Evolved sequence contexts for highly efficient amber suppression with noncanonical amino acids. ACS Chem Biol 9(12):2815–2822. https://doi.org/10.1021/cb5006273

44. Hug N, Longman D, Caceres JF (2016) Mechanism and regulation of the nonsense-mediated decay pathway. Nucleic Acids Res 44(4):1483–1495. https://doi.org/10.1093/nar/gkw010

45. Ori A, Banterle N, Iskar M, Andres-Pons A, Escher C, Khanh Bui H, Sparks L, Solis-Mezarino V, Rinner O, Bork P, Lemke EA, Beck M (2013) Cell type-specific nuclear pores: a case in point for context-dependent stoichiometry of molecular machines. Mol Syst Biol 9:648. https://doi.org/10.1038/msb.2013.4

Chapter 19

MultiBacTAG-Genetic Code Expansion Using the Baculovirus Expression System in Sf21 Cells

Christine Koehler and Edward A. Lemke

Abstract

The combination of genetic code expansion (GCE) and baculovirus-based protein expression in *Spodoptera frugiperda* cells is a powerful tool to express multiprotein complexes with site-specifically introduced non-canonical amino acids. This protocol describes the integration of synthetase and tRNA gene indispensable for GCE into the backbone of the Bacmid, the Tn7-mediated transposition of various genes of interest, as well as the final expression of protein using the MultiBacTAG system with different noncanonical amino acids.

Key words MultiBac, Baculovirus, Genetic code expansion, Noncanonical amino acids

1 Introduction

The expression of eukaryotic proteins and multidomain protein complexes can be challenging, especially in lower organisms like *E. coli*. Therefore, the expression host of choice for such complex proteins are mammalian cultures, yeast, or the Baculovirus-based expression in insect cells (e.g., Sf21 cells), all having their pros and cons. The Baculovirus system has established itself in the last decades in the life sciences, not only because of the easy handling and low running costs, but also because of the very high flexibility of the system and simple cloning to coexpress several proteins of interest [1, 2]. Even large eukaryotic protein complexes can be expressed in high yields via the Baculovirus-based expression system, rendering biochemical and biophysical experiments possible, which require preparative amounts of proteins. Genetic code expansion (GCE), on the other hand, enables the user to site-specifically introduce noncanonical amino acids (ncAA) with a special chemical handle into the protein of choice at any amino acid position. Meanwhile, there are more than 200 ncAAs existing, providing a broad spectrum of possible applications, which can range

Edward A. Lemke (ed.), *Noncanonical Amino Acids: Methods and Protocols*, Methods in Molecular Biology, vol. 1728, https://doi.org/10.1007/978-1-4939-7574-7_19, © Springer Science+Business Media, LLC 2018

from structural biology via the development of new biologics to reprogramming large living organisms (for reviews *see* refs. 3–5).

This protocol describes the entanglement of GCE and the Baculovirus-based expression of proteins in Sf21 cells in a single, robust, and easy-to-handle platform [6]. GCE requires the presence of a tRNA-synthetase and tRNA (RS/tRNA) pair, in this case PylRS/tRNA^{Pyl} from *Methanosarcina mazei*, which has to be orthogonal preventing cross-reactions with the translation machinery of the host cell. One of the two variants of PylRS, WT or a double mutant called AF (Y306A and Y384F, used frequently for encoding a large set of ring strained reactive amino acid side chain) [7–16], is cloned with a tRNA expression cassette into a transfer plasmid (pUCDM), which after integration in the backbone of the Bacmid, yields in two different Baculovirus (MultiBacTAG^{WT} and MultiBacTAG^{AF} respectively). In a second step, different proteins of interest can be ligated into several donor and acceptor plasmids, which get combined to a Multifusion plasmid using Cre recombinase. Three donor plasmids and one acceptor plasmid allow great latitude in various combinations of different proteins in a very handy way [17, 18]. The protein, which should possess the ncAA has to contain an amber mutation (TAG) in the gene sequence at the designated site. Because of the easy combination possibilities, proteins with different amber site mutations can be simply shuffled into the Multifusion plasmid. The Multifusion plasmid will then be incorporated into the Bacmid backbone via a Tn7 site by a standard bacterial transformation. With the resulting DNA of the Multifusion plasmid, the Bacmid DNA, Sf21 cells get infected and produce Virus. The protein expression in presence of the required ncAA occurs by adding the Virus to a fresh batch of Sf21 cells [6].

2 Materials

2.1 Plasmids and Cell Lines

1. pBADZHisCre (available from Imre Berger lab, University of Bristol, Bristol, UK).

2. DH10MultiBac cells (available from Imre Berger lab, University of Bristol, Bristol, UK).

3. pUCDM-PylRS(WT)-U6(Sf21)-2-tRNA^{Pyl}-3term (available from Edward Lemke lab, EMBL, Heidelberg, GE).

4. pUCDM-PylRS(AF)-U6(Sf21)-2-tRNA^{Pyl}-3term (available from Edward Lemke lab, EMBL, Heidelberg, GE).

5. MultiBac plasmids: *see* Table 1 for details.

2.2 Buffers, Chemicals, and Medium

1. 2× YT medium: 1% (w/v) tryptone, 1% (w/v) yeast extract, 86 mM NaCl, pH 7.2.

2. LB-agar plates for blue–white screening (low salt): 1% (w/v) tryptone, 0.5% (w/v) yeast extract, 86 mM NaCl, 1.5% (w/v) agar, 1 mg/ml X-Gal, 1 mM IPTG, pH 7.5 (with NaOH).

Table 1
Details for MultiBac plasmids (type and antibiotic resistance)

Plasmid	Plasmid type	Antibiotic resistance
pAceBac-DUAL	Acceptor plasmid	Ampicillin
pIDC	Donor plasmid	Chloramphenicol
pIDK	Donor plasmid	Kanamycin
pIDS	Donor plasmid	Spectinomycin

3. Antibiotics: *see* Table 2 for abbreviations and concentrations (if not further mentioned, all antibiotics are used in a 1:1000 dilution).

4. LB medium: 1% (w/v) tryptone, 0.5% (w/v) yeast extract, 170 mM NaCl, pH 7.

5. 20% (w/v) arabinose: sterile-filtered through a 0.22 μm filter device.

6. 10% (w/v) glycerol.

7. LB-agar plates for blue–white screening (normal salt concentration): 1% (w/v) tryptone, 0.5% (w/v) yeast extract, 136 mM NaCl, 1.5% (w/v) agar, 1 mg/ml X-Gal, 1 mM IPTG, pH 7.5 (with NaOH).

8. DNA Miniprep Kit (e.g., PureLink Quick Plasmid Miniprep Kit, Thermo Fisher Scientific).

9. Isopropanol.

10. 70% ethanol.

11. Sterile ddH$_2$O.

12. Sf-900 III SFM medium (Thermo Scientific, 12658019).

13. FuGene HD transfection reagent (Promega).

14. 4× PBS (pH 8).

15. 5× SDS loading dye.

16. 10-well NuPAGE Novex 4–12% Bis-Tris Protein Gels (Thermo Fisher Scientific) or other gradient protein gels.

17. NuPAGE MOPS SDS Running Buffer (20×) (Thermo Fisher Scientific): prepare 1× dilution.

18. 5% milk in 1× PBS.

19. Anti-PylRS[AF] (available from Edward Lemke lab).

20. Anti-Rat (Dianova).

21. 1× PBS + 0.2% Tween 20.

22. ECL Kit (GE Healthcare).

Table 2
List of antibiotics

Antibiotic	Abbreviation	Concentration (mg/ml)
Ampicillin	Amp	50
Kanamycin	Kan	50
Tetracycline	Tet	14
Zeocin	Zeo	25
Chloramphenicol	Cm	33
Gentamycin	Gent	10
Spectinomycin	Spec	50

23. 10× Cre reaction buffer: 500 mM Tris–HCl pH 7.5, 330 mM NaCl, 100 mM MgCl$_2$.

24. Cre recombinase (e.g., New England BioLabs).

25. Electrocompetent cells: Top10 (Thermo Scientific, C404050) or alike.

26. SOC medium: 2% (w/v) tryptone, 0.5% (w/v) yeast extract, 10 mM NaCl, 2.5 mM KCl, 10 mM MgCl$_2$, 20 mM glucose.

27. LB-agar plates: 1% (w/v) tryptone, 0.5% (w/v) yeast extract, 136 mM NaCl, 1.5% (w/v) agar, pH 7.5 (with NaOH).

28. ncAA: 1 mM in expression culture.

2.3 Plasticware, Glassware, and Equipment

1. Electroporation cuvettes (0.1 cm) and GenePulser (Bio-Rad) or alike for electroporation.

2. Heat block with shaking function for 1.5 ml tubes.

3. Beckman centrifuge and rotor JLA8.1000 or alike for harvesting up to 1 l cultures.

4. 6-well cell culture plate.

5. Sterile Erlenmeyer flasks in different sizes (100 ml to 5 l).

6. Nitrocellulose membrane for Western blot.

7. Cooling microcentrifuge, e.g., Eppendorf 5427R centrifuge.

8. Benchtop centrifuge, e.g. Eppendorf 5804R centrifuge.

9. Cell culture hood.

10. 27 °C insect cell incubator.

11. Western blot transfer machine, e.g., Trans-Blot Turbo Transfer System (Bio-Rad).

3   Methods

To compose a Baculovirus expression system suitable for GCE, the backbone of the Baculovirus has to be changed. The tRNA-synthetase/tRNA pair (PylRS/tRNA^{Pyl}) has to be integrated into the Bacmid DNA, followed by testing for successful integration. Therefore, Bacmid DNA and Virus are prepared from several colonies and expression of the PylRS is tested by Western blot for all of those colonies. From the colony, which expresses PylRS best, electrocompetent cells have to be prepared. These cells are called DH10MultiBacTAG^{WT}or^{AF}. In parallel, the proteins of interest can be cloned into any of the available three donor and into the acceptor plasmids. The Multifusion plasmid harboring the proteins of interest will be obtained using Cre recombinase. This Multifusion plasmid can be transformed into the freshly prepared DH10MultiBacTAG$^{WT \ or \ AF}$ electrocompetent cells and consequently integrated in the Tn7 site of the Bacmid. After preparing Bacmid DNA, V$_0$-Virus can be harvested and used for protein production. The ncAA will be incorporated into the protein in response to the amber codon, if added to the medium during protein expression.

3.1   Generation of MultiBacTAG$^{WT \& AF}$

3.1.1   Preparation of DH10MultiBac^{Cre} Competent Cells

1. Transform 1 and 4 μl of pBADZHisCre (100 ng/μl) into 100 μl of electrocompetent DH10MultiBac cell using a Gene Pulser at the settings: 1.8 kV, 200 Ω, 25 μF. Include also one transformation with 1 μl H$_2$O as a negative control.

2. Directly after transformation add 800 μl of 2× YT medium and incubate mixture shaking in a heat block at 500 rpm at 37 °C for 4 h.

3. Plate 200 and 400 μl of these cells on plates for blue–white screening (low salt), which contain in addition Kan, Tet, and Zeo.

4. Incubate at 37 °C for 24 h.

5. Inoculate 50 ml LB medium (Kan, Tet, and Zeo) with one blue colony and grow overnight at 37 °C, shaking at 180 rpm.

6. Add 10 ml of this overnight culture into 500 ml LB medium (Kan, Tet, and Zeo) and grow culture at 37 °C shaking until an OD$_{600}$ of 0.25 is reached.

7. Induce expression of Cre recombinase by adding 2.5 ml of 20% arabinose.

8. Reduce temperature to 25 °C and grow the culture up to an OD$_{600}$ of 0.5 and then rapidly cool down cells on ice for 15 min.

9. Centrifuge the cells for 15 min at $3060 \times g$ at 4 °C in a Beckman centrifuge using the JLA8.1000 rotor.

10. Resuspend the cell pellet in 10% glycerol. From this step on, everything should be done at 4 °C or on ice, if possible.

11. Spin down again and take the pellet up in 250 ml 10% glycerol.

12. Harvest cell pellet again and wash it in 10 ml 10% glycerol.

13. After the final centrifugation step, resuspend the pellet in 1 ml 10% glycerol.

14. Prepare 100 µl aliquots, shock-freeze in liquid nitrogen and store them at −80 °C.

3.1.2 Transformation of pUCDM Plasmid into DH10MultiBac^{Cre}

1. Transform 10 ng of the donor plasmid pUCDM, containing the synthetase variant (PylRS WT or PylRS AF) and the tRNA expression cassette (U6(Sf21)-2-tRNA^{Pyl}-3term), into 100 µl freshly prepared DH10MultiBac^{Cre} cells.

2. After adding 800 µl of 2× YT medium to the transformed cells, incubate them over night at 37 °C, shaking at around 500 rpm in a heat block.

3. Plate 200 and 400 µl of the cells on blue–white screening plates (normal salt concentration), containing Amp, Kan, Cm, and Tet as antibiotics.

4. Incubate plates for at least 24 h at 37 °C until blue colonies are visible.

5. Pick eight blue colonies and grow them separately in 2 ml 2× YT medium each (Amp, Kan, Cm, Tet) over night at 37 °C, shaking at 180 rpm and prepare glycerol stocks of each of the overnight cultures (*see* **Note 1**).

3.1.3 Preparation of Bacmid-DNA

1. Prepare Bacmid-DNA of each of the eight colonies starting with a DNA Miniprep Kit. Spin down the overnight cultures, resuspend the cell pellets and lyse the cells. After adding the precipitation buffer, spin down for at least 15 min, collect the supernatants, and spin them again in new tubes for at least 10 min (*see* **Note 2**).

2. Continue with an isopropanol extraction by adding 800 µl isopropanol to the supernatants.

3. After incubation for 5 min at RT, spin down the mixtures for 30 min at $15,700 \times g$ at 4 °C in a microcentrifuge.

4. Remove the supernatants and wash the pellets in 400 µl 70% ethanol.

5. Centrifuge again for 30 min at 4 °C at $15,700 \times g$.

6. Remove the ethanol and dry the pellets in the hood. Resuspend each pellet in 30 µl sterile H_2O 2–3 h before transfection.

3.1.4 Production of V_0-Virus

1. Under the cell culture hood, split Sf21 cells to a cell density of 0.3×10^6 cells/ml and seed 3 ml cells in each well of a 6-well cell culture plate. Seed for each Bacmid-DNA two wells.

2. Mix each 30 µl Bacmid-DNA with 200 µl of Sf-900 III SFM medium.

3. In parallel, mix 10 µl of FuGene HD transfection reagent with 100 µl of Sf-900 III SFM medium for each sample.

4. Mix both solutions together and incubate this mixture for 15 min at room temperature.

5. Use this mixture to transfect two wells by adding it dropwise.

6. Incubate the 6-well cell culture plate at 27 °C for 60 h.

7. Harvest Virus (V_0-Virus) by carefully pipetting off the supernatant from each well. The supernatant of the wells, which have been transfected with the same Bacmid-DNA, can be pooled (*see* **Note 3**).

3.1.5 V_1-Generation

1. Prepare for each V_0-Virus one Erlenmeyer flask with 50 ml cells at a density of 0.6×10^6 cells/ml (*see* **Note 4**) and add 3 ml V_0-Virus into 50 ml cells.

2. Shake flask at 27 °C on a shaker at 110 rpm.

3. Count and split cells each day to 0.6×10^6 cells/ml until cell proliferation arrest (*see* **Note 5**).

4. Incubate cells for additionally 60 h.

5. Harvest cells at $500 \times g$ for 10 min. Keep the pellets for further analysis (*see* **Note 6**).

3.1.6 PylRS Expression Analysis by Western Blot

1. Resuspend pellets of V_1-Generation in 5 ml 4× PBS.

2. Mix 10 µl of the resuspension with 90 µl of 4× PBS, add 100 µl 5× SDS loading dye and heat up the samples for 5 min at 95 °C.

3. Load 30 µl of each V_1 protein expressions per well in a 10-well NuPAGE Novex 4–12% Bis-Tris Protein Gels and run the gel in NuPAGE MOPS SDS running buffer (1×).

4. Transfer on nitrocellulose membrane (using for example Trans-Blot Turbo Transfer System or a standard transfer method).

5. Block the membrane in 5% Milk in 1× PBS for 1 h.

6. Incubate with primary antibody Anti-PylRS[AF] (Edward Lemke lab, 1:400 diluted in 5% Milk, 1× PBS) overnight at 4 °C.

7. Wash 3 × 5 min with 1× PBS + 0.2% Tween 20.

8. Incubate with secondary antibody Anti-Rat (Dianova, 1:5000 in 5% Milk, 1× PBS) for 1 h at RT.

9. Wash 3 × 10 min with 1× PBS, 0.2% Tween 20.

10. Use a ECL Kit (GE Healthcare) to obtain the chemiluminescence signal (*see* Fig. 1).

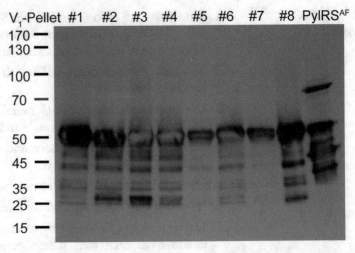

Fig. 1 Western blot against PylRS^{AF} of V$_1$ expressions. The V$_1$ pellet of eight initial colonies is shown. A control of the experiment (PylRS^{AF}, purified from *E. coli* expression) shows a band at the molecular weight of around 50 kDa corresponding to PylRS expression. All eight samples show PylRS expression

3.1.7 Preparation of Electrocompetent DH10MultiBacTAG$^{WT\&AF}$ Cells

1. Of one of the eight colonies that shows highest expression of PylRS in the Western blot analysis (e.g., #1), prepare an overnight culture. Therefore, set up a 10 ml culture with 2× YT medium, containing Amp, Kan, Tet, and Cm as antibiotics.

2. Shake overnight at 37 °C at 180 rpm.

3. Set up a 500 ml culture using 5 ml of the overnight culture and the same antibiotics as before.

4. Grow culture at 37 °C, shaking at 180 rpm, until an OD$_{600}$ of 0.5 is reached.

5. Cool down cells on ice for 15 min.

6. Centrifuge the cells at $3060 \times g$ for 15 min at 4 °C in a Beckman centrifuge using the JLA8.1000 rotor.

7. Resuspend pellet in 500 ml 10% glycerol and spin again. Work on ice or at 4 °C whenever possible.

8. Wash the pellet with 250 ml 10% glycerol and harvest cells again.

9. Take the pellet up in 10 ml 10% glycerol and centrifuge in a Benchtop centrifuge (Eppendorf 5804R) at $2880 \times g$ for 10 min at 4 °C.

10. Add 1 ml 10% glycerol to the pellet and aliquot cells (100 μl/tube).

11. Flash freeze aliquots in liquid nitrogen and store them at −80 °C until further use.

3.2 Preparing the Multifusion Plasmid

3.2.1 Cloning

Plan the cloning of the nucleotide sequences of your protein of choice according to the available donor (pIDC, pIDK and pIDS) plasmids and acceptor plasmid (pAceBac-DUAL, Fig. 2) (*see* **Note 7**). Use the multicloning sites for restriction enzyme-based cloning. All plasmids contain a recombination site (*loxP* site) to combine the plasmid with each other via Cre/loxP recombination and the acceptor plasmid (pAceBac-DUAL) in addition posses a Tn7 transposon for integration in the Bacmid. Verify the resulting plasmids by sequencing.

3.2.2 Cre/loxP-Mediated Recombination of Transfer Plasmids

1. To combine all transfer plasmids into a Multifusion plasmid, set up the Cre/loxP reaction in a 20 µl total volume (follow pipetting scheme, Table 3). The ratio of acceptor:donor:donor:donor plasmid should be 0.8:1:1:1 (*see* **Note 8**).

2. Incubate this reaction at 37 °C for 1 h and heat-inactivate at 70 °C for 10 min (*see* **Note 9**).

3. Transformation into 100 µl electrocompetent cells (e.g., Top10) is carried out with 2 µl of the Cre/loxP reaction, adding 400 µl SOC medium and incubating overnight shaking at 37 °C.

4. Spin down the cells at $2300 \times g$ in a microcentrifuge for 5 min and resuspend in 200 µl SOC medium before plating on LB-agar plates containing the appropriate antibiotics.

5. Grow plates at 37 °C overnight.

6. Pick eight colonies for each Cre/loxP reaction and incubate in 500 µl SOC without any antibiotics for 2 h on a shaker at 37 °C.

7. Meanwhile prepare a 96-well microtiter plate with 150 µl LB medium per well. For each colony prepare as many wells as antibiotics you are using. In each well add only one of the antibiotics.

8. Add 1 µl of the preculture in each well, so that each culture is now growing in different antibiotics separately.

9. Incubate the plate at 37 °C overnight on a shaker.

10. Next day, streak out the cultures on a LB-agar plate containing all the antibiotics in combination. Therefore, divide the plates into as many parts as antibiotics you are using and use approximately 1 µl of each culture for plating.

11. Incubate the plates at 37 °C overnight. The Multifusion plasmid contains all the required transfer plasmids, if the cultures grow in all sections on the plate (*see* **Note 10**).

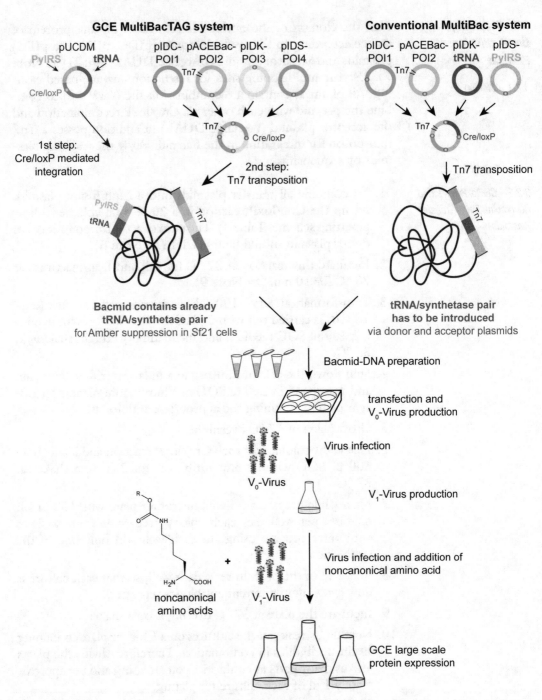

Fig. 2 Schematic overview of the MultiBacTAG system in comparison to the conventional MultiBac system. In the newly established MultiBacTAG system, the PylRS/tRNA^{Pyl} pair gets integrated into the Bacmid backbone by Cre/lox recombination. The Multifusion plasmid can be transformed into the MultiBacTAG Bacmid via the Tn7 site. For the conventional MultiBac system, the PylRS/tRNA^{Pyl} pair has to get introduced into the Bacmid via the Multifusion plasmid. Bacmid preparation is followed by V_0- and V_1-Virus production, before the protein containing a site specific introduced ncAA can get expressed

Table 3
Pipetting scheme for Cre/loxP reaction

	Volume (μl)
ddH$_2$O	8.4
10× Cre reaction buffer	2
pIDC (~500 ng/μl)	2
pIDK (~500 ng/μl)	2
pIDS (~500 ng/μl)	2
pAceBac-DUAL (~500 ng/μl)	1.6
Cre recombinase	2

12. Pick one colony of each Cre/loxP reaction and incubate in 10 ml LB medium with the corresponding combination of antibiotics at 37 °C overnight.

13. Next day, harvest the cell pellet by centrifugation at $2300 \times g$ for 10 min at 4 °C. Prepare Multifusion plasmid DNA using a commercial Miniprep Kit.

3.3 Tn7-Mediated Transposition of Multifusion Plasmid into MultiBacTAG

1. Mix 1 μl of the Multifusion plasmid with 100 μl of the electrocompetent DH10MultiBacTAG$^{WT\ or\ AF}$ cells and incubate on ice for 5 min.

2. Perform electroporation using a Gene Pulser at the settings: 1.8 kV, 200 Ω, 25 μF. Add directly after electroporation 800 μl of SOC medium and incubate shaking at 37 °C overnight.

3. On the following day, plate 200 and 400 μl on a blue–white screening plate (normal salt concentration) containing Amp, Kan, Tet, and Gent as antibiotics.

4. Incubate plate for at least 24 h or until blue and white colonies are visible (*see* **Note 11**).

5. Pick several white colonies for each Multifusion plasmid and grow them overnight in 4 ml 2× YT medium containing the appropriate antibiotics (*see* **Note 12**).

3.4 Expression of Protein in Sf21 Cells Using MultiBacTAG

3.4.1 Preparation of MultiBacTAG- Virus

1. Prepare Bacmid DNA as mentioned in Subheading 3.1.3.

2. Prepare V$_0$-Virus as mentioned in Subheading 3.1.4.

3. Produce V$_1$-Virus following the instructions in Subheading 3.1.5.

3.4.2 Incorporation of ncAAs into Proteins Expressed in Sf21 Cells

1. Set up a 20 ml culture (0.6×10^6 cells/ml) and transfect the cells with different amount of V_1-Virus (from 20 μl to 2 ml) to estimate the MOI (Multiplicity of Infection).

2. Check the cell density after 1 day (*see* **Note 13**).

3. Start a small test expression by adding the right amount of V_1-Virus into 50 ml of cells (0.6×10^6 cells/ml). Prepare one culture for each construct, one for the amber mutant with and one without ncAA, as well as one for the wild-type construct (without amber site) as a positive control. Add the ncAA at 1 mM final concentration (*see* **Note 14**). Incubate cultures for 4 days at 27 °C on a shaker. Harvest cells by centrifugation at $500 \times g$ for 15 min in Beckman centrifuge (JLA8.1000 rotor). Protein expression can be followed up by standard purification methods, such as Immobilized Metal Affinity Chromatography (IMAC) using a 6-his tag or different purification methods in combination with SDS-PAGE analysis (*see* **Note 15**).

4. Large scale protein expressions follow the same protocol as small test expressions by increasing the volume of cells and Virus used. Harvest bigger volumes of cells for 45 min at $500 \times g$, instead of only 15 min (*see* **Note 16**).

4 Notes

1. Preparation of glycerol stocks: 333 μl of 50% glycerol + 666 μl of overnight culture, store at −80 °C. Streak out each overnight culture again on the blue–white screening plates to verify the integration of the plasmid into the Bacmid backbone by the colonies being blue.

2. This is a very critical step. Repeat the centrifugation step as often as needed to ensure there is no precipitation anymore in the supernatant.

3. The supernatant can be centrifuged down at $500 \times g$ for 10 min, to ensure that there are no cells in the V_0-Virus.

4. The flask to culture Sf21 cells should be always five times bigger than the Sf21 cell volume (e.g., 50 ml cells in a 250 ml flask).

5. Virus transfection can be monitored by the cell proliferation and the size of the Sf21 cells. Transfected cells are nearly double the size of nontransfected ones.

6. Pellets can be kept at 4 °C for short-term or at −20 °C for long-term storage.

7. The pAceBacDUAL plasmid contains two promoters each followed by a multicloning site for the cloning of two individual genes of interest. The plasmids pIDK, pIDC and pIDS each

include one promoter and a multicloning site for the cloning of in total three genes of interest. An amber site (TAG) has to be inserted in the gene of protein of interest, which should contain the ncAA. The amber site can be introduced by site-directed mutagenesis into the nucleotide sequence of the gene. The position has to be chosen in a way that the incorporation of the ncAA does not influence the folding or characteristics of the protein. Favorable amino acids to mutate are surface exposed regions of the protein. Add a purification tag, such as a 6-His tag, at the C-terminal domain of the protein, which harbors the amber site, to ensure purification of the full length protein. Prepare also one plasmid with the gene of interest without the amber site as a positive control and to establish the purification protocol. Also make sure, that none of your protein genes end with an amber stop codon.

8. Not all donor plasmids have to be used, only the ones carrying a protein gene of interest. But it is necessary to use the acceptor plasmid, because of the Tn7 transposon site. All donor plasmids have to be propagated in BW23474 cells (Pir+ cells).

9. The reaction can be followed up by agarose gel analysis. Therefore, take 1 µl of the Cre/loxP reaction for a 10 µl analytical digest with a restriction enzyme, which cuts only once and analyze the digest on an agarose gel (1% agarose, TAE buffer). A band of the right height for the Multifusion plasmid should be visible. The Cre/loxP reaction can be optimized by changing incubation time and temperature, e.g., incubation at 30 °C for 2 h.

10. The Multifusion plasmid can be further checked by sequencing using suitable primers, that only anneal once in each transfer plasmid.

11. The colonies are white, if the transposition into the Bacmid at the Tn7 site was successful.

12. Streak out the overnight culture again on a blue–white selection plate to ensure the white color of the colonies and prepare as well glycerol stocks of each culture. In case the Bacmid preparation fails or the experiment has to be repeated, new Bacmid can be prepared starting from the glycerol stock.

13. The MOI should be around 1, meaning that the cell proliferation arrests after 1 day and the cell count after 1 day incubation should not be higher than 1.5×10^6 cells/ml.

14. For the MultiBacTAG[WT] system ncAAs like propargyl-Lysine (PrK) can be used. Several others like cyclooctyne-Lysine (SCO), trans-Cyclooctene-Lysine (TCO), or bicyclo [6.1.0] nonyne-Lysine (BCN) can get incorporated using the MultiBacTAG[AF] system [6–8, 16, 19, 20]. All the ncAAs should be taken up in 2 M NaOH, diluted 1:10 with Sf-900

III SFM medium and pH adjusted to pH 7 before they are added to the Sf21 culture. Make sure to work in sterile conditions.

15. SDS-PAGE analysis should show that only in presence of the ncAA the protein gets expressed. A very low expression of protein even without ncAA in the medium can sometimes be observed. This is due to the leakiness of the system, which is minimal in the case of the *Methanosarcina mazei* PylRS/tRNA^{Pyl} pair. In the presence of ncAA in the medium, the expressed protein will always contain the ncAA and none of the canonical amino acids in response to the amber codon. The expression level of a protein containing a ncAA depends on the ncAA used, but will typically be still lower compared to the expression level of wild-type protein (10–80% of the amount of wild-type protein). The expression level of the protein will also decrease with increasing numbers of amber sites introduced in the gene of interest.

16. Alternatively, the protein can be also expressed using V_2-Virus, which can be obtained by keeping the supernatant of a V_1-generation. This can lead in some cases to a higher expression yield of the protein.

References

1. Bieniossek C, Imasaki T, Takagi Y, Berger I (2012) MultiBac: expanding the research toolbox for multiprotein complexes. Trends Biochem Sci 37(2):49–57. https://doi.org/10.1016/j.tibs.2011.10.005

2. Crepin T, Swale C, Monod A, Garzoni F, Chaillet M, Berger I (2015) Polyproteins in structural biology. Curr Opin Struct Biol 32:139–146. https://doi.org/10.1016/j.sbi.2015.04.007

3. Lemke EA (2014) The exploding genetic code. Chembiochem 15(12):1691–1694. https://doi.org/10.1002/cbic.201402362

4. Liu CC, Schultz PG (2010) Adding new chemistries to the genetic code. Annu Rev Biochem 79:413–444. https://doi.org/10.1146/annurev.biochem.052308.105824

5. Chin JW (2014) Expanding and reprogramming the genetic code of cells and animals. Annu Rev Biochem 83:379–408. https://doi.org/10.1146/annurev-biochem-060713-035737

6. Koehler C, Sauter PF, Wawryszyn M, Girona GE, Gupta K, Landry JJ, Fritz MH, Radic K, Hoffmann JE, Chen ZA, Zou J, Tan PS, Galik B, Junttila S, Stolt-Bergner P, Pruneri G, Gyenesei A, Schultz C, Biskup MB, Besir H, Benes V, Rappsilber J, Jechlinger M, Korbel JO, Berger I, Braese S, Lemke EA (2016) Genetic code expansion for multiprotein complex engineering. Nat Methods 13(12):997–1000. https://doi.org/10.1038/nmeth.4032

7. Plass T, Milles S, Koehler C, Schultz C, Lemke EA (2011) Genetically encoded copper-free click chemistry. Angew Chem 50(17):3878–3881. https://doi.org/10.1002/anie.201008178

8. Plass T, Milles S, Koehler C, Szymanski J, Mueller R, Wiessler M, Schultz C, Lemke EA (2012) Amino acids for Diels-Alder reactions in living cells. Angew Chem 51(17):4166–4170. https://doi.org/10.1002/anie.201108231

9. Nikic I, Plass T, Schraidt O, Szymanski J, Briggs JA, Schultz C, Lemke EA (2014) Minimal tags for rapid dual-color live-cell labeling and super-resolution microscopy. Angew Chem 53(8):2245–2249. https://doi.org/10.1002/anie.201309847

10. Nikic I, Lemke EA (2015) Genetic code expansion enabled site-specific dual-color protein labeling: superresolution microscopy and beyond. Curr Opin Chem Biol 28:164–173. https://doi.org/10.1016/j.cbpa.2015.07.021

11. Nikic I, Kang JH, Girona GE, Aramburu IV, Lemke EA (2015) Labeling proteins on live mammalian cells using click chemistry. Nat Protoc 10(5):780–791. https://doi.org/10.1038/nprot.2015.045

12. Nikic I, Estrada Girona G, Kang JH, Paci G, Mikhaleva S, Koehler C, Shymanska NV, Ventura Santos C, Spitz D, Lemke EA (2016) Debugging eukaryotic genetic code expansion for site-specific click-PAINT super-resolution microscopy. Angew Chem 55(52):16172–16176. https://doi.org/10.1002/anie.201608284

13. Kozma E, Nikic I, Varga BR, Aramburu IV, Kang JH, Fackler OT, Lemke EA, Kele P (2016) Hydrophilic trans-Cyclooctenylated noncanonical amino acids for fast intracellular protein labeling. Chembiochem 17(16):1518–1524. https://doi.org/10.1002/cbic.201600284

14. Hoffmann JE, Plass T, Nikic I, Aramburu IV, Koehler C, Gillandt H, Lemke EA, Schultz C (2015) Highly stable trans-Cyclooctene amino acids for live-cell labeling. Chemistry 21(35): 12266–12270. https://doi.org/10.1002/chem.201501647

15. Lang K, Davis L, Torres-Kolbus J, Chou C, Deiters A, Chin JW (2012) Genetically encoded norbornene directs site-specific cellular protein labelling via a rapid bioorthogonal reaction. Nat Chem 4(4):298–304. https://doi.org/10.1038/nchem.1250

16. Darko A, Wallace S, Dmitrenko O, Machovina MM, Mehl RA, Chin JW, Fox JM (2014) Conformationally strained trans-Cyclooctene with improved stability and excellent reactivity in Tetrazine ligation. Chem Sci 5(10):3770–3776. https://doi.org/10.1039/C4SC01348D

17. Fitzgerald DJ, Berger P, Schaffitzel C, Yamada K, Richmond TJ, Berger I (2006) Protein complex expression by using multigene baculoviral vectors. Nat Methods 3(12):1021–1032. https://doi.org/10.1038/nmeth983

18. Berger I, Fitzgerald DJ, Richmond TJ (2004) Baculovirus expression system for heterologous multiprotein complexes. Nat Biotechnol 22(12):1583–1587. https://doi.org/10.1038/nbt1036

19. Nguyen DP, Lusic H, Neumann H, Kapadnis PB, Deiters A, Chin JW (2009) Genetic encoding and labeling of aliphatic azides and alkynes in recombinant proteins via a pyrrolysyl-tRNA Synthetase/tRNA(CUA) pair and click chemistry. J Am Chem Soc 131(25):8720–8721. https://doi.org/10.1021/ja900553w

20. Borrmann A, Milles S, Plass T, Dommerholt J, Verkade JM, Wiessler M, Schultz C, van Hest JC, van Delft FL, Lemke EA (2012) Genetic encoding of a bicyclo[6.1.0]nonyne-charged amino acid enables fast cellular protein imaging by metal-free ligation. Chembiochem 13(14): 2094–2099. https://doi.org/10.1002/cbic.201200407

Chapter 20

Production and Chemoselective Modification of Adeno-Associated Virus Site-Specifically Incorporating an Unnatural Amino Acid Residue into Its Capsid

Rachel E. Kelemen, Sarah B. Erickson, and Abhishek Chatterjee

Abstract

The ability to modify the capsid proteins of human viruses is desirable both for installing probes to study their structure and function, and to attach retargeting agents to engineer viral infectivity. However, the installation of such capsid modifications currently faces two major challenges: (1) The complex and delicate capsid proteins often do not tolerate large modifications, and (2) capsid proteins are composed of the 20 canonical amino acids, precluding site-specific chemical modification of the virus. Here, we describe a technology for generating adeno-associated virus (AAV) while incorporating an unnatural amino acid (UAA) into specific sites of the virus capsid. Incorporation of this UAA is generally tolerated well, presumably due to its small structural footprint. The resulting virus can be precisely functionalized at the site of UAA incorporation using chemoselective conjugation strategies targeted toward the azido side chain of this UAA. This technology provides a powerful way to modify AAV with unprecedented precision to both probe and engineer its entry process.

Key words Adeno-associated virus, Unnatural amino acid mutagenesis, Genetic code expansion, Bioorthogonal chemistry, Virus labeling, Virus retargeting

1 Introduction

Adeno-associated virus (AAV) is a small non-enveloped human virus belonging to the parvoviridae family. Over the last two decades, it has emerged as one of the most promising vectors for delivering therapeutic genes in vivo, due to its ability to efficiently transduce a broad range of dividing and nondividing cells and facilitate long-term transgene expression, its low immunogenicity, and the lack of any known pathogenicity [1–3]. AAV also represents a model system to study the biology of the parvoviruses. The AAV capsid is composed of 60 total copies of three capsid proteins VP1, VP2, and VP3, which are present in an approximately 1:1:10 ratio [4]. A 533 amino acid C-terminal domain is common to all three capsid proteins, while VP1 and VP2 have additional N-terminal

Edward A. Lemke (ed.), *Noncanonical Amino Acids: Methods and Protocols*, Methods in Molecular Biology, vol. 1728, https://doi.org/10.1007/978-1-4939-7574-7_20, © Springer Science+Business Media, LLC 2018

extensions. They interlock in an intricate manner to create the ico-sahedral AAV capsid [4]. Apart from serving as the protective shell for the viral genome, the capsid proteins also facilitate viral entry into the host cell by sequentially interacting with numerous host factors in a hierarchical manner, a process that involves dynamic structural changes at distinct stages of the entry pathway [4]. Consequently, these complex and delicate capsid proteins often do not tolerate attempts at introducing structural alterations [5].

The ability to modify the virus capsid in a controlled manner is essential to introduce probes (e.g., fluorophores) for investigating how the virus infects cells, or to introduce new binding motifs for engineering the host cell specificity of AAV. Amino acid residues such as lysine and arginine on the virus capsid have been previously targeted to introduce such modifications [6]. However, this strategy offers no site-specificity, given the abundance of these residues on the virus capsid, and can result in the modification of functionally important sites. Another approach involves the insertion of foreign peptide sequences that either bind a desired target, or can be enzymatically modified to introduce various entities [7–9]. Although this approach is site-selective and produces homogeneous preparations of the modified virus, the insertion of such peptides is tolerated only into a very limited number of sites on the virus capsid. Expansion of the mammalian genetic code has recently enabled site-specific incorporation of novel unnatural amino acids (UAAs) into proteins expressed in these cells [10, 11]. This technology uses an engineered aminoacyl-tRNA synthetase (aaRS)/ tRNA pair to site-specifically introduce the UAA in response to a repurposed nonsense codon (usually TAG). Here, we describe a method to use this technology to generate adeno-associated virus serotype 2 (AAV2), site specifically incorporating an UAA into its capsid (Fig. 1) [12]. A *Methanosarcina barkeri*-derived pyrrolysyl-tRNA synthetase (MbPylRS)/tRNA pair is used to deliver the UAA with an azido side chain, whose incorporation is tolerated well at several sites on the virus capsid. We further show the feasibility of specifically functionalizing the azido side chain of the incorporated UAA using the chemoselective strain-promoted azide-alkyne cycloaddition reaction.

2   Materials

Unless otherwise mentioned, all reagents and chemicals were purchased from commercial sources and used without further purification. According to NIH guidelines, AAV2 vectors that do not encode either a potentially tumorigenic or toxic gene product, and are produced in the absence of the helper adenovirus (which is the case with our system) can be handled at biosafety level 1 (BSL-1).

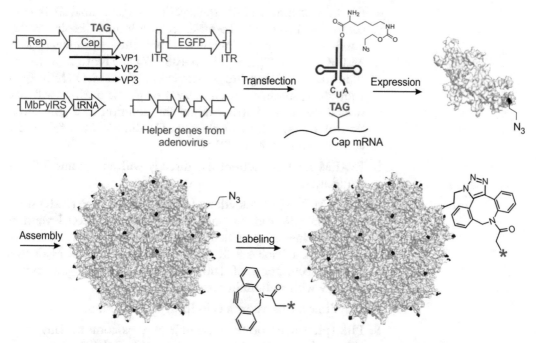

Fig. 1 The schematic description of the strategy to produce UAA-modified AAV and its subsequent chemoselective modification. *Rep* and *cap* genes in the AAV genome encode several replication and capsid proteins, respectively

However, out of an abundance of caution, we perform these experiments under BSL-2 guidelines.

2.1 Cell Culture and Virus Production

1. Low-passage HEK293T cells (ATCC CRL-3216™) maintained in Dulbecco's modified Eagle's medium (DMEM; Thermo Fisher Scientific) supplemented with 10% FBS (fetal bovine serum) and 1× penicillin/streptomycin solution (Thermo Fisher Scientific; final concentration 100 U/mL penicillin and 100 µg/mL streptomycin) at 37 °C and 5% CO_2 (*see* **Note 1**).

2. 12-well and 10 cm cell culture treated dishes, sterile.

3. UAA-modified AAV-production plasmids: Three plasmids are needed to produce AAV2 incorporating UAA. The pHelper plasmid (Cell Biolabs) encodes adenoviral helper genes, pIDTSmart-MbPylRS-RC2 encodes the AAV *rep* and *cap* genes (with stop codon) and the wild-type MbPylRS, and pIDTSmart-8xPytR-ITR-GFP encodes a wild-type EGFP reporter cargo flanked by ITRs for viral packaging and eight tandem copies of the *M. mazei* pyrrolysine tRNA expression cassettes. The latter two plasmids are available upon request from the Chatterjee group. Maps and sequences of these two plasmids are published [12].

4. Polyethyleneimine (PEI) transfection reagent, final stock concentration 1 mg/mL: To make stock solution, dissolve freshly purchased polyethyleneimine (branched, average MW 25,000 g/mol, Sigma-Aldrich 408727) (*see* **Note 2**) to a concentration of 10 mg/mL in sterile deionized H_2O (Milli-Q or equivalent, sterilized by autoclave) by gentle mixing (will take several minutes). Further dilute this solution to a concentration of 1 mg/mL in sterile ddH$_2$O, aliquot, flash-freeze the aliquots, and store at −80 °C.

5. DMEM for transfection: use directly without adding FBS or antibiotics.

6. 100 mM N_6-((2-azidoethoxy) carbonyl)-L-Lysine (AzK) stock solution: AzK can be purchased from commercial vendors (e.g., Sirius Fine Chemicals SiChem GmbH, Bremen, Germany). Dissolve AzK solid in sterile deionized H_2O to a final concentration of 100 mM, adjusting the solution to pH 7–9 with sodium hydroxide.

7. Cell lifter for harvesting cells from 10 cm dishes.

8. PBS (phosphate-buffered saline): used as commercially available, without calcium or magnesium, pH 7.0–7.2, sterile.

2.2 PEG Precipitation

1. Cell lysis solution: PBS (Subheading 2.1) supplemented with 150 mM additional NaCl (final NaCl concentration of approximately 300 mM; filtered through a 0.22 μm sterile filter).

2. Dry ice/ethanol bath: A Dewar flask or suitable vessel containing 95% ethanol cooled with dry ice.

3. 37 °C water bath.

4. 40% polyethylene glycol (PEG) solution: Dissolve polyethylene glycol solid (average MW 8000 g/mol, Fluka) in sterile ddH$_2$O. Heat to 55 °C in a water bath and shake periodically until completely dissolved (several minutes).

2.3 Heparin Sulfate Affinity Chromatography

1. Freeze/thawing baths (*see* Subheading 2.2, **items 2** and **3**).

2. Benzonase (Sigma-Aldrich) or another universal nuclease.

3. 5% sodium deoxycholate solution: Dissolve sodium deoxycholate solid (Sigma-Aldrich) in water to a final concentration of 5% w/v. The solution may be heated to 55 °C.

4. Heparin-agarose beads, such as Sigma-Aldrich H-6508.

5. A narrow-bore gravity column made from disposable plastic 1 mL syringe plugged with cotton wool (*see* **Note 3**).

6. Column elution buffer: PBS (Subheading 2.1) supplemented with 500 mM additional NaCl (sterile filtered).

7. MicroBioSpin P6 columns (Bio-Rad), equilibrated with PBS according to the manufacturer's instructions.

2.4 Visualization and Quantification of Viral Infectivity by Microscopy	1. HEK293T cells, cultured as described in Subheading 2.1. 2. 500 mM sodium butyrate solution: Dissolve sodium butyrate (Sigma-Aldrich) in sterile ddH$_2$O to a final concentration of 500 mM. Filter through a 0.22 μm sterile filter. 3. Fluorescence microscope equipped with a suitable filter set to image EGFP expression.
2.5 Visualization and Quantification of Viral Infectivity by Flow Cytometry	1. All materials described in Subheading 2.4. 2. 0.25% trypsin-EDTA solution (Thermo Fisher Scientific) for cell detachment. 3. DMEM supplemented with FBS as described in Subheading 2.1. 4. 100 μM filters or cell strainer tubes. 5. A flow cytometer capable of analyzing EGFP expressing cell population.
2.6 Titration of Viral Genomes by Quantitative PCR	1. A commercially available AAV2-targeted qPCR kit such as the AAVPro Titration Kit (Clontech). 2. Compatible qPCR instrument and plates.
2.7 Visualization of Virus and Labeling on Protein Gels	1. Polyacrylamide gradient gels. We used 4–15% TGX gels from Bio-Rad. 2. Compatible electrophoresis chamber. 3. 6× SDS gel loading buffer: To make 10 mL total buffer, add 1.5 g sodium dodecyl sulfate (SDS), 3 mL β-mercaptoethanol, 3.5 mL of a 1 M Tris–HCl solution (pH 6.3), 120 μL bromophenol blue, and glycerol to a final volume of 10 mL. For gels to be silver stained, filter the loading buffer through a 0.22 μm filter immediately before mixing with the sample. This removes SDS precipitates which could otherwise cause streaks on the gel. 4. Tris-glycine protein gel running buffer: To make 10× buffer, dissolve 30.3 g Tris base, 114.1 g glycine, and 10 g SDS in 1 L ddH$_2$O. For running gels, dilute the 10× buffer tenfold. For gels to be silver stained, filter the running buffer through a 0.22 μm filter immediately before use to remove SDS precipitates. 5. Molecular weight markers capable of indicating 50–100 kDa, such as Bio-Rad Kaleidescope ladder. 6. A commercial silver staining kit such as the Sigma-Aldrich ProteoSilver staining kit. 7. DBCO-Cy5 (Sigma-Aldrich) or similar fluorophore, dissolved to a concentration of 200 μM in DMSO. 8. Gel imager capable of detecting silver stain and fluorescent label.

3   Methods

3.1   Plasmid System and TAG Mutagenesis for Producing UAA-Containing AAV

The crystal structure of AAV2 capsid only shows the 533 amino acid C-terminal domain, common to all three capsid proteins. The N-terminal extensions associated with rare capsid proteins VP1 and VP2 are believed to be internal. Consequently, sites that are accessible to functionalization upon UAA incorporation reside within the common C-terminal domain. Introducing a TAG mutation within the coding sequence of this common region allows UAA incorporation into all 60 copies of the three capsid proteins. In this protocol, we describe the incorporation of AzK into the T454 site of AAV2 (VP1 numbering). However, simple mutagenesis can be used to introduce the TAG mutation at other sites of the AAV2 capsid for further applications. It is important to note that the site of the TAG mutation can have a significant impact on the efficiency of TAG suppression, virus assembly, and viral infectivity. Additionally, for the incorporation of AzK we have used the wild-type MbPylRS/tRNA pair (*see* **Note 4**). The wild-type MbPylRS is also capable of charging several other UAAs, which may also be introduced onto AAV2 for various applications [10, 13]. Furthermore, an even greater selection of UAAs can be accessed by replacing the wild-type MbPylRS in pIDTSmart-MbPylRS-RC2 (Subheading 2.1, **item 3**) with other engineered variants [10, 11, 14].

3.2   Production of UAA-AAV

1. Seed HEK293T cells in 12-well plates (about 0.7 million cells per well) or 10 cm dishes (about 7–8 million cells per dish) so that they will be mostly (70–80%) confluent in 24 h. One well in a 12-well plate is sufficient for evaluating production efficiency and the infectivity of the resulting virus (*see* **Note 5**). One or more 10 cm dishes are recommended for virus preparations intended for stringent purification, or other large-scale analyses.

2. Set up transfection mixtures in a tissue culture hood. Add ingredients in the following order, and mix the tube by shaking vigorously after the addition of the PEI (*see* **Note 6**).

   (a) UAA-AAV production plasmids. Use 1–1.5 µg of DNA for a single well in a 12-well plate, or 16–24 µg of DNA for one 10 cm dish. An equimolar ratio of the plasmids is a good starting point, but this can be further optimized for modified plasmid systems.

   (b) DMEM media without FBS. Use 17.5 µL of media for every 1 µg of DNA.

   (c) 1 mg/mL polyethyleneimine (PEI). Use 3–4 µL for every 1 µg of DNA—the exact amount can be optimized for your system.

3. Let the transfection mixtures sit in the hood (room temperature) for 10 min.

4. Add transfection mixtures dropwise to the cells, spreading evenly over the entire dish, and mix gently but thoroughly by rocking the plate (*see* **Note 7**).

5. Add required volume of AzK from the 100 mM stock dropwise over the entire dish, to a final concentration of 1 mM. Mix gently but thoroughly by rocking the plate.

6. Optional: gently exchange the media with fresh DMEM (complete) containing 1 mM AzK 24 h after transfection. This step was found to improve yield but is not strictly necessary.

7. Harvest the virus 72 h after transfection. At this point, the cells should be strongly expressing the EGFP reporter (which serves as a facile verification for good transfection efficiency), and may start to appear somewhat rounded, with some detached from the plate. Gently detach the cells from the plate using a 1 mL pipette for a 12-well plate or a cell lifter for a 10 cm dish (*see* **Note 8**). Transfer the resuspended cells in culture media to a centrifuge tube, and use PBS to rinse the plate and combine with the resuspended cells.

8. Pellet the cells by centrifugation at $5000 \times g$ for 5 min. Gently remove the supernatant. The cell pellet can be frozen at −80 °C for long-term storage.

3.3 Semi-purification of AAV by PEG Precipitation

PEG precipitation is a strategy to rapidly concentrate and semi-purify the virus, where small molecules and soluble proteins remain in the solution, while AAV2 (and some components of the cell lysate) will form a precipitate (*see* **Note 9**). This method yields sufficient purity to rapidly verify AzK incorporation into the capsid proteins by fluorophore labeling followed by SDS-PAGE analysis. To more stringently purify the virus by heparin sulfate affinity chromatography, for applications such as visualization by silver-stained SDS-PAGE, skip to Subheading 3.4.

1. Resuspend the cell pellet in lysis solution. Use 1 mL for a 10 cm dish or 100 μL for one well in a 12-well plate.

2. Prepare two freeze/thawing baths: a −80 °C dry ice/ethanol bath and a 37 °C water bath.

3. Lyse cells by freezing and completely thawing the resuspended cell solution twice.

4. Centrifuge at $5000 \times g$ for 5 min to clear cell debris (*see* **Note 10**). Discard the pellet.

5. Add 40% PEG to the lysate to a final concentration of 11% (*see* **Note 11**). Mix very well. The solution will become cloudy. Let it sit overnight at 4 °C.

6. Pellet the precipitate by centrifugation at $\geq 5000 \times g$ for 30 min. A white pellet should be visible, and it will contain the virus.

7. Remove all of the supernatant and discard. Using a pipette, thoroughly resuspend the pellet in PBS—use 1 mL for virus from a 10 cm dish or 100 μL for virus from a well in a 12-well plate. A vortexer can be used on medium speed to break up aggregates and ensure a homogenous solution.

8. This process should yield 1 mL of semi-pure AAV2 from a 10 cm dish. For the resulting virus preparation, the infectivity can be assessed by infecting HEK293T cells (Subheadings 3.5 and 3.6), the genomic titer of the virus can be analyzed by qPCR (Subheading 3.7), and the presence of the azido functionality can be confirmed by fluorophore labeling followed by SDS-PAGE analysis (Subheading 3.8).

3.4 Purification of AAV by Heparin Sulfate Affinity Chromatography

This method relies on the affinity of the assembled AAV2 particles for heparin sulfate. It is important to remember that it cannot be used for mutant viruses which do not bind with the primary receptor heparin sulfate proteoglycan (HSPG). To get high-quality virus from the protocol below, wild-type virus from at least one 10 cm dish or mutant virus from at least four 10 cm dishes must be used. This protocol is adapted and scaled down from a published method [15].

1. Resuspend the cell pellet in DMEM without added FBS. Use 1 mL per 10 cm dish.

2. Prepare two freeze/thawing baths: a −80 °C dry ice/ethanol bath and a 37 °C water bath.

3. Freeze and completely thaw the resuspended cell solution twice.

4. Centrifuge at $5000 \times g$ for 5 min to clear cell debris. Discard the pellet.

5. Treat with universal nuclease (e.g., benzonase) to digest DNA and RNA. We use 1 μL (250 U) of Pierce universal nuclease for the lysate derived from a 10 cm dish, and incubate for 10 min at room temperature.

6. Add 5% (10×) sodium deoxycholate to the supernatant to a final concentration of 0.5%. Mix by inverting the tube and incubate at 37 °C for 30 min. During this time, prepare the column.

7. Add 600 μL heparin-agarose slurry into a narrow-bore gravity column, or a disposable plastic 1 mL syringe plugged with cotton wool. Wash the resin with 3 mL PBS, push carefully through with compressed air to tightly pack the resin bed

(make sure not to run the column dry). The final bed volume should be ~300 µL.

8. After deoxycholate treatment is complete, centrifuge the lysate at 20,000 × g for 5 min to remove any remaining debris which could clog the column.

9. In the tissue culture hood, slowly load the lysate onto the column and let it flow through by gravity (*see* **Note 12**).

10. Wash the resin bed with 3 × 1 mL PBS.

11. Elute virus from the column in four 300 µL fractions using the elution buffer.

12. Concentrate the virus by PEG precipitation by adding 360 µL of 40% PEG to the combined elution fractions in a microcentrifuge tube. Mix thoroughly and incubate at 4 °C overnight.

13. Pellet the precipitate by centrifugation at 12,000 × g for 30 min. The pellet may not be visible.

14. Thoroughly resuspend the precipitated virus in 75 µL PBS by washing the wall of the tube with a pipette.

15. Remove any remaining large aggregates or small-molecule impurities by passing the virus through a Bio-Rad MicroBioSpin P6 column pre-equilibrated with PBS.

16. This process should yield ~75 µL of pure virus. For the resulting virus preparation, the infectivity can be assessed by infecting HEK293T cells (Subheadings 3.5 and 3.6), the genomic titer of the virus can be analyzed by qPCR (Subheading 3.7), the purity can be assessed by SDS-PAGE analysis followed by silver stain (Subheading 3.9), and the presence of the azido functionality can be confirmed by fluorophore labeling followed by SDS-PAGE analysis (Subheading 3.8).

3.5 Visualization of Infectivity by GFP Expression in Cultured Cells

1. Seed HEK293T cells in a 12-well plate so that they will be mostly (70–90%) confluent in 24 h.

2. 24 h after seeding, add 500 mM sodium butyrate (100×) to the cells to a final concentration of 5 mM. Gently mix the plate. Include a control well that receives sodium butyrate but no virus (*see* **Note 13**).

3. Add the PEG-precipitated or heparin column-purified virus to the cells. Use approximately 10 µL of a PEG-precipitated lysate or 2 µL of column-purified virus. Larger or smaller amounts may be needed based on the yield of the virus prep and any mutations which affect infectivity. Mix the plate gently but thoroughly.

4. Return the cells to the incubator for 48 h.

5. 48 h after infection, image the cells using a fluorescent microscope.

3.6 Quantification of Viral Infectivity by Flow Cytometry

This method is best suited for test wells in which <30% of the cells appear infected, and overall fluorescence levels are dim (*see* **Notes 14** and **15**). Adjust the amount of virus added accordingly.

1. Infect and image cells in a 12-well plate as described in Subheading 3.5.

2. Prepare a 1.5 mL microcentrifuge tube for each well and chill in ice. Chill 1 mL DMEM with FBS and 500 μL PBS for each well in ice, and warm 200 μL trypsin per well to 37 °C.

3. Remove the culture media and add 200 μL warm trypsin solution to each well.

4. Incubate the plate at 37 °C for 1–2 min, until the cells are visibly detached.

5. Add 500 μL of chilled media to each well to quench the trypsin, then transfer the contents of each well to a 1.5 mL microcentrifuge tube; use an additional 500 μL of media to rinse the plate and combine in the same microcentrifuge tube.

6. Centrifuge the tubes at 2500 × g for 5 min and remove the supernatant.

7. Add 500 μL chilled PBS to each tube and gently resuspend the cell pellet.

8. Filter the resuspended cells through a 100 μm filter.

9. Analyze the cells by flow cytometry and quantify the fluorescent population. The percentage of infected cells should follow the same trends as the visible fluorescence expectations described in Subheading 3.5.

3.7 Titration of the Virus by Quantitative PCR

Packaged AAV genomes can be titrated using a commercially available qPCR kit like Clontech's AAVPro Titration Kit (*see* **Note 16**).

3.8 Validation of Labeling Chemistry by Protein Gel

Reactivity of the azido side chain on the virus surface can be confirmed by functionalizing intact virus with a cyclooctyne-fluorophore conjugate, then denaturing the virus and visualizing labeled proteins on an SDS-PAGE gel. We recommend using heparin column-purified virus wherever possible for this step. PEG-precipitated lysates may be used for crude characterization, but contain other proteins which can cross-react with the cyclooctyne probe at a low but detectable level (*see* **Note 17**).

1. Standardize the amount of virus to be loaded by infective (Subheadings 3.5 and 3.6) or genomic titer (Subheading 3.7). Start with approximately 10^9 genome copies.

2. Incubate the virus with 10–20 μM cyclooctyne-fluorophore conjugate (such as DBCO-Cy5) for 1 h (*see* **Note 18**).

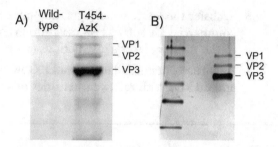

Fig. 2 Chemoselective labeling and SDS-PAGE analysis of AzK-modified AAV2. (**a**) Wild-type and AzK-modified AAV2 were purified by heparin sulfate affinity chromatography and were subjected to labeling with DBCO-Cy5, as described in Subheading 3. Subsequent SDS-PAGE analysis followed by fluorescence imaging shows selective labeling of AzK-modified AAV2. (**b**) Silver stain analysis of AAV2 purified by heparin sulfate affinity chromatography, after resolving the capsid proteins by SDS-PAGE. The left lane shows molecular weight markers

3. Load the virus onto a gradient gel (*see* Subheading 3.9) along with a molecular weight marker which will be visible under fluorescence imaging conditions (e.g., Bio-Rad Kaleidescope ladder), and run until the molecular weight markers in the 50–100 kDa range are well resolved.

4. Detect the presence of fluorophore associated with capsid proteins using a fluorescence gel imager equipped with an appropriate filter set.

5. Lanes containing azido-bearing viruses functionalized with fluorophore should show three bands at 82, 67, and 60 kDa with relative intensities of approximately 1:1:10 (Fig. 2a). These bands should not be observed in lanes containing viruses with no azido side chains (e.g., wild-type).

3.9 Silver Staining of AAV Capsid Proteins in a SDS-PAGE Gel

Virus purity, and the presence of capsid proteins independent of labeling, can be confirmed by silver staining. We recommend using HSPG-purified virus for this step. For best resolution, the AAV capsid proteins should be visualized on a commercially available gradient 4–15% polyacrylamide gradient gel or similar. The sizes of the capsid proteins are 82 kDa for VP1, 67 kDa for VP2, and 60 kDa for VP3, and they should appear at an approximate ratio of 1:1:10 (Fig. 2b).

1. Standardize the amount of virus to be loaded by genomic (Subheading 3.7) or infective titer (Subheadings 3.5 and 3.6). Start with approximately 10^{10} genome copies.

2. Load the virus onto the gel along with a dilute molecular weight marker. The relative amounts for your virus and marker must be empirically optimized for best results—we recommend starting with a 10–100-fold dilution of Bio-Rad Kaleidescope ladder. The electrophoresis should be allowed to continue until 60–80 kDa proteins can be distinguished, approximately 30–60 min at 50 mA.

3. Transfer the gel to a clean staining container and stain using a commercial kit such as Sigma-Aldrich ProteoSilver according to the manufacturer's instructions (*see* **Note 19**).

4. All virus-containing lanes should show three bands at 82, 67, and 60 kDa with relative intensities of approximately 1:1:10.

4 Notes

1. Efficient UAA-modified AAV production requires healthy, low-passage cells.

2. It is important to use the specific kind of PEI mentioned here for UAA-modified AAV production. Use of other kinds of PEIs, which are suitable for transfection in other contexts, led to severely decreased yields. PEI that has been stored for more than 3 months in its solid form was also found to have reduced efficacy. Additionally, vortexing or filtration during the preparation of the PEI solution was found to negatively impact the transfection efficiency.

3. Narrow-bore columns with longer column length led to more efficient purification.

4. The tRNA is the limiting component in UAA-containing protein synthesis. A multi-copy tRNA cassette is used to boost yields [16].

5. It is strongly recommended to test the production of each TAG mutant in the presence or absence of the UAA. Robust production should be observed only in the presence of the UAA. A wild-type (no TAG mutation) positive control should be included as well. Yields of UAA-modified viruses vary between 10% and 30% relative to wild-type virus.

6. Poor mixing will lead to low transfection efficiency. Other potential causes for poor transfection include old or low-quality plasmid stocks, PEI that has been improperly stored, poor cell health, and transfection when cells were less than 50% or more than 90% confluent.

7. HEK293T cells do not adhere strongly to the plate, and care must be taken while mixing or changing media not to dislodge them.

8. PEI transfection makes HEK293T cells adhere to the plate more strongly than usual. Thus, alternative detachment methods such as the addition of EDTA are less effective.

9. PEG precipitation is also a suitable semi-purification method for in vitro retargeting experiments on cultured cells.

10. Higher centrifugation speeds have been observed to lead to lower yields.

11. If a significant amount of virus is found to be released into the culture medium before or during cell harvesting, this can be recovered from the culture medium by PEG precipitation. Combine the cell lysate and culture media in the same centrifuge tube, and add PEG to a final concentration of 11%.

12. Attempts at expediting the flow using compressed air may lead to inefficient binding of the virus to the column, decreasing the yield. Compressed air may be used gently during the wash and elution steps.

13. Sodium butyrate is a histone deacetylase inhibitor which enhances expression of AAV encoded transgenes. The resulting increase in EGFP expression facilitates its visualization and quantification. Sodium butyrate acts as a cell cycle arrestor and can be toxic to cells at higher levels, or upon prolonged exposure.

14. When a small fraction of cells are infected, the number of EGFP-expressing cells more accurately reflects the infective titer of the virus preparation. When the majority of cells are infected, the number of EGFP expressing cells no longer accurately reflects the infective virus titer due to the increased frequency of cells with multiple infections.

15. For AAV preparations infective titer is typically lower than the genomic titer (as determined by qPCR).

16. While we observe no infective virus produced from control wells which were transfected with the required plasmids but not supplemented with the UAA, qPCR titration indicates the production of low levels of packaged AAV genomes (~10% relative to the levels observed for the corresponding +UAA well) from these wells. This may be an artifact of the qPCR assay, due to the presence of a noninfective aggregate between viral genome and truncated capsid proteins.

17. Cyclooctyne probes react at low levels with cysteine-thiols in cell lysates [17]. The appearance of unwanted bands can be minimized by desalting a small amount of PEG-precipitated virus as described in Subheading 3.4, **step 15**, and by optimizing cyclooctyne concentration and reaction time.

18. When DBCO-Cy5 is used at 10–20 μM concentration for 1 h, labeling is only partial (~25%). If complete labeling all azide sites are desired, then higher concentration of the reagent and longer incubation periods should be employed.

19. Silver staining is based upon the reduction of silver nitrate to metallic silver, and unwanted staining may occur due to insufficiently clean running or staining containers or defects in the gel itself. To minimize staining artifacts, sterile-filtration of both the running buffer and sample loading buffer was found to be essential. The electrophoresis chamber and staining con-

tainers should be cleaned thoroughly before use. Lanes with disproportionately high protein concentration (relative to virus proteins) in the gel will also affect the overall sensitivity. Consequently, if a lane of concentrated molecular weight marker is used to monitor the electrophoresis process, it must be cut off before staining.

References

1. Samulski RJ, Muzyczka N (2014) AAV-mediated gene therapy for research and therapeutic purposes. Annu Rev Virol 1:427–451. https://doi.org/10.1146/annurev-virology-031413-085355

2. Mingozzi F, High KA (2011) Therapeutic in vivo gene transfer for genetic disease using AAV: progress and challenges. Nat Rev Genet 12:341–355. https://doi.org/10.1038/nrg2988

3. Kotterman MA, Schaffer DV (2014) Engineering adeno-associated viruses for clinical gene therapy. Nat Rev Genet 15:445–451. https://doi.org/10.1038/nrg3742

4. Agbandje-McKenna M, Kleinschmidt J (2011) AAV capsid structure and cell interactions. Methods Mol Biol 807:47–92. https://doi.org/10.1007/978-1-61779-370-7_3

5. Douar AM, Poulard K, Danos O (2003) Deleterious effect of peptide insertions in a permissive site of the AAV2 capsid. Virology 309:203–208

6. Horowitz ED, Weinberg MS, Asokan A (2011) Glycated AAV vectors: chemical redirection of viral tissue tropism. Bioconjug Chem 22:529–532. https://doi.org/10.1021/bc100477g

7. Liu Y, Fang Y, Zhou Y, Zandi E, Lee CL, Joo KI, Wang P (2013) Site-specific modification of adeno-associated viruses via a genetically engineered aldehyde tag. Small 9:421–429. https://doi.org/10.1002/smll.201201661

8. Shi W, Bartlett JS (2003) RGD inclusion in VP3 provides adeno-associated virus type 2 (AAV2)-based vectors with a heparan sulfate-independent cell entry mechanism. Mol Ther 7:515–525

9. Stachler MD, Chen I, Ting AY, Bartlett JS (2008) Site-specific modification of AAV vector particles with biophysical probes and targeting ligands using biotin ligase. Mol Ther 16:1467–1473. https://doi.org/10.1038/mt.2008.129

10. Dumas A, Lercher L, Spicer CD, Davis BG (2015) Designing logical codon reassignment—expanding the chemistry in biology. Chem Sci 6:50–69

11. Liu CC, Schultz PG (2010) Adding new chemistries to the genetic code. Annu Rev Biochem 79:413–444. https://doi.org/10.1146/annurev.biochem.052308.105824

12. Kelemen RE, Mukherjee R, Cao X, Erickson SB, Zheng Y, Chatterjee A (2016) A precise chemical strategy to alter the receptor specificity of the adeno-associated virus. Angew Chem Int Ed Engl 55:10645–10649. https://doi.org/10.1002/anie.201604067

13. Chatterjee A, Xiao H, Bollong M, Ai HW, Schultz PG (2013) Efficient viral delivery system for unnatural amino acid mutagenesis in mammalian cells. Proc Natl Acad Sci U S A 110:11803–11808. https://doi.org/10.1073/pnas.1309584110

14. Erickson SB, Mukherjee R, Kelemen RE, Wrobel CJJ, Cao X, Chatterjee A (2017) Precise photoremovable perturbation of a virus–host interaction. Angew Chem Int Ed Engl 56:4234–4237. doi: 10.1002/anie.201700683

15. Auricchio A, Hildinger M, O'Connor E, Gao GP, Wilson JM (2001) Isolation of highly infectious and pure adeno-associated virus type 2 vectors with a single-step gravity-flow column. Hum Gene Ther 12:71–76. https://doi.org/10.1089/104303401450988

16. Zheng Y, Lewis TL Jr, Igo P, Polleux F, Chatterjee A (2016) Virus-enabled optimization and delivery of the genetic machinery for efficient unnatural amino acid mutagenesis in mammalian cells and tissues. ACS Synth Biol 6:13. https://doi.org/10.1021/acssynbio.6b00092

17. van Geel R, Pruijn GJ, van Delft FL, Boelens WC (2012) Preventing thiol-yne addition improves the specificity of strain-promoted azide-alkyne cycloaddition. Bioconjug Chem 23:392–398. https://doi.org/10.1021/bc200365k

Chapter 21

Generation of Intramolecular FRET Probes via Noncanonical Amino Acid Mutagenesis

Simone Brand and Yao-Wen Wu

Abstract

Förster resonance energy transfer (FRET) probes are powerful tools to monitor protein–protein interactions and enzyme activities in a spatiotemporal manner in live cells. Using a combination of noncanonical amino acid (ncAA) mutagenesis and bioorthogonal labeling, we have developed intramolecular FRET probes consisting of a fluorescent protein and an organic dye within an individual protein. Herein we present a general approach to establish intramolecular FRET probes for imaging of protein activity in live cells.

Key words Unnatural amino acid mutagenesis, Noncanonical amino acid, Stop codon, Amber suppression, Intracellular labeling, FRET probes, Bioorthogonal labeling

1 Introduction

Research over the last two decades has identified a number of powerful fluorescence-based techniques, allowing for the detection of protein–protein interactions or enzymatic activity in a spatiotemporal manner in live cells [1]. Using fluorescently tagged probes, these techniques have been employed to monitor a variety of biological events including protease or kinase activity, protein–protein interactions, polymerization, conformational changes and small GTPase activity [2–4]. The use of fluorescent proteins in general has several major drawbacks including their large size, limited photostability and the limitation to C- or N-terminal protein tagging. This is especially challenging for proteins whose termini are buried in the structure or are required for posttranslational modifications and protein function. One way to tackle this problem is through insertion of the fluorescent protein into loop regions of the target protein, although this approach requires intense characterization of the insertion site to prevent interference with protein function. Additionally, the properties of the utilized linker can dramatically influence both proper positioning of the fluorescent protein and the dynamic range of the sensor [5].

Edward A. Lemke (ed.), *Noncanonical Amino Acids: Methods and Protocols*, Methods in Molecular Biology, vol. 1728, https://doi.org/10.1007/978-1-4939-7574-7_21, © Springer Science+Business Media, LLC 2018

As an alternative to fluorescent proteins, numerous self-labeling protein tags have been developed (the SNAP tag, eDHFR tag, tetracysteine-based tags, etc.). These tags can be genetically fused to the target protein, utilize synthetic fluorophores exhibiting better biophysical properties than fluorescent proteins and vary in size from 4 to >200 amino acids [6–8].

Chemical approaches have yielded different in vitro strategies to overcome the problem of tag size and the limitation of labeling solely the protein termini. These biocompatible techniques allow for the coupling of any suitably functionalized synthetic dye to the target protein by targeting amino acids residues such as lysine or cysteine moieties [9–11]. However, due to the biological abundance of these amino acids and the fact that all accessible target moieties can potentially react, the labeling specificity of these reactions is limited. To enhance labeling specificity, nontargeted residues could be exchanged if they are not essential for protein folding and/or function. In some cases side reactions can also be avoided by kinetic labeling [9, 12]. However, due to a lack of specificity these labeling strategies are not suitable for live cell applications. In contrast, the ncAA mutagenesis approach enables site-specific incorporation of fluorescent moieties or bioorthogonal handles that allow to specifically address a particular functional group in the target protein [13–15]. This strategy utilizes ncAA as the "minimalist" tag and allows introduction of organic dyes with minimal perturbance to a target protein [16–18].

We have used this approach to generate a number of intramolecular FRET probes with a fluorescent protein as the donor and an organic dye as the acceptor. Some of them have been used as conformational sensors for small GTPases activity (COSGAs) to spatiotemporally visualize Ras and Rab activity in live cells [12]. Herein we present a general protocol to generate the FRET probes using stop-codon amber suppression and bioorthogonal labeling in the live cell. Many cell-permeable organic dyes are known to accumulate in endomembranes, resulting in high background signals within the cell and thereby hindering detection of specific labeling of intracellular proteins. To avoid this we have equipped the previously described reporter construct eGFP182TAG with a flexible linker and a CaaX box at its C-terminus [19]. When expressed in cells, the CaaX box gets prenylated and thereby targets the fusion protein to the cytosolic side of the plasma membrane [20] (Fig. 1). The stop codon at position 182 in the eGFP prevents maturation of the fluorophore in the absence of the noncanonical amino acid, facilitating verification of the ncAA incorporation. In the second step, the bioorthogonal group of the ncAA can be targeted with cell-permeable tetrazine dyes, resulting in rapid and highly specific labeling of the target protein (Fig. 2). The intramolecular FRET is monitored by fluorescence lifetime imaging microscopy (FLIM).

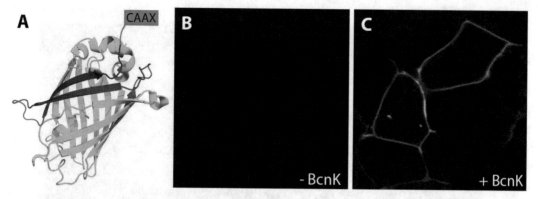

Fig. 1 Composition of the reporter gene eGFP182TAG-CaaX. (**A**) Structure of eGFP, the N-terminal sequence to the amber stop codon is colored green, while the sequence downstream is colored red. In the absence of noncanonical amino acid the amber codon is not suppressed, preventing the eGFP fluorophore to mature and fluoresce (**B**). Presence of noncanonical amino acid, here BCN, leads to full length protein expression, maturation of the fluorophore, and prenylation of the CaaX box, targeting the fusion protein to the cytosolic side of the cytoplasma membrane (**C**)

2 Materials

2.1 Construction of the Sensor

1. cDNA of the protein of interest (POI).
2. Citrine or GFP encoding plasmid for expression in mammalian cells (e.g. pCitrine from Clonentech).
3. Competent *E. coli* suitable for cloning.
4. Primers for insertion of the target gene into the expression vector.
5. Primers for site-directed mutagenesis to introduce amber codons into the target gene.

2.2 Mammalian Cell Culture Consumables and Reagents

1. Tissue culture vessels, e.g. 10 cm diameter dishes (Sarstedt).
2. Sterile single packed pipets (5 and 10 mL).
3. Pipette controller.
4. Sterile microliter pipets and pipet tips (10, 200, and 1000 μL).
5. Glass bottom imaging vessels such as 35 mm MaTek dishes (MaTek) or x-well chambers (Sarstedt).
6. 50 mL conical tubes, sterile.
7. 1.5 mL reaction tubes, sterile.
8. Inverted light microscope.
9. HEK293T cells (ATCC® CRL-3216™).
10. Cell culture medium: DMEM supplemented with 10 % (v/v) FBS, 1 % (v/v) NEAA, and 1 % (v/v) sodium pyruvate.

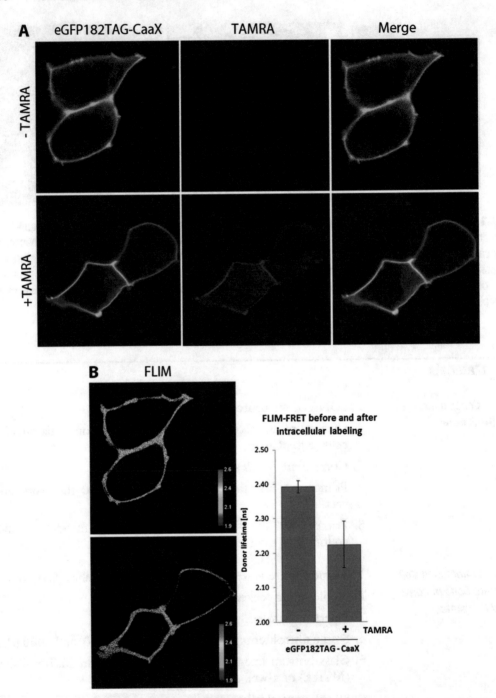

Fig. 2 Intracellular labeling of eGFP182TAG-CaaX with TAMRA. (**A**) Confocal fluorescence images of HEK293T cells expressing amber suppressed eGFP182TAG-CaaX. The highly specific labeling is clearly visible in the red channel only in presence of the dye. (**B**) Verification of the labeling reaction by FLIM-FRET: The fluorescence lifetime of the donor eGFP in the absence of an acceptor is 2.4 ns. After successful labeling, TAMRA acts as a FRET acceptor and the donor fluorescence lifetime drops to 2.2 ns, verifying successful labeling

11. Opti-MEM, serum-free.

12. XtremeGene HP transfection reagent (Roche).

13. Phosphate-buffered saline (PBS).

14. 0.01% Poly-L-lysine solution (mol wt 15,000–300,000).

2.3 Site-Specific BCN Incorporation and Protein Labeling in Mammalian Cells

1. PylRS AF plasmid [21].

2. endo BCN-L-Lysine (SiChem, SC-8014).

3. 0.1 M NaOH, sterile filtered.

4. 1 M HEPES buffer pH 7.0, sterile filtered.

5. H-tetrazine-5-TAMRA (Jena Bioscience, CLK-017-05).

6. DMSO, cell culture grade.

7. Inverted fluorescence microscope such as JuLI™-Smart Fluorescent Cell Analyzer (NanoEnTek).

8. Confocal fluorescent microscope such as a Leica TCS SP5, equipped with a White Light Laser (Koheras) and a multidimensional TCSPC module for FLIM.

3 Methods

3.1 Design and Cloning of the Target Construct

1. Determination of the incorporation site: To generate an intramolecular FRET probe, the ncAA incorporation site, and thereby the labeling site within the protein has to fulfil several criteria: (a) It should be within 10 nm distance to the FRET-partner in at least one conformational state of the probe. (b) The labeling site must not be critical for any interaction or function of the protein. (c) It needs to be sterically accessible for chemical labeling. These parameters can be assessed by rational design based on the crystal structure of the target protein (*see* **Note 1**).

2. When N-terminally attached to the POI, the fluorescent protein is translated first and can mature even if the POI is not properly expressed. In this case, fluorescence is not an indicator for successful incorporation of the ncAA into the target protein. Thus, it is recommended to position the fluorescent protein downstream of the target protein by fusing it to its C-terminus. If the C-terminus is essential for protein function, subcellular localization can serve as an indicator for successful amber suppression and thereby successful expression of the POI.

3. Clone the cDNA of the target protein into a Citrine or eGFP containing plasmid by restriction enzyme digestion and subsequent DNA ligation (*see* **Note 2**). Transform the ligation into

competent *E. coli* and screen for positive clones by PCR. Verify the sequence of the positive clones.

4. Mutate each of the ncAA incorporation sites to a TAG codon by single site-directed mutagenesis and verify the sequence of the final construct.

3.2 Site-Specific Incorporation of BCN into Proteins in Mammalian Cells

All solutions used for cell culture should be cell culture grade, sterile, and prewarmed in a water bath before use. To avoid unwanted cell detachment, tilt the vessel slightly and add/remove solutions by pipetting to/from the vessel wall. Unless indicated otherwise, all incubation steps are carried out in a humidified cell culture incubator at 37 °C and 10 % CO_2.

1. To enhance cell adherence to the culture vessel, coat the bottom with Poly-L-Lysine according to the manufacturer's protocol.

2. Remove the growth medium from HEK293T cells and wash once with PBS. Add 1 mL Trypsin solution, incubate for several minutes at room temperature and verify cell detachment under a light microscope. Add 9 mL of DMEM and gently pipet to disperse cell aggregates. Transfer the cell suspension to a 50 mL tube and count the cell number.

3. Seed 1.8×10^4 HEK293T cells per cm^2 growth area in the precoated glass bottom dish or chamber objective slide (*see* **Note 3**). Grow the cells overnight.

4. Prepare the transfection mix per 15,000 cells as follows: Dilute 150 ng of PylRS AF plasmid and 150 ng of the POI plasmid in 40 μL Opti-MEM in a 1.5 mL reaction tube (*see* **Note 4**). Mix gently and add 0.4 μL XtremeGene HP reagent; do not mix after addition of the transfection reagent. Incubate for at least 15 min at room temperature to allow complex formation. Mix the transfection mix gently and add 40 μL of the mix to each well. Rock the wells gently to achieve equal distribution of the complex in the culture media.

5. Dissolve BCN in 0.1 M NaOH to a final concentration of 250 mM. Prepare a 1:5 dilution of BCN in 1 M HEPES pH 7.0 buffer freshly before each experiment. Add the diluted BCN solution directly after transfection to the samples to a final concentration of 250 μM and rock gently to obtain optimal BCN distribution (*see* **Note 5**). Incubate the transfected samples overnight (*see* **Note 6**).

6. The following day exchange media in all samples to remove free BCN from the growth medium. Incubate overnight (*see* **Note 7**).

7. Expression of amber suppressed POI can be monitored by fluorescence microscopy.

3.3 Generating Intramolecular FRET-Probes via Intracellular Protein Labeling with Tetrazine Dyes

1. Prepare a 10 mM stock solution of H-tetrazine-5-TAMRA by solubilizing 100 µg of TAMRA powder in 16.68 µL DMSO (*see* **Note 8**). Mix vigorously.

2. Dilute tetrazine-TAMRA to a final concentration of 0.5 µM into prewarmed DMEM and mix vigorously (*see* **Note 9**). Remove the growth medium from the transfected samples and add the dye solution. Incubate for 30–45 min (*see* **Note 10**).

3. Remove the dye solution from the cells and carefully rinse at least once with fresh medium. After labeling, incubate the samples for at least 2 h in growth medium to remove residual free dye (*see* **Note 11**).

4. Exchange media to phenol red-free medium before imaging the samples under a fluorescent microscope. Successful labeling leads to strong colocalization of the TAMRA signal with the donor fluorescence signal and can be validated by sensitized emission imaging and/or FLIM-FRET.

5. Covalent labeling can also be verified by in-gel fluorescence analysis: lyse samples before and after labeling, and run the lysates on a SDS-PAGE. Scan the gel with a Typhoon Scanner to monitor specific fluorescence upon labeling and verify the identity of the band by Western blot.

4   Notes

1. The length and flexibility of the linker between the POI and the fluorescent protein can affect the distance and orientation of the FRET-partners. We found that varying both linker length and backbone flexibility can increase the FRET-signal up to 20 % [12]. Besides optimizing the linker, the orientation of the FRET-partners can be altered by using circularly permuted fluorescent proteins [22, 23].

2. The choice of donor fluorophore depends on the FRET-pair, as well as on the detection method: due to its higher photostability in vivo, its monoexponential decay curve and higher fluorescent lifetime, Citrine is superior to eGFP for FLIM-FRET measurements [24].

3. We use x-well objective slides on cover glass (Sarstedt, order number 94.6190.802) with 0.8 cm^2 growth area per well and 300–400 µL total volume.

4. Optimal plasmid ratio should be determined empirically for the protein of interest. Higher PylRS AF:POI plasmid ratios may increase incorporation efficiency.

5. The concentration of BCN in the growth medium is critical for optimal amber suppression and can vary for different cell lines,

target constructs, and/or incorporation sites. The optimal working concentration can be determined by titration of BCN and subsequent quantification of the yield of full-length amber suppressed POI by Western blot.

6. The expression time in the presence of BCN in the growth medium can affect the overall yield of amber suppressed POI. It can vary between few hours up to several days and should be determined empirically for each POI and incorporation site. Some POIs may have a high cellular turnover and/or may be degraded or cytotoxic during long expression periods.

7. High amounts of BCN may require longer wash-out time before labeling to reduce unspecific background signals. In our hands, around 30 h expression time in the presence of BCN, followed by 8 h incubation without BCN and subsequent labeling, results in highly specific labeling and low background signal.

8. Aliquot the stock solution and store aliquots at −20 °C. Frequent freeze–thaw cycles may lead to precipitation of TAMRA and can reduce labeling efficiencies.

9. Avoid final concentrations of DMSO higher than 0.1 % (v/v) as this may cause altered intracellular signaling and cytotoxicity.

10. Although the inverse electron demand Diels–Alder reaction in live cells has been reported to be complete within 20 min, we have observed that the labeling yield can be improved by prolonging the incubation time up to 45 min. Longer incubation periods, even up to several hours, did not further increase the overall labeling yield.

11. We found that the washing time does not only depend on the concentration of the dye but also on its biophysical properties and the used cell line. On one hand, prolonging the washing time increases the signal-to-noise ratio for imaging, while on the other hand, depending on the cellular half-life of the POI, the overall amount of labeled POI may be reduced due to protein turnover.

Acknowledgments

This work is supported by the Deutsche Forschungsgemeinschaft, DFG (grant No.: SPP 1623), European Research Council, ERC (ChemBioAP), and Behrens-Weise-Stifung. S.B. acknowledges support from the IMPRS-CMB. We thank Edward Lemke and Carsten Schultz for the kind gift of PylRS AF plasmid and BCN.

References

1. Rodriguez EA, Campbell RE, Lin JY et al (2017) The growing and glowing toolbox of fluorescent and photoactive proteins. Trends Biochem Sci 42(2):111–129

2. Pertz O, Hahn KM (2004) Designing biosensors for Rho family proteins--deciphering the dynamics of Rho family GTPase activation in living cells. J Cell Sci 117:1313–1318

3. Komatsu N, Aoki K, Yamada M et al (2011) Development of an optimized backbone of FRET biosensors for kinases and GTPases. Mol Biol Cell 22:4647–4656

4. Ueda Y, Kwok S, Hayashi Y (2013) Application of FRET probes in the analysis of neuronal plasticity. Front Neural Circuits 7:163

5. Sabet O, Stockert R, Xouri G et al (2015) Ubiquitination switches EphA2 vesicular traffic from a continuous safeguard to a finite signalling mode. Nat Commun 6:8047

6. Chen X, Li F, Wu YW (2015) Chemical labeling of intracellular proteins via affinity conjugation and strain-promoted cycloadditions in live cells. Chem Commun (Camb) 51:16537–16540

7. Chen X, Wu YW (2016) Selective chemical labeling of proteins. Org Biomol Chem 14:5417–5439

8. Liu W, Li F, Chen X et al (2014) A rapid and fluorogenic TMP-AcBOPDIPY probe for covalent labeling of proteins in live cells. J Am Chem Soc 136:4468–4471

9. Voss S, Zhao L, Chen X et al (2014) Generation of an intramolecular three-color fluorescence resonance energy transfer probe by site-specific protein labeling. J Pept Sci 20:115–120

10. Sletten EM, Bertozzi CR (2009) Bioorthogonal chemistry: fishing for selectivity in a sea of functionality. Angew Chem Int Ed Engl 48:6974–6998

11. Wu YW, Goody RS (2010) Probing protein function by chemical modification. J Pept Sci 16:514–523

12. Voss S, Kruger DM, Koch O et al (2016) Spatiotemporal imaging of small GTPases activity in live cells. Proc Natl Acad Sci U S A 113:14348–14353

13. Lang K, Chin JW (2014) Cellular incorporation of unnatural amino acids and bioorthogonal labeling of proteins. Chem Rev 114:4764–4806

14. Lang K, Chin JW (2014) Bioorthogonal reactions for labeling proteins. ACS Chem Biol 9:16–20

15. Nikic I, Plass T, Schraidt O et al (2014) Minimal tags for rapid dual-color live-cell labeling and super-resolution microscopy. Angew Chem Int Ed Engl 53:2245–2249

16. Lang K, Davis L, Wallace S et al (2012) Genetic encoding of bicyclononynes and transcyclooctenes for site-specific protein labeling in vitro and in live mammalian cells via rapid fluorogenic Diels-Alder reactions. J Am Chem Soc 134:10317–10320

17. Liu DS, Tangpeerachaikul A, Selvaraj R et al (2012) Diels-Alder cycloaddition for fluorophore targeting to specific proteins inside living cells. J Am Chem Soc 134:792–795

18. Uttamapinant C, Howe JD, Lang K et al (2015) Genetic code expansion enables live-cell and super-resolution imaging of site-specifically labeled cellular proteins. J Am Chem Soc 137:4602–4605

19. Wang W, Takimoto JK, Louie GV et al (2007) Genetically encoding unnatural amino acids for cellular and neuronal studies. Nat Neurosci 10:1063–1072

20. Zhang H, Constantine R, Frederick JM et al (2012) The prenyl-binding protein PrBP/delta: a chaperone participating in intracellular trafficking. Vis Res 75:19–25

21. Plass T, Milles S, Koehler C et al (2012) Amino acids for Diels-Alder reactions in living cells. Angew Chem Int Ed Engl 51:4166–4170

22. Baird GS, Zacharias DA, Tsien RY (1999) Circular permutation and receptor insertion within green fluorescent proteins. Proc Natl Acad Sci U S A 96:11241–11246

23. Nagai T, Sawano A, Park ES et al (2001) Circularly permuted green fluorescent proteins engineered to sense Ca2+. Proc Natl Acad Sci U S A 98:3197–3202

24. Walther KA, Papke B, Sinn MB et al (2011) Precise measurement of protein interacting fractions with fluorescence lifetime imaging microscopy. Mol BioSyst 7:322–336

Fluorogenic Tetrazine-Siliconrhodamine Probe for the Labeling of Noncanonical Amino Acid Tagged Proteins

Eszter Kozma, Giulia Paci, Gemma Estrada Girona, Edward A. Lemke, and Péter Kele

Abstract

Tetrazine-bearing fluorescent labels enable site-specific tagging of proteins that are genetically manipulated with dienophile modified noncanonical amino acids. The inverse electron demand Diels-Alder reaction between the tetrazine and the dienophile fulfills the criteria of bioorthogonality allowing fluorescent labeling schemes of live cells. Here, we describe the detailed synthetic and labeling protocols of a near infrared emitting siliconrhodamine-tetrazine probe suitable for super-resolution imaging of residue-specifically engineered proteins in mammalian cells.

Key words Bioorthogonality, Tetrazine, Fluorogenicity, NIR emission, Super-resolution microscopy

1 Introduction

Emerging super-resolution microscopy (SRM) techniques have brought substantial progress in the exploration of biomolecular processes in the sub-diffraction range [1–3]. Live organisms can now be studied in fine details, yet, improvements can result in further increase of resolution and enable new biological insights. To address current limitations that impede such improvements, small synthetic dyes with suitable spectral characteristics that allow site-specific tagging of intracellular structures even under in vivo conditions are needed [4, 5]. Preferred synthetically tailored, small-sized organic fluorophores are membrane permeant, photostable, brightly fluorescent, and allow minimal background labeling and autofluorescence in order to result in a high signal-to-noise ratio. The means by which such ideal probes are installed onto the biomolecule of interest is also crucial. The applied chemistry should be biocompatible and highly selective. Such chemical transformations are termed bioorthogonal [6–8]. Most of the time, fast

Edward A. Lemke (ed.), *Noncanonical Amino Acids: Methods and Protocols*, Methods in Molecular Biology, vol. 1728,
https://doi.org/10.1007/978-1-4939-7574-7_22, © Springer Science+Business Media, LLC 2018

kinetics are also required. Inverse electron demand Diels-Alder (IEDDA) cycloaddition of tetrazines and strained unsaturated ring systems enable fast and highly selective reactions [9, 10]. Most IEDDA labeling schemes rely on the use of tetrazine bearing fluorescent probes in combination with cyclooctyne or *trans*-cyclooctene modified biomolecules (e.g., by means of genetically encoded noncanonical amino acids, ncAAs). Furthermore, tetrazine scaffolds can efficiently quench fluorescence of dyes giving rise to fluorogenic scaffolds [11–15].

In live cell imaging applications, phototoxicity and autofluorescence can be minimized if the spectral characteristics of the applied probe allow far-red/near-infrared (NIR) excitation/emission. Since labeling schemes often apply large excess of the labeling species, background fluorescence of unreacted probes bound nonspecifically to hydrophobic surfaces is often encountered, and several washing cycles are needed before imaging, which excludes labeling of, e.g., proteins with rapid turnovers. So-called fluorogenic probes efficiently reduce background fluorescence as they are minimally fluorescent when bound nonspecifically but become intensely emitting upon the particular specific chemical reaction [16, 17].

Siliconrhodamines (SiRs) are widely used membrane-permeable NIR dyes suitable for SRM applications [18–23]. Besides their high photostability and brightness, the unique environment-dependent fluorescence of carboxy-SiRs due to a polarity dependent lactone-formation offers an appealing opportunity to distinguish between specific and nonspecific labeling (polarity-based fluorogenicity) [18]. Carboxyl-SiRs exist in a fluorescent zwitterionic form when they are in polar environment such as near protein surfaces. In non-polar microenvironments, however, they isomerize to their respective, non-fluorescent spirolactone form. These characteristics render carboxy-SiRs minimally emitting when bound nonspecifically onto nonpolar surfaces. Despite their wide applications [21], there are only a few examples of SiR-based probes for site-specific labeling of proteins [18, 22, 23].

Herein, we describe the detailed synthesis of a bioorthogonally applicable, NIR-emitting, membrane permeable, double fluorogenic carboxy-SiR suitable for SRM imaging, as well as the labeling scheme for genetically modified proteins in living cells (Fig. 1) [24]. We show the applicability of the method by labeling site-specifically an intracellular skeletal protein, vimentin, engineered with an IEDDA reactive ringstrained ncAA by means of Amber suppression using an orthogonal tRNA/tRNA synthetase system derived from *Methanosarcia mazei* as described previously [18, 25–30]. We demonstrate the power of the developed SiR-dye in subsequent site-specific SRM imaging applications.

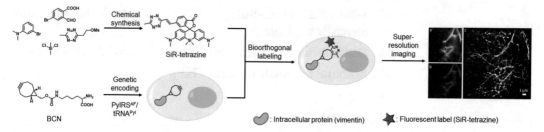

Fig. 1 Illustration of chemical synthesis, biorthogonal labeling, and subsequent imaging steps described in the protocol

2   Materials

The laboratory should be equipped to perform basic chemical synthesis. All synthetic steps shall be performed in a chemical hood equipped with magnetic stirrer, nitrogen-line, and vacuum line. The laboratory should be further equipped with a rotary evaporator (with a high vacuum pump), and a standard S1 cell culture. In this protocol specific examples of cell line, protein of interest, and microscopy setup are given, but the user should keep in mind that the method can easily be applied to other biological systems and that imaging can be performed on any fluorescence microscope.

2.1   Synthesis of 3,3′-(Dimethylsilanediyl) bis(N,N-dimethylaniline) (2)

1. Magnetic stirrer with a stand clamp.
2. 250 mL round bottom flask with a conical joint NS 29/32.
3. 500 mL round bottom flask with a conical joint NS 29/32.
4. 250 mL separatory funnel.
5. 500 mL Erlenmeyer flask.
6. Magnetic stir bar.
7. Bubbler filled with mineral oil.
8. Nitrogen gas.
9. Crystallizing dish.
10. Rotary evaporator.
11. High vacuum pump (like Vacuubrand Vario Pro PC3001 chemistry pumping unit).
12. Automated flash chromatographer (such as Teledyne Isco CombiFlash Rf+).
13. Empty solid load cartridge 65 g with three pieces of frit and a loading rod (like RedisepRf).
14. Silica gel (25–40 μm).
15. Analytical thin-layer chromatography (TLC) plates (such as silica gel 60 F_{254} precoated aluminum TLC plates from Merck).
16. 3-bromo-N,N-dimetylaniline (**1**, commercially available).

17. *n*-Butyllithium (*n*-BuLi, in 1.6 M solution in hexane, commercially available).

18. Dichlorodimethylsilane ($SiCl_2Me_2$, commercially available).

19. Absolute tetrahydrofurane (THF).

20. Dry ice.

21. Acetone.

22. Water (distilled or purified).

23. Ethyl acetate (EtOAc).

24. Saturated NaCl solution.

25. Anhydrous magnesium sulfate ($MgSO_4$).

26. Dichloromethane (DCM).

27. Celite.

28. Hexane.

2.2 Synthesis of 4-bromo-2-formylbenzoic acid (3)

1. Magnetic stirrer with stand clamp and oil bath.

2. 250 mL round bottom flask with a conical joint NS 29/32.

3. Reflux condenser for with a conical joint NS 29/32.

4. Magnetic stir bar.

5. Rotary evaporator.

6. High vacuum pump (like Vacuubrand Vario Pro PC3001 chemistry pumping unit).

7. Fritted funnel.

8. Desiccator.

9. 5-bromophtalide (commercially available).

10. *N*-bromosuccinimide (NBS) (commercially available).

11. Azobisisobutyronitrile (AIBN) (commercially available).

12. 1,2-dichloroethane (DCE).

13. Water (distilled or purified).

2.3 Synthesis of 6'-bromo-3,7-bis(dimethylamino)-5,5-dimethyl-3'H,5H-spiro[dibenzo[b,e]siline-10,1'-isobenzofuran]-3'-one (SiR-Br, 4)

1. Magnetic stirrer with stand clamp and oil bath.

2. 2–6 mL borosilicate microwave vial (like Anton Paar G10) with snap cap and PTFE-coated silicone septa.

3. Magnetic stir bar.

4. Sonicator (such as Selecta ultrasons 6.5 L).

5. Rotary evaporator.

6. High vacuum pump (like Vacuubrand Vario Pro PC3001 chemistry pumping unit).

7. Chromatography column with fused-in-frit and PTFE stopcock.

8. Silica gel (60–200 μM).

9. Preparative TLC plate (glass-backed such as Kiesegel 60 F$_{254}$ 20 × 20 cm, 2 mm).

10. Glass chamber for preparative TLC purification.

11. Analytical TLC plates (such as silica gel 60 F$_{254}$ precoated aluminum TLC plates from Merck).

12. Compound **2** (for preparation *see* Subheading 3.1).

13. Copper(II) bromide (CuBr$_2$) (commercially available).

14. Compound **3** (for preparation *see* Subheading 3.2).

15. DCM.

16. Celite.

17. Hexane.

18. EtOAc.

19. Triethylamine (Et$_3$N).

20. Methanol (MeOH).

2.4 Synthesis of ethyl 3-hydroxypropionimidate hydrochloride (9)

1. Magnetic stirrer with stand clamp.

2. 250 mL round bottom flask with a conical joint NS 29/32.

3. 500 mL two-neck round bottom flask.

4. Magnetic stir bar.

5. Dropping funnel.

6. Analytical TLC plates (such as silica gel 60 F$_{254}$ precoated aluminum TLC plates from Merck).

7. Rotary evaporator.

8. High vacuum pump (like Vacuubrand Vario Pro PC3001 chemistry pumping unit).

9. 3-Hydroxypropionitrile (commercially available).

10. Acetyl chloride (commercially available).

11. Absolute ethanol (EtOH).

12. NaCl.

13. cc. H$_2$SO$_4$.

14. Diethyl ether (Et$_2$O).

2.5 Synthesis of 2-(6-methyl-1,2,4,5-tetrazin-3-yl)ethyl methanesulfonate (OMs-tet, 5)

1. Magnetic stirrer with stand clamp.

2. 250 mL round bottom flask with a conical joint NS 29/32.

3. 100 mL round bottom flask with a conical joint NS 29/32.

4. 500 mL Erlenmeyer flask.

5. 250 mL separatory funnel.

6. Vacuum line (with vacuum pump like Vacuubrand rotary vane pump RZ 2.5).

7. Rotary evaporator.

8. High vacuum pump (like Vacuubrand Vario Pro PC3001 chemistry pumping unit).

9. Magnetic stir bar.

10. Bubbler filled with mineral oil.

11. Nitrogen gas.

12. Crystallizing dish for ice/water bath.

13. Automated flash chromatographer (such as Teledyne Isco CombiFlash Rf+).

14. Empty solid load cartridge 65 g with three pieces of frit and a loading rod (like RedisepRf).

15. Silica gel (25–40 μm).

16. Analytical TLC plates (such as silica gel 60 F_{254} precoated aluminum TLC plates from Merck).

17. Compound **9** (for preparation *see* Subheading 3.4).

18. Acetonitrile (MeCN).

19. Hydrazine hydrate (N_2H_4 50–60%) (commercially available).

20. Sodium nitrite ($NaNO_2$) (commercially available).

21. Distilled water.

22. Ice.

23. cc. HCl.

24. EtOAc.

25. Saturated NaCl solution.

26. Anhydrous $MgSO_4$ (commercially available).

27. DCM.

28. Celite.

29. Hexane.

30. Methanesulfonyl chloride (MsCl, commercially available).

31. Et_3N.

2.6 Synthesis of (E)-3,7-bis(dimethylamino)-5,5-dimethyl-6'-(2-(6-methyl-1,2,4,5-tetrazin-3-yl)vinyl)-3'H,5H-spiro[dibenzo[b,e]siline-10,1'-isobenzofuran]-3'-one (SiR-tetrazine, 6)

1. 2–6 mL borosilicate microwave vial (like Anton Paar G10) with snap cap and PTFE-coated silicone septa.

2. 50 mL round bottom flask with conical joint NS 29/32.

3. Magnetic stir bar.

4. Microwave reactor (such as Anton Paar Monowave 300).

5. Analytical TLC plates (such as silica gel 60 F_{254} precoated aluminum TLC plates from Merck).

6. Rotary evaporator.

7. High vacuum pump (like Vacuubrand Vario Pro PC3001 chemistry pumping unit).

8. Preparative TLC plate (such as glass-backed Kiesegel 60 F_{254} 20 × 20 cm, 2 mm).

9. Glass chamber for preparative TLC.

10. Preparative HPLC system.

11. Preparative C18 column (such as Gemini 5 μm C18 110 Å LC column 150 × 21.2 mm).

12. Round bottom flask freeze-drier for lyophilization.

13. SiR-Br (compound 4, for preparation see Subheading 3.3).

14. OMs-tet (compound 5, for preparation see Subheading 3.5).

15. Tris(dibenzylideneacetone)dipalladium(0) ($Pd_2(dba)_3$, commercially available).

16. 1,2,3,4,5-Pentaphenyl-1'-(di-tert-butylphosphino)ferrocene (QPhos, commercially available).

17. Anhydrous dimethylformamide (DMF).

18. N,N-dicyclohexylmethylamine ($(Cy)_2NMe$, commercially available).

19. DCM.

20. Hexane.

21. EtOAc.

22. MeCN.

23. Distilled water.

2.7 Site-Specific Protein Engineering with the Ringstrained ncAA Bicyclo[6.1.0] nonyne-Lysine (BCN^{endo}): Cell Culture and Transfections

Note that many steps are similar, if not even identical to procedures described in Chapter 18, by Nikic et al. [31].

1. COS-7 (Sigma 87021302) cell line or any other cell line of interest.

2. 1× PBS (phosphate-buffered saline).

3. Dulbeccos's modified Eagle medium (DMEM, Gibco 41965039) supplemented with 10% (v/v) fetal bovine serum (FBS, Sigma F7524), 1% (v/v) penicillin-streptomycin 10,000 U/mL (Gibco 15140122), 1% (v/v) 100 mM sodium pyruvate (Gibco 11360070), and 1% (v/v) 200 mM L-glutamine (Sigma G7513). Store at 4 °C (see Note 1). Other manufacturers providing similar formulations could also be used.

4. Trypsin-EDTA (Gibco 25300054 or similar). Long-term storage at −20 °C. Once thawed, keep at 4 °C.

5. Hemocytometer.

6. Cell culture incubator at 37 °C with a humidified 5% CO_2 atmosphere.

7. Cell culture hood.

8. Water bath at 37 °C.

9. Sterile plastic ware (Falcon and microcentrifuge tubes, 100 mm cell culture dishes, serological pipettes).

10. 4-well Lab-Tek™ II Chambered Coverglass (Nunc™).

11. JetPrime transfection reagent (Polyplus, or other transfection reagents appropriate for the cell line of choice).

12. Eukaryotic expression vector for the Amber suppression machinery. In this protocol a vector with *M. mazei* NESPylRS^{AF}/ tRNA^{Pyl} was used (available from Dr. Edward Lemke, EMBL, Heidelberg) (*see* **Note 2**).

13. Eukaryotic expression vector for the Amber mutant of the protein of interest. In this protocol pVimentin^{N116TAG}-PSmOrange was used (*see* **Notes 3** and **4**).

14. Plasmid Maxiprep kit (low or endotoxin-free recommended, e.g., Qiagen 12362 or Invitrogen K210007).

15. Noncanonical amino acid stock: 100 mM endo BCN-L-Lysine (BCN^{endo}, SiChem SC8014) in 15% (v/v) DMSO, 0.2 M NaOH. Store at −20 °C.

16. Sterile 1 M HEPES. Store at 4 °C.

17. Vortex.

18. Benchtop mini-centrifuge.

19. SiR-tetrazine dye of choice. Stock in DMSO at 0.5 mM.

20. 2% PFA: 2% paraformaldehyde in 1× PBS (*see* **Note 5**).

2.8 Localization-Based Super-Resolution Imaging and Analysis

1. 20× TN buffer: 1 M Trizma base, 0.2 M NaCl, adjust pH to 8.0 with HCl and filter. Store at 4 °C.

2. 20% Glucose. Store at 4 °C for maximum 2 weeks.

3. 20× Glucose Oxidase/Catalase (GO/C) stock: 51% glycerol, 50 mM Tris–HCl pH 8.0, 800 μg/mL Catalase (Sigma C3155), 10 mg/mL Glucose Oxidase (Sigma G0543). Store at −20 °C.

4. MEA (Sigma 411000). Store at −20 °C, aliquotes can be reused if kept cold all the time.

5. GLOX-MEA buffer: 1× TN buffer, 10% glucose, 10 mM MEA, and 1× GO/C mix) in water (*see* **Notes 6** and **7**).

6. TIRF-microscope (e.g., Leica GSDIM) with appropriate laser and filter cube for dyes of choice (*see* **Note 8**) and software for data analysis (*see* **Note 9**).

3 Methods

Perform all synthesis steps in a chemical hood (for safety instructions, *see* **Note 10**) (*see* Scheme 1).

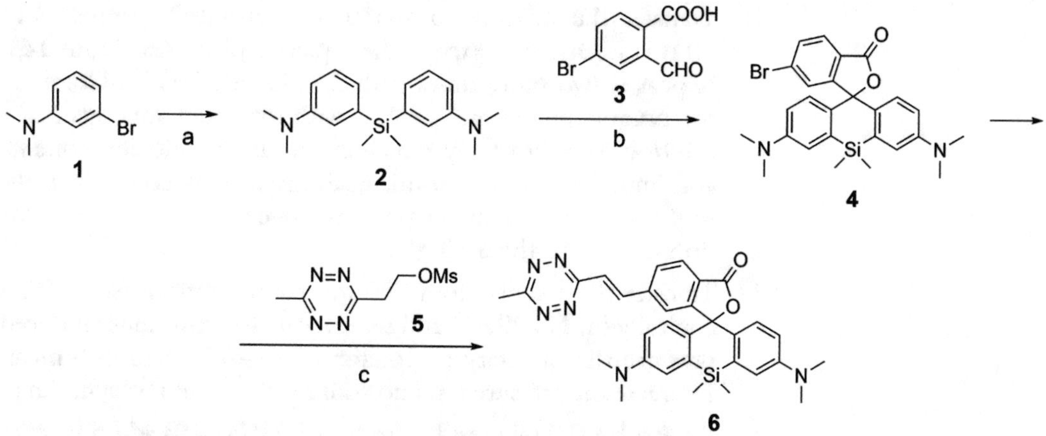

Scheme 1 Synthesis of SiR-tetrazine (**6**). Reaction conditions: (a) (1) *n*-BuLi, THF, −78 °C, 2 h; (2) SiCl$_2$Me$_2$, RT, 16 h, 71%; (b) CuBr$_2$, 140 °C, 16 h, 21%; (c) Pd$_2$(dba)$_3$, QPhos, (Cy)$_2$NMe, DMF, 50 °C, 40 min, MW, 48%

3.1 Synthesis of 3,3'-(dimethyl-silanediyl)bis(N,N-dimethylaniline) (2)

1. Weigh 5.0 g 3-bromo-*N*,*N*-dimethylaniline (**1**) with a glass pipette into a well-dried 250 mL round bottom flask with conical joint NS 29/32 equipped with a magnetic stir bar (*see* **Note 11**).

2. Close the flask with a fold over rubber septum, mount it above a magnetic stirrer using a stand clamp, and flush the flask carefully with nitrogen using a bubbler. Make sure the reaction is kept under nitrogen atmosphere until otherwise stated (*see* **Note 12**).

3. Transfer 100 mL absolute THF to the flask while stirring the solution.

4. Fill a large crystallizing dish with dry ice and add acetone very carefully until the temperature reaches −78 °C. Place the dish under the flask and make sure the temperature is kept constant during the reaction. Refill dry ice if needed.

5. When the temperature of the reaction mixture reaches −78 °C, add 17 mL *n*-butyllithium (in 1.6 M solution in hexane) dropwise to the reaction mixture over the course of 15 min.

6. Stir the reaction at −78 °C under nitrogen atmosphere for 2 h.

7. Add 1.8 mL dichlorodimethylsilane dropwise.

8. Remove the cooling dish, and let the reaction mixture warm to room temperature. Keep stirring the reaction mixture for an additional 16 h (*see* **Note 13**).

9. Remove the septum and very carefully quench the reaction with 30 mL of distilled water. Add only small portions of the water at once and wait until the bubbling stops before adding another portion.

10. Transfer the mixture to a 250 mL separatory funnel. Add 30 mL EtOAc and extract the aqueous phase (*see* **Note 14**). Repeat it two more times with 2 × 30 mL EtOAc. Combine the organic phases and wash it with saturated aqueous NaCl solution to remove any remaining water soluble components and impurities. In a 500 mL Erlenmeyer flask add anhydrous $MgSO_4$ to the organic phase to remove water. Filter the solution to remove the drying agent.

11. Transfer the solution to a 500 mL round bottom flask with a conical joint NS 29/32 and remove the solvents under reduced pressure with a rotary evaporator set to 40 °C and 400 mbar. Decrease the pressure until no additional solvent is evaporating.

12. Redissolve the oily residue in 20 mL DCM and add 5 g celite to the solution. Remove the solvent under reduced pressure set to 40 °C and 850 mbar on a rotary evaporator slowly decreasing the pressure until the celite is completely dry (*see* **Note 15**).

13. Fill a 65 g cartridge with silica gel up to 2/3, add a frit using a loading rod, transfer the celite from the previous step onto the top, place a second a frit on top of it, and place the column into an automated flash chromatographer. Start the chromatographic separation using a 1–15% EtOAc gradient in hexane over 25 min. Use 254 and 280 nm detection wavelengths for fraction collection. Check the collected fractions by TLC (*see* **Notes 13** and **16**).

14. Combine the fractions containing purely the product in a round bottom flask, remove the solvent under reduced pressure with a rotary evaporator set to 40 °C and 300 mbar. Further decrease the pressure until no additional solvent is evaporating.

15. Collect the light yellow oil formed and store it at 4 °C until further use (*see* **Note 17**).

3.2 Synthesis of 4-bromo-2-formylbenzoic acid (3) (Scheme 2)

1. Weigh 4.582 g 5-bromophtalide and add it to a 250 mL round bottom flask with a conical joint NS 29/32 equipped with a magnetic stir bar.

2. Weigh 4.209 g NBS and add it to the 250 mL round bottom flask.

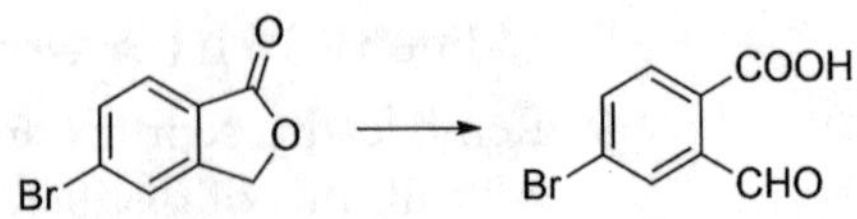

Scheme 2 Synthesis of 4-bromo-2-formylbenzoic acid (**3**). Reaction conditions: (1) NBS, AIBN, DCE, reflux, 2 h; (2) water, reflux, 2 h; 90%

3. Weigh 177 mg AIBN and mix it with the solid compounds in the 250 mL round bottom flask (*see* **Notes 18** and **19**).

4. Dissolve the solid in 100 mL dichloroethane (DCE) and mount the flask using a stand clamp and immerse it in an oil bath on a magnetic stirrer. Adjust a reflux condenser with a conical joint NS 29/32.

5. While continuously stirring, heat the reaction mixture to 88 °C until the solvent starts to reflux and keep the temperature for an additional 2 h (*see* **Note 20**).

6. Let the reaction mixture cool to room temperature. Close the flask with a fold over rubber septum and keep it at −20 °C for 2 h.

7. Filter the precipitate and collect the filtrate.

8. Using a 250 mL round bottom flask, remove the solvent under reduced pressure with a rotary evaporator set to 40 °C and 200 mbar. Further decrease the pressure until no additional solvent is evaporating. A white crystalline residue forms.

9. Suspend the crystals in 50 mL water in a 250 mL round bottom flask with a conical joint NS 29/32. Adjust a condenser with a conical joint NS 29/32 and mount the flask using a stand clamp and immerse it in an oil bath on a magnetic stirrer. Start stirring the suspension.

10. Adjust the temperature to 100 °C and keep the suspension refluxed for 2 h under continuous stirring (*see* **Note 21**).

11. Let the reaction mixture cool to room temperature. Close the flask with a fold over rubber septum and keep it at 4 °C for 16 h.

12. Collect the precipitate using a fritted funnel by applying vacuum. Wash the precipitate with 2 × 10 mL ice-cold water. Remove as much water as possible from the precipitate and then place the funnel into a desiccator, supplied with a drying agent, under vacuum for 24 h to remove any residual water.

13. Remove the funnel from the desiccator and collect the white crystalline powder. Store it at 4 °C until further use (*see* **Note 17**).

3.3 Synthesis of 6'-bromo-3,7-bis(dimethylamino)-5,5-dimethyl-3'H,5H-spiro[dibenzo[b,e]siline-10,1'-isobenzofuran]-3'-one (SiR-Br, 4)

1. Weigh 641 mg compound **3** and transfer it to a microwave borosilicate glass vial (2–6 mL) equipped with a magnetic stir bar.

2. Weigh 13 mg $CuBr_2$ and add it to the microwave vial.

3. Weigh 167 mg of compound **2** and add it to the vial using a pipette. Slightly mix the mixture.

4. Close the tube with a snap cap containing a PTFE-coated silicone septum and mount it using a stand clamp and immerse it in an oil bath (covered 2/3 in oil) on a magnetic stirrer. Start stirring.

5. Heat the reaction to 140 °C and keep stirring for 16 h at this temperature (*see* **Note 22**).

6. Let the reaction mixture cool to room temperature.

7. Add 3 mL DCM and dissolve the reaction mixture by applying sonication (*see* **Note 23**).

8. Transfer the solution to a round bottom flask and add 4 g celite. Remove the solvent under reduced pressure with a rotary evaporator set to 40 °C and 850 mbar. Further decrease the pressure until no solvent is evaporating (*see* **Note 15**).

9. Purify the product by column chromatography. For this, suspend silica gel in a 4:1 mixture of hexane and EtOAc with 1% (v/v) Et₃N and transfer to a chromatography column with a fused-in frit and PTFE stopcock. Cover the silica gel with sand and then place the celite on the top. Perform chromatographic purification using a 4:1 mixture of hexane and EtOAc with 1% (v/v) Et₃N and collect the eluting fractions in test tubes. Check the presence of the product by TLC (R_f = 0.4) (*see* **Note 24**).

10. Combine fractions containing the product and remove the solvent under reduced pressure with a rotary evaporator set to 40 °C and 200 mbar. Further decrease the pressure once no additional solvent is evaporating. A yellow solid forms.

11. Perform preparative thin-layer chromatography purification (*see* **Note 25**). For this, preincubate a large glass chamber with 50:1 mixture of DCM and MeOH. Dissolve the yellow residue in 2 mL of DCM and slowly apply the solution onto glass-backed preparative TLC plates using a pipette near one edge of the plate. Let the DCM evaporate before placing the plate into the chamber to develop (*see* **Note 26**).

12. Remove the plate and scrape the silica gel containing the blue product. Suspend the silica gel in DCM to dissolve the product from the silica. Filter the silica gel and collect the solution. In order to dissolve as much product from the silica as possible, repeat this step twice.

13. Collect the filtrates in a round bottom flask and remove the solvent under reduced pressure with a rotary evaporator set to 40 °C and 850 mbar. A white crystalline residue forms. Store at 4 °C until further use (*see* **Note 17**).

3.4 Synthesis of ethyl 3-hydroxypro-pionimidate hydrochloride (9) (Scheme 3)

1. To synthesize OMs-tet (**5**), start preparing compound **9** from commercially available 3-hydroxypropionitrile (**8**).

2. For this, weigh 1.4 mL 3-hydroxypropionitrile (**8**) and transfer it to a 250 mL round bottom flask with a conical joint NS 29/32 charged with a magnetic stirring bar. Dissolve the compound in 14 mL abs. EtOH and mount the round bottom flask above a magnetic stirrer using a stand clamp.

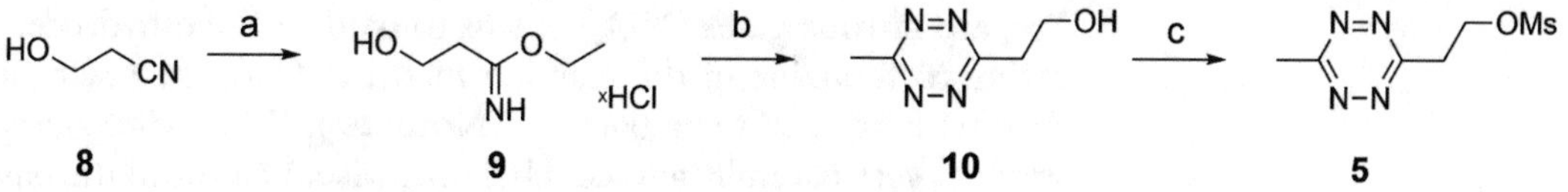

Scheme 3 Synthesis of mesyl-tetrazine (OMs-tet) (**5**). Reaction conditions: (a) EtOH, HCl gas, acetyl chloride, RT, 2 h, 83%; (b) (1) MeCN, hydrazine hydrate, N_2, RT, 2 h; (2) NaNO$_2$, cc. HCl; 30%; (c) MsCl, Et$_3$N, DCM, 91%

3. While continuously stirring, add 3 mL acetyl chloride dropwise.

4. Prepare HCl gas in situ. To do this, fill a 500 mL two-necked round bottom flask with NaCl and a magnetic stir bar. Adjust a dropping funnel onto one neck and fill it with cc. H_2SO_4. Adjust a teflon tube to the other neck using a connecting adapter. Insert a glass pipette at the other end of the teflon tube and insert it into the carefully stirred EtOH solution. Start dropping the cc. H_2SO_4 while continuously stirring the solution. (For safety instructions during HCl gas evolution *see* **Note 27**).

5. Continue purging the reaction with HCl gas for 2 h at room temperature. Check for completion (disappearance of starting material) by TLC (starting material R_f = 0.3 in DCM:MeOH 30:1, the product is at R_f = 0).

6. Purge the solution with nitrogen gas to remove residual HCl gas and remove the solvent under reduced pressure with a rotary evaporator set to 40 °C and 150 mbar. Further decrease the pressure until no additional solvent is evaporating. Close the flask with a fold over rubber septum and keep it at −20 °C for 16 h.

7. Collect the formed off-white crystals on a vacuum-filter using a fritted funnel and wash the crystals with Et$_2$O (2 × 10 mL). Store at 4 °C until further use (*see* **Note 17**).

3.5 Synthesis of 2-(6-methyl-1,2,4,5-tetrazin-3-yl)ethyl methanesulfonate (OMs-tet, 5)

1. First, prepare the OH-tetrazine (**10**). For this, weigh 6.14 g compound **9** and transfer it to a 250 mL round bottom flask with a conical joint NS 29/32.

2. Suspend compound **9** in 24 mL acetonitrile (MeCN) and 40 mL hydrazine hydrate and start stirring.

3. Close the flask with a fold over rubber septum, mount it above a magnetic stirrer using a stand clamp and flush the flask carefully with nitrogen using a bubbler. Keep the nitrogen flow throughout the reaction (*see* **Note 12**).

4. Keep stirring the solution for 2 h at room temperature (*see* **Note 28**).

5. Remove the septum. Weigh 34.5 g NaNO$_2$ and dissolve it in 50 mL water. Add the solution to the reaction mixture carefully while continuously stirring.

6. Prepare nitrous gases (NO_x) in situ to oxidize the dihydrotetrazine to tetrazine in the reaction mixture: Cool the reaction flask with an ice/water bath (*see* **Note 29**). While vigorously stirring, very carefully add cc. HCl dropwise. Wait until the gas evolution stops and add more cc. HCl dropwise. Repeat it until pH reaches 3 (check it with universal pH paper) and no more gas is evolved (*see* **Notes 27** and **30**).

7. Add 50 mL EtOAc to the mixture and transfer to a separatory funnel. Separate the organic phase. Add 50 mL EtOAc to the aqueous phase (*see* **Note 14**) and extract it. Repeat the extraction of the aqueous phase three more times each time with 50 mL EtOAc.

8. Combine the organic phases and extract it with 50 mL saturated NaCl solution to further remove water soluble components and impurities. Place the organic phase in an Erlenmeyer flask and add anhydrous $MgSO_4$ to the organic phase to remove residual water. Filter the solution to remove the drying agent.

9. Transfer the solution to a round bottom flask and remove solvents under reduced pressure with a rotary evaporator set to 40 °C and 240 mbar. Further decrease the pressure until no additional solvent is evaporating (*see* **Note 31**).

10. To remove the volatile pink 3,6-dimethyl-1,2,4,5-tetrazine side-product, place the round bottom flask to a high vacuum line and apply vacuum for at least 10 h, or until no further pink residue is removed.

11. Dissolve the remaining pink product in 20 mL DCM and add 5 g celite. Remove the solvent under reduced pressure with a rotary evaporator set to 40 °C and 850 mbar. Further decrease the pressure until no solvent is evaporating (*see* **Note 15**).

12. Fill a 65 g cartridge with silica gel up to 2/3, add a frit using a loading rod, transfer the celite from the previous step to the top, place a second frit on top of it, and place the column into an automated flash chromatographer. Start the chromatography separation using a 0–70% EtOAc gradient in hexane over 25 min. Use 280 and 524 nm for fraction collection. Check the collected fractions using TLC (*see* **Notes 16** and **32**).

13. Combine the fractions containing the product and remove the solvent under reduced pressure with a rotary evaporator set to 40 °C and 300 mbar. Further decrease the pressure until no additional solvent is evaporating. Collect the resulting OH-tetrazine as pink oil. Keep it at 4 °C until further use.

14. Dissolve the pink oil in 20 mL DCM and transfer to a 100 mL round bottom flask with a conical joint NS 29/32 equipped with a magnetic stir bar. Mount the reaction flask above a magnetic stirrer using a stand clamp and cool it with an ice/water bath. Start stirring the reaction mixture.

15. Measure 1.81 mL MsCl and add it to the reaction mixture dropwise.

16. Measure 3.26 mL Et_3N and add it to the reaction mixture dropwise.

17. Keep stirring the solution for 10 min at room temperature. Check for completion by TLC (*see* **Note 33**).

18. Transfer the reaction mixture into a separatory funnel and wash it with 50 mL water. Collect the organic phase. Extract the aqueous phase with an additional 3 × 30 mL DCM (*see* **Note 34**). Combine the organic phases in an Erlenmeyer flask and add anhydrous $MgSO_4$ to the organic phase to remove residual water. Filter the solution to remove the drying agent.

19. Transfer the solution to a round bottom flask, add 3 g celite, and remove solvents under reduced pressure with a rotary evaporator set to 40 °C and 850 mbar. Further decrease the pressure until no additional solvent is evaporating.

20. Perform flash chromatography purification. Fill a 65 g cartridge with silica gel up to 2/3, add a frit using a loading rod, transfer the celite from the previous step to the top, add a second frit on top of it and place the column into an automated flash chromatographer. Start chromatography using a 5–55% EtOAc gradient in hexane over 20 min. Use 280 and 524 nm for fraction collection. Check the collected fractions using TLC (*see* **Notes 16** and **33**).

21. Combine the fractions containing the product in a round bottom flask and remove the solvent under reduced pressure with a rotary evaporator set to 40 °C and 300 mbar. Further decrease the pressure until no more solvent is evaporating. Collect the resulting OMs-tetrazine as pink crystals. Keep it at −20 °C until further use (*see* **Notes 17** and **35**).

3.6 Synthesis of (E)-3,7-bis(dimethylamino)-5,5-dimethyl-6'-(2-(6-methyl-1,2,4,5-tetrazin-3-yl)vinyl)-3'H,5H-spiro[dibenzo[b,e]siline-10,1'-isobenzofuran]-3'-one (SiR-tetrazine, 6)

1. Weigh 61 mg SiR-Br (**4**) and transfer it to a dry microwave borosilicate glass vial (2–6 mL) equipped with a magnetic stir bar.

2. Weigh 24 mg OMs-tet (**5**) and add it to the vial.

3. Weigh 10 mg $Pd_2(dba)_3$ and transfer it to the vial.

4. Weigh 32 mg QPhos and add it to the solids in the vial.

5. Mix all the solids and dissolve them in 3 mL anhydrous DMF.

6. Add 94 μL $(Cy)_2NMe$ to the solution, purge the vial with nitrogen, and close it with a snap cap containing a PTFE-coated silicon septum.

7. Transfer the vial to a microwave reactor and start the reaction at 50 °C for 40 min. Check for completion by TLC (*see* **Note 36**).

8. Remove the reaction mixture from the vial and transfer it to a 50 mL round bottom flask. Remove the solvent under reduced pressure with a rotary evaporator set to 45 °C and 30 mbar. Further decrease the pressure until no additional solvent is evaporating.

9. Dissolve the residue in 2 mL DCM and perform preparative TLC purification. For this, preincubate a large glass chamber with 2:1 mixture of hexane and EtOAc. Slowly apply the solution onto glass-backed preparative TLC plates using a pipette near one edge of the plate. Let the DCM evaporate before placing the plate into the chamber to develop.

10. Remove the plate and scrape the silica gel containing the product (*see* **Note 36**). Suspend the silica gel in MeCN to dissolve the product (*see* **Note 37**). Filter the silica gel and collect the solution. In order to dissolve as much product from the silica as possible, repeat this step two more times with fresh MeCN portions.

11. Collect the filtrates in a round bottom flask and remove the solvent under reduced pressure with a rotary evaporator set to 40 °C and 200 mbar.

12. Perform preparative HPLC purification (*see* **Note 25**). For this, dissolve the residue in MeCN:H_2O 1:1. Purify the product using a C18 column with the following gradient: A = H_2O B = MeCN, flow rate: 15 mL/min, 0 min 70% A, 70 min 0% A. Use 220, 296, 520 and 620 nm detection wavelengths. Analyze the fractions with MS (or LC-MS). Combine pure fractions and remove the solvent using a round bottom flask freeze-drier for lyophylization. Store the blue crystalline product at −20 °C until further use (*see* **Note 17**).

3.7 Cell Seeding for Labeling Experiments

Note that many steps are similar, if not even identical to procedures described in Chapter 18, by Nikic et al. [31].

Perform cell seeding under aseptic conditions in a cell culture hood.

1. Warm up PBS and growth medium in a water bath at 37 °C.

2. Warm up trypsin-EDTA to room temperature (RT).

3. Take cells out of the cell culture incubator.

4. Aspirate off growth medium.

5. Rinse the 100 mm cell culture plate with 5–10 mL of PBS.

6. Aspirate off PBS.

7. Add 2 mL of trypsin-EDTA to one 100 mm plate.

8. Put the plate back to the incubator for 3–5 min.

9. Check if the cells are detached from the plate surface (*see* **Note 38**). When detached, inactivate trypsin-EDTA by adding 8 mL of growth medium.

10. Pipette up and down a few times, rinsing the entire plate and homogeneously resuspend the trypsinized cells.

11. Transfer the cell suspension to a 15 mL falcon tube.

12. Count the number of cells with a hemocytometer.

13. Seed the appropriate number of cells required for the chosen culture surface. In this case, COS-7 cells are seeded in 4-well Lab-Teks at a density of 35,000 cells/well (*see* **Note 39**). Add the required volume of cell suspension to each well and add fresh medium to a total of 500 μL per well. Rock the Lab-Tek to distribute evenly. For multiple well seeding, prepare a master mix.

14. Incubate the cells in the cell culture incubator overnight.

3.8 Transfections of the Amber Suppression System and the Protein of Interest

Perform transfections under aseptic conditions in a cell culture hood. Note that many steps are similar, if not even identical to procedures described in Chapter 18, by Nikic et al.) [31].

1. Transfections are performed on the following day (15–20 h after the seeding).

2. Prepare the transfection mix according to the manufacturer's recommendations. In this protocol, we used 0.5 μg of plasmid coding for NESPylRS^{AF}/tRNA^{Pyl} and 0.5 μg of plasmid coding for vimentin Amber mutant per well (*see* **Note 40**). For each well 50 μL of JetPrime buffer are mixed with the DNAs in a microcentrifuge tube. Please note that you can prepare a master mix by multiplying this amount with the number of wells that you want to transfect.

3. Vortex the tube for 10 s at maximum speed and then briefly spin it down using a mini-centrifuge.

4. Add JetPrime reagent to the tube using a 1:2 DNA to JetPrime ratio (w/v). Each well contains a total of 1 μg of the total DNA and therefore 2 μL of JetPrime reagent are added to the tube.

5. Vortex the tube for 10 s at maximum speed and then briefly spin it down using a mini-centrifuge.

6. Incubate the transfection mix for 10 min at RT.

7. After the incubation time is over, take the Lab-Tek with cells out of the cell culture incubator.

8. Add the transfection mix dropwise to the well.

9. Return the Lab-Tek to the incubator.

10. After 4–6 h the ncAA is added. First, prepare the ncAA working solution. For each well of the Lab-Tek, mix 1.25 μL of ncAA stock and 3.75 μL of 1 M HEPES in a tube (*see* **Note 41**). Prepare a master mix if working with several wells at the same time.

11. Aspirate off the medium containing the transfection mix from the Lab-Tek.

12. Add 500 μL of fresh, pre-warmed (37 °C) growth medium to the well.

13. Add 5 μL of the ncAA working solution per well.

14. Gently rock the Lab-Tek back and forth and from side to side.

15. Return the cells to the incubator and keep for 24 h.

16. After 24 h aspirate off the growth medium and add fresh pre-warmed medium to the Lab-Tek.

17. Incubate the cells in fresh medium without ncAA overnight in the cell culture incubator (*see* **Note 42**).

3.9 IEDDA-Click Chemistry-Based Live Cell Labeling

Perform the labeling under aseptic conditions in a cell culture hood. When handling the dye stock and solution, it is recommended to turn off the light of the cell culture hood and not expose the labeled sample to light. Please note that the steps during and after cell fixation do not require aseptic conditions. Note that many steps are similar, if not even identical to procedures described in Chapter 18, by Nikic et al. [31].

1. On the following day, proceed to label the transfected cells.

2. Pre-warm growth medium at 37 °C.

3. Take the cells out of the cell culture incubator.

4. Aspirate off the medium.

5. Rinse with growth medium once.

6. Prepare the dye solution by diluting the dye stock to 3 μM in growth medium.

7. Add 500 μL of dye solution to each well.

8. Return the Lab-Tek to the incubator and keep for 10 min (*see* **Note 43**).

9. Aspirate off the dye solution.

10. Rinse the well twice with fresh growth medium.

11. Return the Lab-Tek to the incubator and keep for 2 h for additional washing and better image contrast (*see* **Note 43**).

12. Before imaging, fix the cells.

13. Aspirate off the medium and rinse with PBS.

14. Add 500 μL of 2% PFA per well and incubate for 10 min at RT.

15. Aspirate off PFA.

16. Rinse with PBS twice.

17. Leave the cells in 500 μL/well PBS.

18. Proceed with imaging or keep the cells in the fridge (up to 2 days prior to imaging).

3.10 Super-Resolution Imaging and Data Processing

1. Once you are ready to image the cells, change the medium of the well you want to image to freshly prepared GLOX-MEA buffer (*see* **Note 44**).

2. Take the cells to the microscope.

3. Use mOrange laser and excitation/emission filters to identify transfected cells. For optimal results, look for bright cells showing high expression levels and characteristic expression pattern of the target protein (*see* **Note 45**, Figs. 2 and 3).

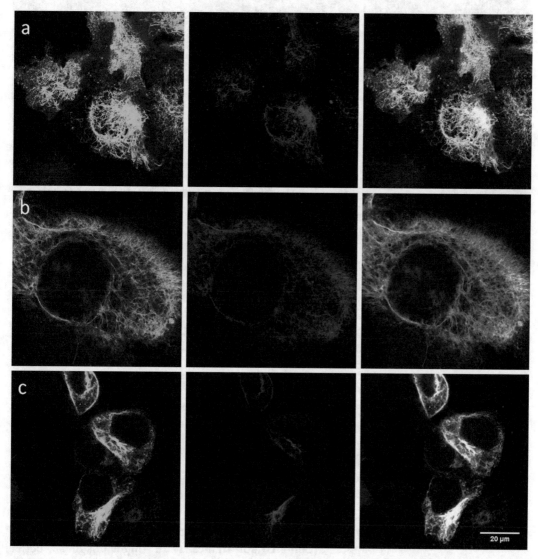

Fig. 2 Representative confocal images of live cell SiR labeling of vimentin[BCNendo]–mOrange with SiR-tetrazine (dye **6**). Left to right: reference channel (mOrange, in cyan), labeling channel (SiR, in magenta), and merge. The labeling was performed in all cases at 37 °C with a dye concentration and reaction time of 1.5 μM for 10 min (**a**), 3 μM for 10 min (**b**), and 3 μM for 30 min (**c**—images scaled differently). Reprinted with permission from [24]

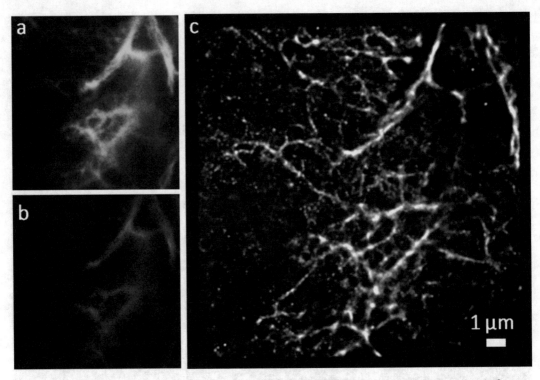

Fig. 3 TIRF SRM imaging of vimentin[BCNendo]-mOrange labeled with SiR-tetrazine (dye **6**). Panels **a** (mOrange, cyan) and **b** (SiR labeling, magenta) are used as a reference for protein expression and expected structure/pattern. Corresponding SRM image from dye **6** labeling (3 μM for 30 min at 37 °C) is shown in panel **c**, with a resolution of 35 nm as determined by Fourier ring correlation (FRC) [32]. Reprinted with permission from [24]

4. Change to the laser of the dye used for the labeling. Select the appropriate filter set and check that the labeling has been successful (you should see signal colocalizing with the reference mOrange image).

5. Adjust the TIRF illumination angle (*see* **Note 46**).

6. Switch the laser to maximum power to bring the fluorophores to a dark state. You should see the signal becoming very bright at first and gradually bleach until individual blinking molecules appear (*see* **Note 47**).

7. Lower the laser power to an appropriate value (*see* **Note 48**), set the exposure time to 30 ms, and start the acquisition.

8. Acquire 10,000–30,000 frames. The optimal length depends on the sample quality, but recognizable features should already appear around 10,000 frames. With longer imaging a better signal-to-noise ratio could be achieved (*see* **Notes 49** and **50**).

9. Do further image processing in appropriate software (*see* **Note 9**).

10. To localize the spots, apply a threshold based on the maximum likelihood ratio and perform fitting with a symmetrical 2D Gaussian function.

11. If desired, consolidate identical emitters (falling within one standard deviation of the spot fit) into a single intensity-weighed localization.

12. Reconstruct a final super-resolved image from binning all the detected events and convolving the resulting image with a Gaussian width according to the resolution determined by the Fourier ring correlation criterion [32] (Fig. 3).

4 Notes

1. Once the growth medium is prepared with all supplements, we do not recommend using it for longer than a month.

2. Several Amber suppression expression systems for eukaryotes exist. We use the NESPylRS^{AF}/tRNA^{Pyl} system because of its enhanced efficiency and reduced background in imaging experiments [25].

3. When testing new reagents or labeling methods/conditions, we recommend using target proteins with very characteristic features, e.g., cytoskeletal proteins. However, any other protein of interest can be used.

4. When testing new reagents or labeling methods/conditions, we recommend using a fusion of the target protein with a C-terminally installed fluorescent protein. This provides a direct readout obtained after a successful transfection and ncAA incorporation (only with successful Amber suppression full-length protein will be generated) that can be later on used as a reference for labeling.

5. PFA is a toxic reagent. Avoid inhalation or contact with skin and eyes. Wear protective gear while handling and follow the relevant institutional rules for using chemicals and discarding waste material.

6. The buffer composition is based on [33] and frequently used for blinking (localization-based) super-resolution microscopy.

7. We recommend always preparing the buffer freshly before starting the imaging experiment.

8. We used a commercial Leica GSDIM microscope (based on ground-state depletion and single molecule localization [34]) but any other TIRF microscope with appropriate lasers, cameras, and filter cubes can be used. Other localization-based microscopy techniques (such as STORM) would also be suitable [2].

9. We used the Localizer package for IgorPro but any other software for localization-based microscopy, such as Leica's GSDIM tools and various ImageJ plugins can be used. The following webpage (http://bigwww.epfl.ch/smlm) provides a benchmarking tool for developers to test different localization-based image analysis algorithms and provides an extensive list of tools available.

10. While performing chemical synthesis, follow general and institutional safety rules. Perform all steps in a well-ventilating chemical hood. Always wear safety glasses, lab coat, protective gloves, and proper clothing. The laboratory has to be equipped with a fire extinguisher, safety shower, and eye wash device. If you do get a chemical in your eye rinse immediately with large quantities of water using the eye-wash station. If possible, collect halogenated and non-halogenated chemical waste separately. Specific instructions on highly hazardous steps are specified at each step.

11. Dry the 250 mL round bottom flask in an oven at 110 °C and let it cool to room temperature before reaction. Make sure that there is no water remaining in the flask before performing the reaction as it can destroy n-butyllithium.

12. Turn on the nitrogen flow so that a reasonably rapid stream of bubbles passes through the mineral oil in the bubbler. Flush the apparatus with a gentle flow of nitrogen delivered through a needle; another needle in the top serves as the gas outlet during purging. When adding reagents to the mixture under inert atmosphere, use a syringe and a needle and add it through the septum. Argon can be used instead of nitrogen if needed.

13. The reaction can be followed using thin-layer chromatography. In hexane:EtOAc 10:1 R_f(starting material) = 0.7, R_f(product) = 0.4.

14. The organic phase is the upper phase.

15. When transferring compound mixtures onto celite for chromatography purification, make sure that the mixture is uniformly distributed on the celite powder. If the celite is still oily or cannot be dried completely, resuspend it in an organic solvent (DCM or EtOAc for example), add more celite and remove the solvent under reduced pressure.

16. Here, flash chromatography is used to enhance separation by enabling gradient elution. Alternatively, you can use the classic column chromatography technique.

17. Compounds can be checked by nuclear magnetic resonance (NMR) or MS. For reference spectra *see* ref. 24.

18. AIBN is an explosive compound; handle with care, use an eye protector and protective gloves.

19. The reaction can be followed by thin-layer chromatography. In hexane:EtOAc 3:1 R_f(starting material) = 0.55, R_f(first step product) = 0.76. In hexane:EtOAc 1:1 R_f(second step product) = 0.3.

20. The suspension transforms into a brown solution over the course of 2 h.

21. The white suspension becomes thick after 10–15 min when stirring may be challenging, and then a smooth white suspension again. Check the reaction frequently and make sure that the stirring is continuous.

22. Once it reached 140 °C, the mixture starts to turn blue and it develops a dark blue color by the end of the reaction.

23. The mixture can be challenging to remove from the vial. Use prolonged (15–30 min) sonication on a high-performance sonicator to dissolve the blue residue. Methanol can be used as a co-solvent.

24. The product is blue when on silica gel (column and TLC), but colorless in solution (hexane:EtOAc 4:1 with 1% (v/v) Et_3N) and forms white crystals as a solid.

25. The second purification step is optional. If the compound is pure after the first purification, omit this step.

26. The side-product to be separated is colorless on silica gel and runs just above the product.

27. Take all safety precautions for this step: wear gloves, safety glasses and the reaction must be performed in a ventilation hood. Concentrated H_2SO_4 is seriously corrosive. HCl is a pungent, irritating gas that can cause severe damage to the eyes, skin, lungs, and upper respiratory tract. NO_x is harmful for the lung when inhaled.

28. The reaction can be exothermic. In that case, cool the reaction flask with ice/water bath.

29. Use fresh ice/water bath if the ice melted completely.

30. The orange suspension will turn to magenta.

31. The removed solvent may contain pink 3,6-dimethyl-1,2,4,5-tetrazine side product.

32. R_f(product) = 0.25 in hexane:EtOAc 1:1.

33. R_f(product) = 0.38 in hexane:EtOAc 1:1.

34. The organic phase is the bottom phase under the aqueous solution.

35. The pink crystals can be kept at −20 °C without any degradation up to 9 months.

36. In hexane:EtOAc 1:1 R_f(SiR-Br) = 0.9, R_f(product) = 0.7, R_f(OMs-tet) = 0.4.

37. The product has a blue color on silica gel, but turns light rose in MeCN solution.

38. Under the microscope, check for cell detachment. When detachment is observed, it is important to not leave the cells in trypsin for much longer since this can have toxic effects. Note that trypsinization time is dependent on the cell line and that proper PBS rinse before the addition of the trypsin is required in order to avoid inactivation by any remaining medium.

39. For imaging, very high confluency is not usually desired, but too low confluency might be insufficient for proper transfection. Appropriate densities might need to be optimized first, given that the number of seeded cells will depend on the cell line, the seeding surface, and the transfection reagent.

40. Transfection conditions might need to be optimized for each protein, cell line, and transfection reagent.

41. HEPES is used to buffer the ncAA stock. This step is not necessary but recommended since it helps in maintaining the pH of the medium and will avoid the impact that the direct addition of the basic ncAA stock into the well has on the cell monolayer. Dilution of the ncAA stock with HEPES is always done fresh prior to addition to the medium.

42. Total expression time will depend on the protein of interest as well as the cell line. Longer incubation without ncAA will help in reducing the background during the labeling.

43. A good labeling efficiency is observed when using 3 μM dye for 10 min. However, dye concentration and labeling time can be adapted according to the user/experiment needs. Similar labeling efficiencies (Fig. 2) were observed for lower concentrations (1.5 μM) as well as longer labeling times (up to 30 min). In addition, the user can also adapt the washing time after the labeling reaction: a low background signal was observed with washes as short as 45 min, however, the longer the wash, the better the final contrast on the image becomes. Conditions where the sample was only quickly rinsed showed also specific labeling; nonetheless, here one might suffer from higher background and might need to optimize further the labeling reaction conditions.

44. Leave cells in the GLOX-MEA buffer only when necessary during the SRM imaging: we have observed a detrimental effect of the buffer on the cells; we recommend always changing the buffer back to PBS if you plan to reuse the same cells in a further experiment.

45. Before performing super-resolution imaging, we recommend a first round of imaging experiments at a confocal microscope (for example Leica SP8, Fig. 2) to establish the method, in order to optimize both expression levels and labeling conditions.

46. E.g. TIRF or HILO [35] can be used and it needs to be adapted to the needs of your experiment, according to the cells used and the protein being imaged. The illumination angle needs to be adjusted to optimize the signal-to-background ratio (Fig. 3).

47. In this step it is quite critical to obtain good blinking in order to achieve an optimal super-resolution image. Issues with the blinking (for example, too long waiting time before the blinking appears or overall insufficient blinking) might be caused among other reasons by: unsuccessful labeling, unsuitable conditions for the dye being used, and old GLOX-MEA buffer. After approximately 30 min we recommend changing the GLOX-MEA buffer with fresh one, to ensure optimal blinking.

48. Laser power should be adjusted to optimize the blinking: if too many events are detected (overlapping blinking particles), try increasing the laser power.

49. If the blinking decreases after some time, back-pumping can be applied by switching on the 405 laser to increase the number of blinking events.

50. Note that longer acquisition time might result in a worse image quality in case of significant drift in the microscopy setup used.

Acknowledgments

Present work was supported by the Hungarian Scientific Research Fund (OTKA, NN-116265) and the "Lendület" Program of the Hungarian Academy of Sciences (LP2013-55/2013). E.K. is grateful for the support of The New National Excellence Program of The Ministry of Human Capacities (Hungary). EAL acknowledges the SPP1623 and SFB1129 for funding.

References

1. Heilemann M (2010) Fluorescence microscopy beyond the diffraction limit. J Biotechnol 149(4):243–251. https://doi.org/10.1016/j.jbiotec.2010.03.012

2. Huang B, Babcock H, Zhuang X (2010) Breaking the diffraction barrier: super-resolution imaging of cells. Cell 143(7):1047–1058. https://doi.org/10.1016/j.cell.2010.12.002

3. Hell SW (2007) Far-field optical nanoscopy. Science 316(5828):1153–1158. https://doi.org/10.1126/science.1137395

4. Fernandez-Suarez M, Ting AY (2008) Fluorescent probes for super-resolution imaging in living cells. Nat Rev Mol Cell Biol 9(12):929–943. https://doi.org/10.1038/nrm2531

5. Chozinski TJ, Gagnon LA, Vaughan JC (2014) Twinkle, twinkle little star: photoswitchable fluorophores for super-resolution imaging. FEBS Lett 588(19):3603–3612. https://doi.org/10.1016/j.febslet.2014.06.043

6. Ramil CP, Lin Q (2013) Bioorthogonal chemistry: strategies and recent developments. Chem Commun 49(94):11007–11022. https://doi.org/10.1039/c3cc44272a

7. Prescher JA, Bertozzi CR (2005) Chemistry in living systems. Nat Chem Biol 1(1):13–21. https://doi.org/10.1038/nchembio0605-13

8. Sletten EM, Bertozzi CR (2009) Bioorthogonal chemistry: fishing for selectivity in a sea of functionality. Angew Chem 48(38):6974–6998. https://doi.org/10.1002/anie.200900942

9. Knall AC, Slugovc C (2013) Inverse electron demand Diels-Alder (iEDDA)-initiated conjugation: a (high) potential click chemistry scheme. Chem Soc Rev 42(12):5131–5142. https://doi.org/10.1039/c3cs60049a

10. Kozma E, Demeter O, Kele P (2017) Bioorthogonal fluorescent labelling of biopolymers through inverse-electron-demand Diels-Alder reactions. Chembiochem 18(6):486–501. https://doi.org/10.1002/cbic.201600607

11. Knorr G, Kozma E, Herner A, Lemke EA, Kele P (2016) New red-emitting tetrazine-phenoxazine fluorogenic labels for live-cell intracellular bioorthogonal labeling schemes. Chemistry 22(26):8972–8979. https://doi.org/10.1002/chem.201600590

12. Meimetis LG, Carlson JC, Giedt RJ, Kohler RH, Weissleder R (2014) Ultrafluorogenic coumarin-tetrazine probes for real-time biological imaging. Angew Chem 53(29):7531–7534. https://doi.org/10.1002/anie.201403890

13. Wu H, Yang J, Seckute J, Devaraj NK (2014) In situ synthesis of alkenyl tetrazines for highly fluorogenic bioorthogonal live-cell imaging probes. Angew Chem 53(23):5805–5809. https://doi.org/10.1002/anie.201400135

14. Wieczorek A, Werther P, Euchner J, Wombacher R (2017) Green- to far-red-emitting fluorogenic tetrazine probes - synthetic access and no-wash protein imaging inside living cells. Chem Sci 8(2):1506–1510. https://doi.org/10.1039/c6sc03879d

15. Agarwal P, Beahm BJ, Shieh P, Bertozzi CR (2015) Systemic fluorescence imaging of zebrafish glycans with bioorthogonal chemistry. Angew Chem 54(39):11504–11510. https://doi.org/10.1002/anie.201504249

16. Cserep GB, Herner A, Kele P (2015) Bioorthogonal fluorescent labels: a review on combined forces. Methods Appl Fluoresc 3(4). https://doi.org/10.1088/2050-6120/3/4/042001. Artn 042001

17. Nadler A, Schultz C (2013) The power of fluorogenic probes. Angew Chem 52(9):2408–2410. https://doi.org/10.1002/anie.201209733

18. Lukinavicius G, Umezawa K, Olivier N, Honigmann A, Yang G, Plass T, Mueller V, Reymond L, Correa IR Jr, Luo ZG, Schultz C, Lemke EA, Heppenstall P, Eggeling C, Manley S, Johnsson K (2013) A near-infrared fluorophore for live-cell super-resolution microscopy of cellular proteins. Nat Chem 5(2):132–139. https://doi.org/10.1038/nchem.1546

19. Uno SN, Kamiya M, Yoshihara T, Sugawara K, Okabe K, Tarhan MC, Fujita H, Funatsu T, Okada Y, Tobita S, Urano Y (2014) A spontaneously blinking fluorophore based on intramolecular spirocyclization for live-cell super-resolution imaging. Nat Chem 6(8):681–689. https://doi.org/10.1038/nchem.2002

20. Shieh P, Siegrist MS, Cullen AJ, Bertozzi CR (2014) Imaging bacterial peptidoglycan with near-infrared fluorogenic azide probes. Proc Natl Acad Sci U S A 111(15):5456–5461. https://doi.org/10.1073/pnas.1322727111

21. Kushida Y, Nagano T, Hanaoka K (2015) Silicon-substituted xanthene dyes and their applications in bioimaging. Analyst 140(3):685–695. https://doi.org/10.1039/c4an01172d

22. Peng T, Hang HC (2016) Site-specific bioorthogonal labeling for fluorescence imaging of intracellular proteins in living cells. J Am Chem Soc 138(43):14423–14433. https://doi.org/10.1021/jacs.6b08733

23. Uttamapinant C, Howe JD, Lang K, Beranek V, Davis L, Mahesh M, Barry NP, Chin JW (2015) Genetic code expansion enables live-cell and super-resolution imaging of site-specifically labeled cellular proteins. J Am Chem Soc 137(14):4602–4605. https://doi.org/10.1021/ja512838z

24. Kozma E, Girona GE, Paci G, Lemke EA, Kele P (2017) Bioorthogonal double-fluorogenic siliconrhodamine probes for intracellular super-resolution microscopy. Chem Commun. https://doi.org/10.1039/C7CC02212C

25. Nikic I, Estrada Girona G, Kang JH, Paci G, Mikhaleva S, Koehler C, Shymanska NV, Ventura Santos C, Spitz D, Lemke EA (2016) Debugging eukaryotic genetic code expansion for site-specific click-PAINT super-resolution microscopy. Angew Chem 55(52):16172–16176. https://doi.org/10.1002/anie.201608284

26. Plass T, Milles S, Koehler C, Szymanski J, Mueller R, Wiessler M, Schultz C, Lemke EA (2012) Amino acids for Diels-Alder reactions in living cells. Angew Chem 51(17):4166–4170. https://doi.org/10.1002/anie.201108231

27. Plass T, Milles S, Koehler C, Schultz C, Lemke EA (2011) Genetically encoded copper-free click chemistry. Angew Chem 50(17):3878–3881. https://doi.org/10.1002/anie.201008178

28. Nikic I, Plass T, Schraidt O, Szymanski J, Briggs JA, Schultz C, Lemke EA (2014) Minimal tags for rapid dual-color live-cell labeling and super-resolution microscopy. Angew Chem 53(8):2245–2249. https://doi.org/10.1002/anie.201309847

29. Borrmann A, Milles S, Plass T, Dommerholt J, Verkade JM, Wiessler M, Schultz C, van Hest

JC, van Delft FL, Lemke EA (2012) Genetic encoding of a bicyclo[6.1.0]nonyne-charged amino acid enables fast cellular protein imaging by metal-free ligation. Chembiochem 13(14): 2094–2099. https://doi.org/10.1002/cbic.201200407

30. Lang K, Davis L, Torres-Kolbus J, Chou C, Deiters A, Chin JW (2012) Genetically encoded norbornene directs site-specific cellular protein labelling via a rapid bioorthogonal reaction. Nat Chem 4(4):298–304. https://doi.org/10.1038/nchem.1250

31. Nikic-Spiegel I (2017) Genetic code expansion- and click chemistry-based site-specific protein labelling for intracellular DNA-PAINT imaging. Methods Mol Biol

32. Banterle N, Bui KH, Lemke EA, Beck M (2013) Fourier ring correlation as a resolution criterion for super-resolution microscopy.

J Struct Biol 183(3):363–367. https://doi.org/10.1016/j.jsb.2013.05.004

33. Dempsey GT, Vaughan JC, Chen KH, Bates M, Zhuang X (2011) Evaluation of fluorophores for optimal performance in localization-based super-resolution imaging. Nat Methods 8(12):1027–1036. https://doi.org/10.1038/nmeth.1768

34. Folling J, Bossi M, Bock H, Medda R, Wurm CA, Hein B, Jakobs S, Eggeling C, Hell SW (2008) Fluorescence nanoscopy by ground-state depletion and single-molecule return. Nat Methods 5(11):943–945. https://doi.org/10.1038/nmeth.1257

35. Tokunaga M, Imamoto N, Sakata-Sogawa K (2008) Highly inclined thin illumination enables clear single-molecule imaging in cells. Nat Methods 5(2):159–161. https://doi.org/10.1038/nmeth1171. nmeth1171 [pii]

Chapter 23

Site-Specific Protein Labeling Utilizing Lipoic Acid Ligase (LplA) and Bioorthogonal Inverse Electron Demand Diels-Alder Reaction

Mathis Baalmann, Marcel Best, and Richard Wombacher

Abstract

Here, we describe a two-step protocol for selective protein labeling based on enzyme-mediated peptide labeling utilizing lipoic acid ligase (LplA) and bioorthogonal chemistry. The method can be applied to purified proteins, protein in cell lysates, as well as living cells. In a first step a W37V mutant of the lipoic acid ligase (LplA^{W37V}) from *Escherichia coli* is utilized to ligate a synthetic chemical handle site-specifically to a lysine residue in a 13 amino acid peptide motif—a short sequence that can be genetically expressed as a fusion with any protein of interest. In a second step, a molecular probe can be attached to the chemical handle in a bioorthogonal Diels-Alder reaction with inverse electron demand (DA$_{inv}$). This method is a complementary approach to protein labeling using genetic code expansion and circumvents larger protein tags while maintaining label specificity, providing experimental flexibility and straightforwardness.

Key words Site-specific protein labeling, Bioorthogonal reactions, Fluorescent probes, Inverse-electron-demand Diels–Alder cycloaddition, Ligases, Tetrazines

1 Introduction

In the last decade, bioorthogonal chemistry has attracted a lot of attention for the selective modification and labeling of biomolecules. While classic approaches for bioconjugation focus on the nucleophilicity of amino acid residues for nucleophilic substitution (i.e., cysteine, lysine, tryptophane) or nucleophilic addition reactions, bioorthogonal chemistry depends on methods to selectively introduce unnatural chemical functions to a biomolecule.

Most protein labeling protocols in this compendium describe the incorporation of unnatural amino acids [1] that either carry the desired probe functionality by themselves or contain a chemical function that can be modified in a second step using bioorthogonal chemistry [2, 3]. Although these labeling methods are minimal in size and have an extraordinary high spatial precision, nonsense suppressor mutagenesis strategies cannot be easily adapted to every

Edward A. Lemke (ed.), *Noncanonical Amino Acids: Methods and Protocols*, Methods in Molecular Biology, vol. 1728, https://doi.org/10.1007/978-1-4939-7574-7_23, © Springer Science+Business Media, LLC 2018

organism. The adaption of protocols for other organisms is often-times experimentally demanding and requires specially modified and optimized strains [4]. The herein reported protocol represents an alternative strategy to introduce a chemical function to a protein for subsequent modification by bioorthogonal chemistry. In general, the described method can be easily applied to most biological organisms.

The protocol is based on enzyme-mediated peptide labeling utilizing the biological specificity of *E. coli* lipoic acid ligase A (LplA) [5, 6] which naturally catalyzes the covalent attachment of lipoic acid to the ε-amino moiety of lysine within an amino acid sequence, a specific lipoylation motif. It has been found that LplA also accepts other organic molecules with a terminal carboxylic function as a substrate [7, 8] (Fig. 1). However, the incorporation of other functionalized molecules requires a mutation in the active center of the enzyme. Uttamapinant et al. developed a W37V mutant (LplA^{W37V}) to site specifically attach reactive molecules for subsequent CuAAC- [9, 10], SPAAC- [11], hydrazine-coupling [12], or DA$_{inv}$ [13], the latter of which is based on the attachment of *trans*-cyclooctene (TCO). Despite showing fast kinetics in DA$_{inv}$ (k values up to $3 \times 10^5/M/$ s) [14], TCO derivatives lack long-term stability through *trans*-to-*cis* isomerization in the presence of nucleophiles [15]. Here, we report a protocol for the site specific incorporation of a chemically easily accessible and stable substrate for LplA^{W37V} bearing a norbornene moiety. First, the norbornene substrate is site-specifically attached to a peptide motif using the LplA^{W37V} followed by subsequent labeling with a tetrazine fluorophore via DA$_{inv}$. We demonstrate that the protocol is suitable for highly efficient and scalable in vitro labeling of proteins (purified protein as well as in eukaryotic or mammalian cell lysates) and show that an adapted protocol can be used for selective labeling of cell surface proteins on living mammalian cells.

1.1 The Lipoic Acid Protein Ligase

Lipoic acid is an essential cofactor for oxidative metabolism and plays an important role in the pyruvate dehydrogenase complex (PDH). Lipoic acid ligase (EC 6.3.1.20) is the enzyme that attaches lipoic acid to the lipoyl domains of certain key enzymes involved in oxidative metabolism, including pyruvate dehydrogenase. The lipoylation of the lipoyl domains is catalyzed in a magnesium(II)-dependent manner under consumption of ATP. The cofactor α-lipoate is activated via ATP as a lipoyl-AMP and transferred to the ε-amino moiety of an amino acid sequence motif within specialized lipoylation domains (lipoate acceptor peptide, LAP, *see* Fig. 1).

Substrates and reactions according to Cronan et al. [5], Uttamapinant et al. [16], Puthenveetil et al. [17], Fernandez-Suarez et al. [7], Liu et al. [13], Hauke et al. [18], Best et al. [19]).

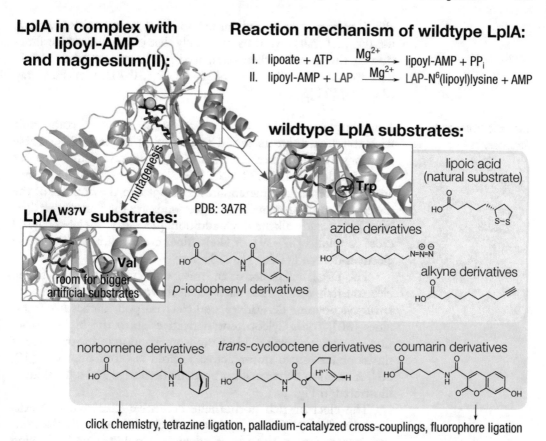

Fig. 1 Native and mutated *E. coli* K12 lipoate-protein ligase A (LplA) is suitable for site-specific installation of chemical handles. Sphere represents Mg^{2+}, LAP = lipoate acceptor peptide (also known as LplA acceptor peptide)

The LplA from *E. coli* exhibits a narrow degree of promiscuity for other lipoate-like substrates, such as selenolipoate, 6-thiooctanoate [8], azide alkylcarboxylic acids and alkine carboxylic acids [7]. In a study by the Ting group, the necessary recognition sequence for lipoylation was reduced to 13 amino acids using yeast display evolution [17]. It was further demonstrated that *E. coli* LplB-deficiency mutants with spontaneous mutations in the *lplA* gene acquired capability of ligating octanoate to LAPs [20], a step that normally required the function of LplB. Structure-guided mutagenesis experiments by the Ting group gave rise to the LplA[W37V] mutant that was now able to accept different non-natural substrates (Fig. 1) priming the protein of interest for several bioorthogonal chemical reactions or for biophysical analyses in in vitro or ex vivo settings.

This great versatility of the LplA[W37V] makes it a useful tool for the chemical biologist to site-specifically introduce chemical handles on proteins of interest in vitro and ex vivo. Usage of LplA[W37V] allows maintaining a minimal tag size (13 amino acids/1.5 kDa for the LAP tag) and thus minimally disturbing the native function of the protein when compared to commonly used protein tags for labeling like the SNAP/CLIP (20 kDa) or Halo tag (33 kDa) [21].

1.2 Bioorthogonal Reactions

Bioorthogonal reactions are chemical reactions that are compatible with the environments found in biological systems [22]. These reactions ideally do not interfere with the functionalities that are found inside or outside of a cell [23], like in proteins, DNA, sugars, lipids, or cellular metabolites. Examples in the context of the modification of proteins and nucleic acids are the Staudinger ligation [24], azide-alkyne cycloaddition [25, 26], metal-catalyzed cross couplings [27–29], or Diels-Alder cycloaddition with inverse electron-demand (DA_{inv}) [30, 31].

The DA_{inv} makes use of an irreversible cycloaddition between electron-rich dienophils, such as norbornene, cyclopropene- or *trans*-cylcooctene derivatives, and electron-poor dienes like tetrazines [30]. *trans*-Cylcooctene derivatives show the highest reaction rates with tetrazines [32], but isomerization in the presence of nucleophiles into a non-reactive scaffold is often observed [19, 33]. A reaction between a norbornene derivative and a tetrazine is illustrated in Fig. 2.

The electron-rich norbornene derivative (depicted in black) acts as a dienophile and undergoes cycloaddition with electron-deficient tetrazine derivatives (depicted in blue) while leaving nitrogen makes the reaction irreversible. Modified according to Blackman et al. [30].

Among all bioorthogonal transformations, DA_{inv} belongs to the fastest bioorthogonal reactions that are reported [23]. It is therefore not surprising that DA_{inv} has found broad application in the life sciences as a convenient tool for the introduction of fluorophores [1, 3, 13, 31] or radiotracers [33–35] in in vitro settings, living cells, or even whole animals.

1.3 Summary

Here, we describe a protocol (Fig. 3) for selective labeling of proteins. The first step comprises the enzyme-mediated and site-specific introduction of a chemical norbornene handle to the protein of interest (POI). This chemical handle is stable under physiologi-

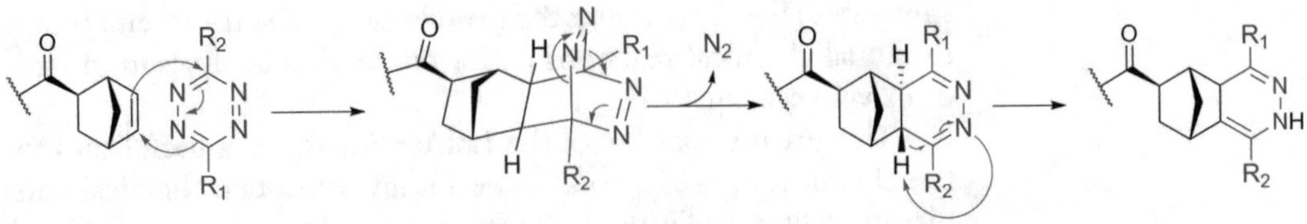

Fig. 2 Diels-Alder cycloaddition with inverse electron demand using norbornene derivatives and tetrazines

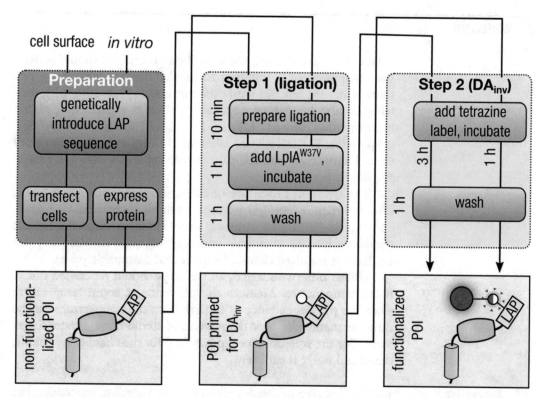

Fig. 3 Workflow of the two-step labeling procedure for both in vitro and ex vivo application

cal conditions both in isolated form and attached to a protein and forms a stable covalent linkage after the bioorthogonal reaction with tetrazine derivatives. The strength of this approach is that the method is fully scalable, flexible, and modular since it is adaptable for a variety of different tetrazine labels that are commercially available. The method does not require special strains, meaning that the labeling is independent of the protein expression system, and maintains minimal size perturbation of the protein.

The preparation for the labeling procedure is briefly discussed in the materials section. For in vitro and ex vivo application it is necessary to genetically introduce a LAP sequence to a DNA template encoding the POI. After expression of the POI, incubation is performed with the norbornene substrate and LplA^{W37V}. It is necessary to remove the excess norbornene substrate through a washing step. Labeling is then performed via DA$_{inv}$ by adding a tetrazine fluorophore of choice. Note that the second step of the procedure accounts for 1 h for in vitro and 3 h for cell surface labeling experiments.

2 Materials

A laboratory conducting the described procedures should possess the equipment of a standard molecular biology laboratory, like thermoshakers, pipettes with different ranges (0.1–1000 µL), centrifuges, and laboratory disposables, such as pipette tips, plastic reaction tubes (250 µL, 1.5 mL, 15 mL, and 50 mL), and serological pipettes. For cell surface labeling of living cells, the laboratory should have access to a cell culture with cell incubators and a laminar flow cabinet.

The synthesis procedure of the norbornene substrate is facile, does not require inert gas atmosphere or highly harmful chemicals, and can be performed in nearly every standard organic synthesis chemistry lab. All chemicals and reagents that are necessary are available at standard chemical suppliers at reasonable prices.

Unless otherwise stated, all the steps should be carried out at room temperature. Measure the pH value at room temperature since the pH dependency of buffers from the temperature can be quite dramatic. Use pure deionized and sterile-filtered water for all steps. For the aqueous workup required for the chemical synthesis, deionized water is sufficient.

2.1 Generation of an LAP-Tagged Protein of Interest

For the first step of the described two-step labeling procedure, the protein of interest (POI) needs to be equipped with the necessary amino acid motif for lipoylation (or in this case for the attachment of the norbornene substrate to the internal lysine residue of the LAP). The amino acid sequence of the most efficient and smallest LAP tag is GFEID**K**VWYDLDA (with the lysine in boldface) (known as LAP2 from Puthenveetil et al. [17]). This 13 amino acid peptide motif is ideally introduced by molecular cloning, such as classical restriction enzyme cloning, or cloning procedures like ligation independent cloning (LIC) or Gibson assembly (*see* **Note 1**).

2.2 Synthesis of the Norbornene Handle

1. Solvents: acetonitrile (MeCN), ethanol (EtOH), ethyl acetate (EtOAc), cyclohexane (CH), *N*,*N*-dimethylformamide (DMF), dimethylsulfoxide (DMSO), acetic acid (AcOH).

2. Reactants/reagents: exo-5-norbornene carboxylic acid (1 g, CAS-#: 934-30-5), *N*-hydroxysuccinimide (NHS, 1 g, CAS-#: 6066-82-6), *N*-(3-dimethylaminopropyl)-*N'*-ethylcarbodiimide hydrochloride (EDC·HCl, 1.5 g, CAS-#: 25952-53-8), 7-aminoheptanoic acid (1.85 g, CAS-#: 929-17-9), triethylamine (NEt₃), anhydrous MgSO₄ (30 g).

3. Solutions for aqueous workup: saturated aqueous NaCl solution (brine), saturated aqueous NH₄Cl solution, saturated aqueous NaHCO₃ solution, 1 M aqueous HCl solution.

4. Glassware and supplies: Round bottom flasks (2 × 100 mL with ground joints) with compatible rubber septa, column cartridge (>4 cm inner diameter), test tubes, rack for test tubes, separatory funnel (250 mL), Erlenmeyer flasks (100 mL and 250 mL), measuring cylinder (100 mL), spatulas.

5. Disposables: plastic syringes, cannulae.

6. Magnetic stirrer and magnetic stirbars.

7. Column purification and reaction monitoring: Silica gel for normal phase flash chromatography (50 g; high purity silica gel, 60 Å pore size, 230–400 mesh particle size), silica gel normal phase thin-layer chromatography plates, chamber for TLC plates, blue-shift reagent (10 g of phosphomolybdic acid in 100 mL of EtOH).

8. Evaporation and concentration: Schlenk line, rotary film evaporator, oil pump connected to the Schlenk line.

9. Analytics: NMR device, NMR tubes and deuterated chloroform ($CDCl_3$), mass spectrometer (preferably HR-ESI).

2.3 LplA^{W37V}-Mediated Ligation In Vitro

1. LplA^{W37V} solution: 0.5–10 mg/mL of purified LplA^{W37V} in PBS (*see* **Notes 2** and **3**).

2. ATP solution (20×):100 mM ATP solution in water.

3. Mg(OAc)$_2$ solution (100×): 500 mM Mg(OAc)$_2$ solution in water.

4. Sodium phosphate LplA buffer (10×) (*see* **Note 4**): Prepare a 250 mM solution of Na_2HPO_4 and a 250 mM solution of NaH_2PO_4. Add a fraction of the basic Na_2HPO_4 solution to a beaker and carefully add the acidic NaH_2PO_4 solution to adjust the pH to 7.0.

5. 200 mM norbornene substrate solution in DMSO.

6. Purified LAP-tagged protein (or MBP-LAP-HA) in PBS (*see* **Note 1**).

7. Amicon Ultra-0.5 mL filter unit with Ultracel 10 Membrane (10,000 MWCO, Merck Millipore).

8. Temperature-controlled microcentrifuge with rotation speeds up to 14,000 rcf.

9. Tabletop mini centrifuge.

10. Temperature-controlled shaker.

11. Microliter photospectrometer.

12. Vortex mixer.

2.4 Diels-Alder Cycloaddition with Inverse Electron Demand (DA$_{inv}$) In Vitro

1. Methyl-tetrazine-label stock solution in DMSO in a concentration between 10–200 mM (*see* **Notes 5** and **6**).

2. Phosphate-buffered saline (PBS, 10× stock solution, *see* **Note 7**): Dissolve the following salts for a final volume of 1000 mL in 800 mL of water: 1.37 M NaCl, 27 mM KCl, 100 mM Na$_2$HPO$_4$, 18 mM KH$_2$PO4. Adjust the pH to 7.4 with HCl and fill up to 1 L in a volumetric flask with water. Sterile-filter the 10× PBS concentrate through a vacuum-driven sterile filter and store at RT.

3. Amicon Ultra-0.5 mL filter unit with Ultracel 10 Membrane (10,000 MWCO, Merck Millipore).

4. Temperature-controlled shaker.

5. Temperature-controlled microcentrifuge with rotation speeds up to 14,000 rcf.

6. Tabletop mini centrifuge.

7. Microliter photospectrometer.

8. Vortex mixer.

2.5 LplA^{W37V}-Mediated Labeling at the Cell Surface

1. LplA^{W37V} solution: 2–10 mg/mL of purified LplA^{W37V} in PBS (*see* **Notes 2** and **3**).

2. Appropriate adherent cell lines for your experiments that can be transfected with plasmids (we recommend cell lines such as human embryonic kidney cells 293, HEK 293, or 3T3 cells).

3. Laminar flow cabinet.

4. Cell incubator (37 °C, 5% CO$_2$).

5. Water bath (filled with autoclaved, deionized water, preheated to 37 °C).

6. Cell culture flasks for adherent cells (flasks with a vented cap and a surface area of 25 cm^2 are sufficient).

7. Cell media: DMEM (GlutaMAX®, 4500 mg/L glucose, with Phenol Red), DMEM (GlutaMAX®, 4500 mg/L glucose, without Phenol Red), Opti-MEM I Reduced Serum Medium, DPBS (*see* **Note 8**). Store every medium at 4 °C. DPBS may be stored at room temperature. Both media have to be supplemented with Fetal Bovine Serum (50 mL per 500 mL medium) and Penicillin/Streptomycin (5 mL of a commercially available 100× concentrated solution per 500 mL medium).

8. Sterilized glass pipettes or similar devices to remove medium by aspiration. The glass pipettes have to be attached to a plastic pipe (connected either to a water-jet pump or to an electric pump).

9. Trypsin-EDTA (0.05%) solution, store at −20 °C.

10. Cell chambers: IBIDI μ-slides VI$^{0.4}$ (ibidi GmbH).

11. Centrifuge for 15 mL tubes.

12. pDisplay-LAP2-CFP-TM (from Alice Ting, Addgene plasmid # 34842) as a 0.2 µg/µL solution in PBS.

13. Transfection reagent of choice.

14. Ringers buffer (150 mM NaCl, 1 mM $CaCl_2$, 1 mM $MgCl_2$, 5 mM KCl, 20 mM HEPES (pH = 7.2), 2 g/L glucose) (*see* **Note 9**).

15. A conventional light microscope.

16. Neubauer counting chamber or other cell counter.

17. Buffer for in vivo ligation: 10 µM $LplA^{W37V}$, 5 mM ATP, 5 mM $Mg(OAc)_2$ and 1 mM norbornene substrate in DMEM (GlutaMAX®, 4500 mg/L glucose, without Phenol Red). Prepare 1 mL (6 × 120 mL) for every IBIDI µ-slide $VI^{0.4}$. Add the enzyme in the last step.

18. A 0.2 µM solution of tetrazine-TAMRA [36] in DMEM (GlutaMAX®, 4500 mg/L glucose, without Phenol Red); prepare 1 mL for every IBIDI µ-slide $VI^{0.4}$. Use a 200 mM stock solution of the Tetrazine dye in DMSO.

19. A fluorescence microscope (*see* **Note 10**).

3 Methods

3.1 Synthesis of N-Hydroxysuccinimidyl Exo-Nor-born-5-Ene-Carboxylate (Norbornene NHS Ester)

It should be noted that all the procedures should be conducted under a well-ventilated fume hood. Before the synthesis, make sure that you considered all of the necessary protective measures and wear the appropriate protective clothes (lab coat, protective eyewear, gloves, if necessary). The reaction scheme for the synthesis of the norbornene NHS ester is shown in Scheme 1. The NHS ester is used for the synthesis of the final norbornene LplA substrate (Scheme 2).

1. Put 25 mL MeCN in a 50 mL round bottom flask and dissolve 691 mg of exo-5-norbornene carboxylic acid (5.00 mmol, 1.00 eq.), 719 mg of N-hydroxysuccinimide (6.25 mmol, 1.25 eq.), and 1200 mg of EDC·HCl (6.26 mmol, 0.1.25 eq.). Close the round bottom flask with a septum and stir it overnight at room temperature by using a magnetic stirrer and magnetic stir bar.

2. Remove the septum and the stir bar and connect the flask to a rotatory film evaporator. Remove the volatile compounds at 40 °C and 170 mbar under constant rotation.

3. Take up the residue with 60 mL of a 1:1 mixture of ethyl acetate and water. You should observe the formation of two layers (*see* **Note 11**).

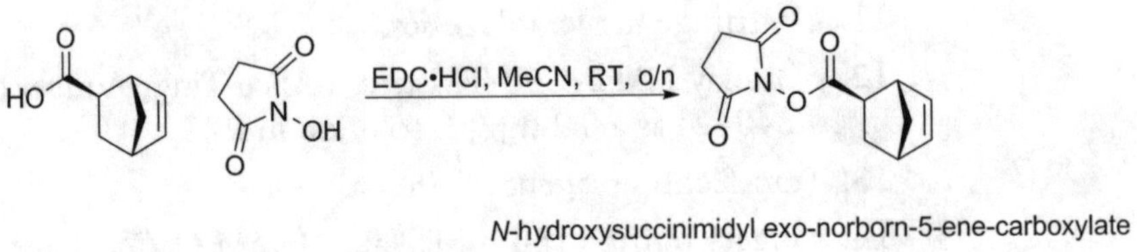

N-hydroxysuccinimidyl exo-norborn-5-ene-carboxylate

Scheme 1 Synthesis of norbornene NHS ester

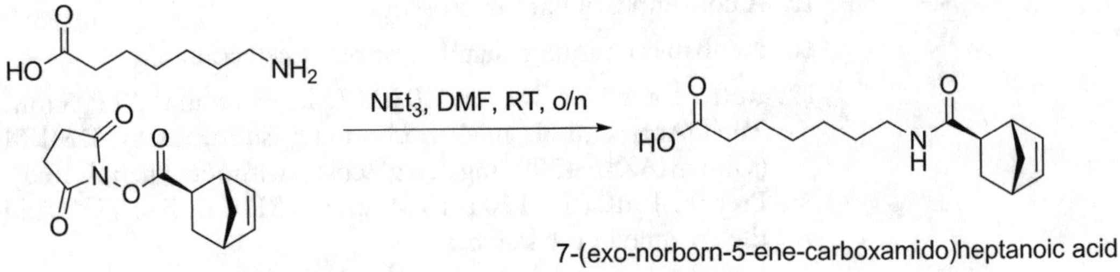

7-(exo-norborn-5-ene-carboxamido)heptanoic acid

Scheme 2 Synthesis of norbornene substrate

4. Separate the aqueous layer (the lower layer) from the ethyl acetate layer by using a separatory funnel. Extract the aqueous layer with 10 mL ethyl acetate twice. Discard the aqueous layer.

5. Combine the organic layers and wash them three times with 20 mL saturated aqueous NH_4Cl solution using a separatory funnel. Discard the aqueous solution.

6. Wash the organic layer three times with 20 mL of a saturated $NaHCO_3$. Discard the aqueous solution.

7. Wash the organic layer once with brine. Discard the aqueous solution.

8. Transfer the organic layer in a 100 mL Erlenmeyer flask.

9. Dry the organic layer over $MgSO_4$: Add $MgSO_4$ slowly and shake the flask vigorously (*see* **Note 12**).

10. Filter the organic solution through a glass funnel equipped with a pleated filter into a 100 mL round bottom flask.

11. Remove the volatile compounds at 40 °C and 185 mbar by using a rotary evaporator until the residue is sufficiently dry.

12. Dry the residue overnight at <1 mbar (*see* **Note 13**).

13. You should receive your final compound as a colorless solid in quantitative yield. Verify successful synthesis by ^{1}H-NMR, ^{13}C-NMR and mass spectrometry.

3.1.1 Chemical Characterization

^{1}H NMR (300 MHz, $CDCl_3$) δ 6.23–6.11 (m, 2H), 3.33–3.17 (m, 1H), 3.05–2.94 (m, 1H), 2.83 (d, J = 1.7 Hz, 4H), 2.54–2.46 (m, 1H), 2.11–1.98 (m, 1H), 1.64–1.41 (m, 4H). ^{13}C NMR

(75 MHz, CDCl$_3$) δ 171.8, 169.4, 138.7, 135.4, 47.3, 46.5, 41.9, 40.4, 31.1, 25.8. HR-ESI, calculated for [C$_{12}$H$_{13}$NO$_4$ + Na]$^+$: 258.0737, found: 258.0742.

3.2 Synthesis of 7-(Exo-Norborn-5-Ene-Carboxamido) Heptanoic Acid (Norbornene Substrate)

1. Dissolve 1 g of norbornene NHS ester (4.26 mmol, 1 eq.) in 65 mL of DMF. Add 1.85 g of 7-aminoheptanoic acid (12.77 mmol, 3 eq.) and 1.17 mL NEt$_3$. Close the round bottom flask with a septum and stir it overnight at room temperature by using a magnetic stirrer and magnetic stir bar.

2. Add 65 mL of an aqueous 1 M HCl solution and extract the mixture three times with 20 mL EtOAc in a separatory funnel (*see* **Note 11**).

3. Wash the combined organic layers with brine in a separatory funnel.

4. Put the organic layer in a 250 mL Erlenmeyer flask.

5. Dry the organic layer over MgSO$_4$: Add MgSO$_4$ slowly and shake the flask vigorously (*see* **Note 12**).

6. Filter the organic solution through a glass funnel equipped with a pleated filter into a 100 mL round bottom flask.

7. Remove the organic solvents of the filtrate at 40 °C and 180 mbar using a rotary evaporator.

8. Perform normal phase flash chromatography purification (1:1 EtOAc:CH, supplemented with 1% v/v acetic acid).

9. Remove the volatile compounds at 40 °C and 180 mbar by using a rotary evaporator and remove the residual solvent at <1 mbar (*see* **Note 13**).

10. You should receive your final compound as a colorless solid in about 75% yield. Verify successful synthesis by ^{1}H-NMR, ^{13}C-NMR and mass spectrometry (preferably HR-ESI).

3.2.1 Chemical Characterization

^{1}H NMR (300 MHz, CDCl$_3$): δ 6.12 (dd, J = 5.52, 2.87 Hz, 1H), 6.10–6.05 (m, 1H), 5.66 (s, 1H), 3.23 (dd, J = 13.39, 6.67 Hz, 2H), 2.89 (d, J = 0.46 Hz, 2H), 2.33 (t, J = 7.40 Hz, 2H), 2.01–1.93 (m, 1H), 1.88 (td, J = 11.19, 3.82 Hz, 1H), 1.64 (td, J = 14.56, 7.79 Hz, 3H), 1.56–1.42 (m, 2H), 1.32 (m, 6H). ^{13}C NMR (75 MHz, CDCl$_3$): δ 179.0, 175.9, 138.2, 136.0, 47.1, 46.3, 44.7, 41.5, 39.5, 33.9, 30.5, 29.4, 28.6, 26.5, 24.5. HR-ESI, calculated for [C$_{15}$H$_{23}$NO$_3$ + Na]$^+$: 288.1570, found: 288.1545.

The norbornene substrate is stable in a dry form at −20 °C for several years. As a 200 mM DMSO stock solution, the substance is also stable for more than 2 years when stored at −20 °C.

3.3 LplA^W37V-Mediated In Vitro Ligation of the Norbornene Substrate

1. Prepare the ligation reaction as follows on ice for a volume of 100 μL (*see* **Notes 14–16**) in 1.5 mL reaction tubes: Mix 1 μL of the $Mg(OAc)_2$ (100×) solution, 10 μL of the sodium phosphate LplA buffer (10×), 5 μL of the ATP (20×), and 0.5 μL of the norbornene substrate. Add the LAP-tagged protein of interest in a concentration between 1 μM and 50 μM to the ligation mixture (*see* **Note 17**).

2. Calculate the volume of the LplA^W37V to add and fill up to 100 μL water minus the volume of the LplA^W37V solution that is added in the next step. Spin down, vortex the samples, and spin down again. Estimate the protein concentration of each sample with a microliter photospectrometer at 280 nm.

3. Use the LplA^W37V in a concentration between 0.2 μM and 1 μM for labeling within 1 h at 37 °C (*see* **Note 18**): Add the LplA^W37V solution to the ligation mixture on ice, spin down, vortex, spin down again, and transfer the tubes for the ligation mixture to a preheated (37 °C) thermoshaker.

4. Incubate the ligation reaction with the MBP-LAP-HA at 37 °C under shaking for 1 h (*see* **Note 19**).

5. Centrifugal filtration (*see* **Note 20**): Transfer the ligation reaction mixture to an Amicon Ultra-0.5 mL filter unit with Ultracel 10 Membrane (10,000 MWCO) inserted into the filtrate collection tube. Fill up the ligation reaction mixture up to the maximum level of 500 μL with PBS. Centrifuge the assembled filter unit at 14,000 rcf for 15 min at 4 °C and discard the flow-through (*see* **Note 21**). Repeat this step twice. Label new concentration tubes and weigh each tube. Turn the cartridge upside down, put it into the new concentrate collection tubes, and centrifuge at 4 °C and 1000 rcf for 2 min. Discard the filter unit.

6. Measure the obtained volume by weight difference from the initial tube weight and add PBS until the original volume of 100 μL is reached again (*see* **Notes 22** and **23**). Spin down, vortex, spin down again, and transfer the samples to fresh 1.5 mL reaction tubes.

7. Store the samples at 4 °C or directly proceed with the chemical functionalization step (*see* **Note 24**).

3.4 In Vitro Functionalization (Diels-Alder Cycloaddition with Inverse Electron Demand) Using Tetrazine Labels

1. Prepare the chemical functionalization using Diels-Alder cycloaddition as follows: To more than 2.5 μM of the norbornene-functionalized POI from the last step, you add a 25× (concentration > 10 μM) to 100× fold molar excess of the tetrazine-label stock solution (in DMSO) over the initial protein concentration that needs to be labeled (*see* **Notes 25** and **26**).

2. Incubate the reaction at 37 °C under shaking for 1 h (*see* **Note 19**) in a thermoshaker.

3. Transfer the functionalization reaction mixture to an Amicon Ultra-0.5 mL filter unit with Ultracel 10 Membrane (10,000 MWCO) inserted into the filtrate collection tube. Fill up the ligation reaction mixture up to the maximum level of 500 μL with PBS. Centrifuge the assembled filter unit at 14,000 rcf for 15 min at 4 °C and discard the flow-through (*see* **Note 21**). Repeat this step twice. Label new concentration tubes and weigh each tube. Turn the cartridge upside down, put it into the new concentrate collection tubes, and centrifuge at 4 °C and 1000 rcf for 2 min. Discard the filter unit.

4. Measure the obtained volume by weight difference from the initial tube weight and add PBS until the original volume of 100 μL is reached again (*see* **Note 22**).

5. Estimate the protein concentration/absorbance with a microliter photospectrometer at 280 nm and calculate the protein recovery by comparison to the initial absorbance before LplA^{W37V} addition.

For the analysis of the ligation reaction, we recommend using the LAP-tagged model protein MBP-LAP-HA as a positive control and performing the Diels-Alder cycloaddition with a tetrazine fluorophore derivative bearing the same (or similar) tetrazine moiety. The labeling protocol can be conducted in a shortened version without the final washing step (*see* **Note 27**). The reaction is followed by an SDS-PAGE, in-gel fluorescence measurement (fluorescence laser scanner), and a protein staining procedure (*see* Fig. 4).

The model protein MBP-LAP-HA was first labeled enzymatically with the norbornene substrate in a purified form or in a human embryonic kidney (HEK) cell lysate and washed with PBS using centrifugal filtration. For a negative control at the enzymatic labeling stage, LplA^{W37V} was omitted. The labeling reactions on the purified MBP-LAP-HA were split into two parts and spiked into a HEK lysate. The samples were incubated with 125 μM of methyltetrazine-fluorescein (5/6 regioisomer mixture). For a negative control, the tetrazine-fluorophore was omitted. After this, the reaction was quenched with 5 mM of TCO-OH, incubated for 5 min at RT, heated in reducing Laemmli Buffer at 95 °C for 5 min, and run on an SDS-PAGE (10% of acrylamide:bisacrylamide in a ratio of 19:1) at 150 V for 80 min. The gel was imaged using a fluorescence laser scanner (Typhoon FLA 9500, 473 nm excitation, LBP filter) and then stained via colloidal Coomassie Blue G-250 [37]. The fluorescent dye was synthesized according to Liu et al. [13] (in contrast to the paper from Liu et al., the methyl-substituted aminobenzyltetrazine derivative was used for the synthesis of the fluorophore).

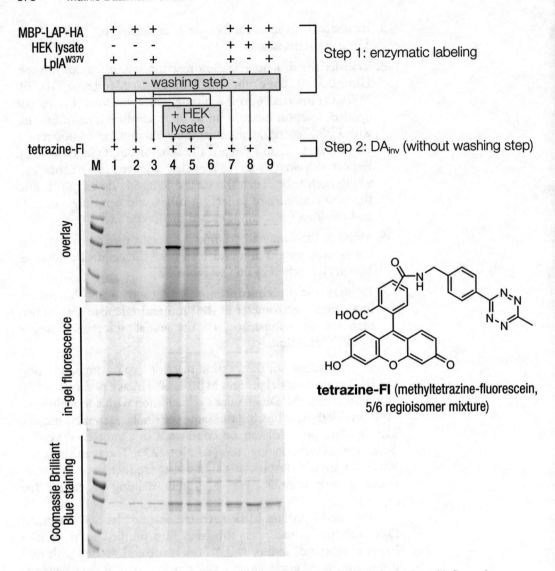

Fig. 4 The two-step labeling procedure is suitable for the functionalization of proteins with fluorophores

3.5 LplA^W37V-Mediated Imaging at the Cell Surface

All the following steps (except from the steps using a microscope or centrifuge) have to be performed in a sterilized, ventilated hood suitable for cell culture to minimize external sources of contamination.

According to our experience, we recommend using IBIDI μ-slide VI^0.4 as they allow experiments in a minimal scale in order to save compounds such as the tetrazine fluorophore or enzyme. Certainly, these experiments can also be performed in well-plates or other cell chambers that are suitable for fluorescence microscopy. Fluorescence imaging of live cell labeled pDisplay-LAP2-CFP-TM, a transmembrane protein bearing the cyan fluorescent protein and the LAP displayed on the cell surface, is shown in Fig. 5.

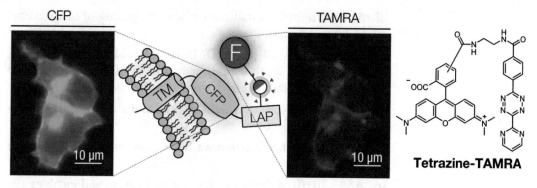

Fig. 5 Cell surface labeling in living HEK 293 cells in a two-step procedure

1. Thaw cells and cultivate them to at least the third passage.

2. Trypsinise cells at a confluency of 90% with 1 mL Trypsin-EDTA (0.05%) solution and add 5 mL of DMEM (GlutaMAX®, 4500 mg/L glucose, with Phenol Red). Carefully mix the suspension and incubate in a cell incubator for the detachment of the adherent cells.

3. Take an aliquot of the cell suspension and determine the number of cells, *i.e.* with a Neubauer chamber.

4. Adjust the number of cells in the suspension to 100 cells/µL by diluting with DMEM (GlutaMAX®, 4500 mg/L glucose, with Phenol Red) (*see* **Note 28**).

5. Add 30 µL of the cell suspension into each channel of the IBIDI µ-slide VI$^{0.4}$: Slightly tilt the slide and carefully inject the suspension into the opening of one side per channel. Avoid air bubbles.

6. Incubate the cells for 30 min in a cell incubator.

7. Add 100 µL of DMEM (GlutaMAX®, 4500 mg/L glucose, with Phenol Red) to each channel (*see* **Note 29**).

8. Incubate the cells overnight in the cell incubator.

9. Prepare six plastic tubes (1.5 mL) for every IBIDI µ-slide VI$^{0.4}$.

10. For transfection (here described exemplarily for FuGENE® HD, Promega) put 10 µL of Opti-MEM I into each plastic tube (*see* **Note 30**). Add 0.2 µg of plasmid (pDisplay-LAP2-CFP-TM; 1 µL from the stock solution) into each tube, mix by frequent pipetting, and incubate the transfection mix for 5 min at room temperature. Add 0.6 µL of transfection reagent into each transfection tube and mix by frequent pipetting. Incubate for 10 min at room temperature.

11. Remove the medium from each channel and wash the cells twice with 120 µL of DPBS.

12. Add the transfection mixture into one opening of each channel to be transfected.

13. Carefully add 60 μL of DMEM (GlutaMAX®, 4500 mg/L glucose, with Phenol Red) at each side of each channel. Do not flush the channel with the medium.

14. Incubate the cells overnight in the incubator for the transfection to take place.

15. Remove the medium and wash the cells twice with 120 μL of DPBS.

16. Add 120 μL of the buffer for ex vivo ligation and incubate the cells for 60 min in a cell incubator.

17. Remove the buffer for in vivo ligation, wash the cells twice with 120 μL DBPS, and incubate the cells with 120 μL DMEM (GlutaMAX®, 4500 mg/L glucose, without phenol red) for 30 min in the incubator.

18. Remove the medium and add 120 μL of a 0.2 μM solution of the tetrazine dye solution in DMEM (GlutaMAX®, 4500 mg/L glucose, without phenol red) and incubate for 3 h in the cell incubator.

19. Remove the medium, wash the cells twice with DPBS, and add 120 μL of Ringers buffer into each channel.

20. Perform live cell imaging in an epifluorescence microscope with $\lambda_{ex} = 472$ nm und $\lambda_{em} = 520$ nm (or similar channel settings) for CFP and $\lambda_{ex} = 543$ nm, $\lambda_{em} = 593$ nm (or similar) for Tetrazin-TAMRA, respectively.

Cells were transiently transfected with pDisplay-LAP2-CFP-TM, a transmembrane protein bearing the cyan fluorescent protein and the LAP displayed on the cell surface. In a first step, the cells were incubated with the norbornene substrate and LplA^{W37V}. After washing, labeling was performed with tetrazine TAMRA (TAMRA = carboxy tetramethyl-rhodamine). Epifluorescence microscopy was then performed at $\lambda_{ex} = 472$ nm, $\lambda_{em} = 520$ nm for CFP and $\lambda_{ex} = 543$ nm, $\lambda_{em} = 593$ nm for TAMRA, respectively. The tetrazine dye Tetrazine-TAMRA that was used here was made according to Schoch et al. [36].

4   Notes

1. We have functionalized various proteins using the two-step labeling procedure with a number of different tetrazine probes using this method. The LAP tag should be accessible by the LplA^{W37V} and can be N-terminal, C-terminal, or internal. Internal LAP tags between protein domains are more critical but possible to functionalize. Make sure that the LAP tag is

accessible by the LplA^{W37V} and is not buried inside of a protein. To ensure maximum flexibility and accessibility, we often include a GGGGS linker between the tag and the POI or between the different protein domains of the POI.

For positive controls, an LAP-containing fusion protein should be designed. We recommend using the maltose binding protein (MBP) as a fusion partner of the LAP for high yield expression and an immunotag (such as the human influenza hemagglutinin immunotag, HA) for the possible detection over western blots. The design and the construction of this MBP-LAP-HA protein is shortly sketched and can easily be adapted for your POI: To generate the MBP-LAP-HA protein (~47 kDa), perform a Gibson assembly (Or ligation independent cloning) with SspI-linearized pET His6 MBP TEV LIC cloning vector (2M-T) (plasmid from Scott Gradia, Addgene plasmid # 29708) and a synthetic double-stranded DNA molecule containing the sequence for LAP-HA. Transform the expression plasmid into *E. coli* BL21(DE3)-pLysS, express the protein, and purify it with standard Ni^{2+}-nitrilotriacetic acid (NTA) immobilized metal ion chromatography (IMAC). Using this system, MBP-LAP-HA can be obtained in a total yield of higher than 0.2 g from a 1 L cultivation in a terrific broth medium. Dialyze the purified protein fractions against PBS and concentrate it using centrifugal filtration. Then, aliquot the protein, shock freeze in liquid nitrogen, and store it at −80 °C for later use.

Double-stranded synthetic DNA (also known as synthetic genes) can be purchased by different manufacturers worldwide. A possible sequence tailored for ligation independent cloning or Gibson assembly in combination with the 2M-T plasmid is provided as follows:

5' -

TACTTCCAATCCAATGCAggtggtggcggtagcGGTTTCGAAATTGATAAAGTTTGGTATGA

CCTGGACGCG

ggcggtggcggttccTACCCGTACGATGTTCCGGACTACGCCtaataacattggaagtggat

aa-3'

Legend (5' → 3'): LIC v1 tag (left), GGGGS linker 1, LAP sequence, GGGGS linker 2, HA immunotag, LIC v1 tag (right).

The sequence of the model protein we used for the experiment is as follows:

```
>sp|MBP-LAP-HA|MBP-LAP-HA fusion protein

MKSSHHHHHHGSSMKIEEGKLVIWINGDKGYNGLAEVGKKFEKDTGIKVTVEHPDKLEEKFP

QVAATGDGPDIIFWAHDRFGGYAQSGLLAEITPDKAFQDKLYPFTWDAVRYNGKLIAYPIAV

EALSLIYNKDLLPNPPKTWEEIPALDKELKAKGKSALMFNLQEPYFTWPLIAADGGYAFKYE

NGKYDIKDVGVDNAGAKAGLTFLVDLIKNKHMNADTDYSIAEAAFNKGETAMTINGPWAWSN

IDTSKVNYGVTVLPTFKGQPSKPFVGVLSAGINAASPNKELAKEFLENYLLTDEGLEAVNKD

KPLGAVALKSYEEELAKDPRIAATMENAQKGEIMPNIPQMSAFWYAVRTAVINAASGRQTVD

EALKDAQTNGIEENLYFQSNAGGGGSGFEIDKVWYDLDAGGGGSYPYDVPDYA*
```

Legend (N- to C-terminus): His$_6$-tag, MBP tag, TEV protease recognition site, GGGGS linker 1, LAP sequence, GGGGS linker 2, HA immunotag.

2. Express His-tagged LplA^{W37V} (~39 kDa) in *E. coli* and purify the enzyme with standard expression and Ni^{2+}-NTA-IMAC protocols. The plasmid for His$_6$-tagged LplA^{W37V} expression in *E. coli* is available at the Addgene platform (pYFJ16-LplA(W37V), plasmid from Alice Ting, Addgene plasmid # 34838). Make sure that you work at lowered temperatures (4 °C) during and after the lysis. Between the lysis of the bacteria and final storage should only be 24 h to minimize loss of activity. Concentrate the enzyme to 0.5–10 mg/mL using centrifugal filtration. Aliquot the enzyme and shock freeze at −80 °C. For 100 μL labeling reactions (labeling of 0.1–50 nmol of LAP-tagged protein), aliquots of 5 μL with LplA^{W37V} concentrations around 1 mg/mL (~20 μM) are convenient. For cell surface labeling, the concentration should be higher (>2 mg/mL or >40 μM). We routinely use plain PBS as the storage buffer for LplA^{W37V}. If prolonged storage above 0 °C or prolonged incubation is desired, DTT/DTE addition is possible since LplA is known to form intramolecular disulfide bonds by reaction with molecular oxygen over the time [8]. We never had issues with oxidative loss of activity because we use fresh aliquots for each experiment and short incubation times.

3. The enzyme retains activity for at least 2 years at −80 °C in the shock-frozen state in PBS. Once an aliquot of LplA^{W37V} is needed, it should be thawed and handled on ice. For reproduc-

ibility, we recommend using a fresh aliquot for each experiment. Any remaining enzyme should be discarded after the experiment. Avoid repeated freeze-thaw cycles since the activity of LplA^{W37V} will be lost rapidly.

4. We recommend using a phosphate buffer with a pH near to the pH optimum (pH 6.8–7.0) [8] of the native *LplA* from *Escherichia coli* for the ligation reaction. The ligation reaction also takes place at physiological pH (pH 7.4) in buffered media such as DMEM used for cell culture. Usage of other buffers is possible, but try to avoid buffers containing nucleophiles (such as the primary amine of Tris) in high concentrations. Use inorganic buffers or organic non-nucleophilic buffers instead (for example MOPS- or HEPES-buffers; Good's buffers).

5. Methyl tetrazine-fluorophores are commercially available (Sulfo-Cy5-Tetrazine, 6-Methyl-Tetrazine-5-TAMRA, 6-Methyl-Tetrazine-5-FAM) as well as methyl tetrazine biotin conjugates (6-Methyl-Tetrazine-PEG4-Biotin) and bifunctional methyl-tetrazine compounds for prior conjugation to a small molecule of interest (i.e., 6-Methyl-Tetrazine-PEG5-NHS ester).

6. It should be noted that there are more reactive tetrazine compounds available, for instance such with an H-substituted tetrazine moiety. Although they are around 20 times faster than the methyl-substituted versions, their stability in aqueous buffers and sera is impaired [14]. In terms of mild chemical derivatization reactions utilizing peptide chemistry, methyl-substituted tetrazines are more suitable because of their improved resistance toward nucleophiles which is helpful in multi-step syntheses [38]. The choice of the tetrazine is thus guided by the experimental requirements, for instance by-reactions vs. clean reactions or stability vs. reactivity. Also consider that an organic chemistry laboratory might be able to synthesize the desired tetrazine-containing label of interest with tailored properties for your experiments quite easily.

7. 10× PBS concentrates are convenient in both handling and storage.

8. Always warm the temperature up to 37 °C before using both media and buffer.

9. Store at 4 °C and pre-warm to 37 °C before using.

10. It is sufficient to use a standard epifluorescence microscope.

11. If you do not see the formation of two layers, either add more ethyl acetate or brine.

12. Do not add too much MgSO$_4$. Finish the addition after no more clumping of MgSO$_4$ can be observed.

13. We recommend using an oil pump connected to a Schlenk line with a Dewar flask filled with liquid nitrogen.

14. We do not recommend reactions in volumes smaller than 50 µL; however, upscaling is no problem.

15. The labeling reaction also works in the presence of impurities or even in whole cell lysates. Make sure that no phosphatases are present in the labeling reaction that might degrade ATP which is a required substrate of the enzymatic reaction (*see* Fig. 1). If there are phosphatases present, use an appropriate phosphatase inhibitor.

16. Ethylenediaminetetraacetic acid (EDTA) or other chelators will inhibit the enzymatic labeling reaction by the reduction of free Mg^{2+} levels. Mg^{2+} is a compulsory cofactor of the LplA^{W37V}. Desalt the protein sample or dilute it before performing the enzymatic functionalization.

17. We recommend a concentration of the LAP-tagged protein in 2.5 µM or higher. Do not exceed a concentration of higher than 250 µM.

18. Too much enzyme, i.e., >5 µM or longer incubation times, i.e., >4 h at 37 °C will lead to excessive formation of the mixed anhydride intermediate between the norbornene substrate and AMP (*see* Fig. 1). This activated carboxylic acid is rather weakly stable in the bulk buffer, but in high concentrations or when non-modified LAP-tagged protein is already depleted, it reacts nonspecifically with nucleophiles that are present in the ligation reaction mix, i.e., lysine residues in proteins thus leading to background functionalization.

19. Alternatively incubate at 30 °C for 2 h or at 4 °C overnight.

20. We found that the centrifugal filtration has a higher protein recovery rate compared to gel filtration units and provides more flexibility, such as sample concentration, for a similar price.

21. When you use centrifugal filters with a lower molecular cutoff or a higher volume, please consider the manual from the manufacturer first to adjust centrifugal force and filtration time.

22. The volume to add can be reduced to make the protein solution more concentrated. This way, a twofold concentration from an initial volume of 100–50 µL is easily feasible.

23. The concentration of the LAP-tagged protein of interest for the next step (DA$_{inv}$) should be equal to or higher than 2.5 µM.

24. For each new protein, we recommend performing an analysis reaction although our optimized reaction conditions worked for every LAP-tagged protein we have in our laboratory.

25. Please consider that centrifugal filter units that are used in a later step in this protocol have a limited scope of DMSO toler-

ance. A total DMSO concentration of 1% did not impair the centrifugal concentration in our applications. When the DMSO concentration is too high, dilute the sample to a concentration suitable for the filter device.

26. This reaction is optimized for methyl-tetrazine-phenyl compounds. If you plan to use faster tetrazine derivatives for the DA_{inv} [14], the reaction conditions need to be optimized carefully.

27. After the DA_{inv} reaction, excess tetrazine-fluorophore is quenched with a 10× molar excess of *trans*-cyclooctenol (TCO-OH, prepare a 100–200 mM stock solution in DMSO, prepare freshly and do not store at RT for prolonged periods, CAS-#: 85081-69-2) over the fluorophore and incubated for 5 min at RT. This step is necessary before the treatment of the protein sample with SDS-sample buffer at elevated temperatures that will otherwise facilitate side-reactions of the tetrazine moiety with reactive residues of the proteins. Tetrazine derivatives are prone to react with other nucleophiles such as cysteine and amines [38]. For SDS-PAGE, the removal of basic TCO-OH-tetrazine fluorophore conjugates (like Cy dyes, fluorescein, rhodamines, fluoresceins) is not necessary since it will run with the bromophenol blue front and does not impair further analysis. For the generation of fluorescently labeled proteins that are not analyzed in SDS-PAGE and are supposed to be clean and used for further experiments, the fluorophore has to be removed by centrifugal filtration (see full protocol).

28. It is sufficient to prepare a volume of at least 200 μL cell suspension for each IBIDI μ-slide VI$^{0.4}$ used in your experiment.

29. Do not inject the medium into one side of the channel. Carefully add 50 μL of DMEM (GlutaMAX®, 4500 mg/L glucose, with Phenol Red) at each side of each channel. Do not flush the channel with the medium to ensure maximum cell attachment in the imaging channel.

30. We recommend setting up one tube of transfection mixture for every channel, do not prepare master mixes.

Acknowledgment

This work was supported by funding from the Deutsche Forschungsgemeinschaft DFG (SPP1623, WO 1888/1-2). Furthermore, Mathis Baalmann acknowledges support from the Landesgradiuertenförderung Baden-Württemberg (LGF BW). The authors thank Stefanie Kühn, Hagen Sparka, and Tobias T. Schmidt for technical and experimental support.

References

1. Plass T, Milles S, Koehler C, Szymański J, Mueller R, Wießler M, Schultz C, Lemke EA (2012) Amino acids for Diels–Alder reactions in living cells. Angew Chem Int Ed 51:4166–4170

2. Kaya E, Vrabel M, Deiml C, Prill S, Fluxa VS, Carell T (2012) A genetically encoded norbornene amino acid for the mild and selective modification of proteins in a copper-free click reaction. Angew Chem Int Ed 51:4466–4469

3. Lang K, Davis L, Torres-Kolbus J, Chou C, Deiters A, Chin JW (2012) Genetically encoded norbornene directs site-specific cellular protein labelling via a rapid bioorthogonal reaction. Nat Chem 4:298–304

4. Nakamura Y, Ito K, Isaksson LA (1996) Emerging understanding of translation termination. Cell 87:147–150

5. Cronan JE, Zhao X, Jiang Y (2005) Function, attachment and synthesis of lipoic acid in Escherichia coli. In: Poole RK (ed) Advances in microbial physiology, vol 50. Academic Press, Cambridge, pp 103–146

6. Fujiwara K, Toma S, Okamura-Ikeda K, Motokawa Y, Nakagawa A, Taniguchi H (2005) Crystal structure of lipoate-protein ligase a from Escherichia coli determination of the lipoic acid-binding site. J Biol Chem 280:33645–33651

7. Fernandez-Suarez M, Baruah H, Martinez-Hernandez L, Xie KT, Baskin JM, Bertozzi CR, Ting AY (2007) Redirecting lipoic acid ligase for cell surface protein labeling with small-molecule probes. Nat Biotech 25:1483–1487

8. Green D, Morris T, Green J, Cronan J, Guest J (1995) Purification and properties of the lipoate protein ligase of Escherichia coli. Biochem J 309:853–862

9. Uttamapinant C, Sanchez MI, Liu DS, Yao JZ, White KA, Grecian S, Clark S, Gee KR, Ting AY (2013) Site-specific protein labeling using PRIME and chelation-assisted click chemistry. Nat Protoc 8:1620–1634

10. Uttamapinant C, Tangpeerachaikul A, Grecian S, Clarke S, Singh U, Slade P, Gee KR, Ting AY (2012) Fast, cell-compatible click chemistry with copper-chelating azides for biomolecular labeling. Angew Chem Int Ed 51:5852–5856

11. Yao JZ, Uttamapinant C, Poloukhtine A, Baskin JM, Codelli JA, Sletten EM, Bertozzi CR, Popik VV, Ting AY (2012) Fluorophore targeting to cellular proteins via enzyme-mediated azide ligation and strain-promoted cycloaddition. J Am Chem Soc 134:3720–3728

12. Cohen JD, Zou P, Ting AY (2012) Site-specific protein modification using lipoic acid ligase and bis-aryl hydrazone formation. Chembiochem 13:888–894

13. Liu DS, Tangpeerachaikul A, Selvaraj R, Taylor MT, Fox JM, Ting AY (2012) Diels–Alder cycloaddition for fluorophore targeting to specific proteins inside living cells. J Am Chem Soc 134:792–795

14. Karver MR, Weissleder R, Hilderbrand SA (2011) Synthesis and evaluation of a series of 1,2,4,5-tetrazines for bioorthogonal conjugation. Bioconjug Chem 22:2263–2270

15. Yang J, Šečkutė J, Cole CM, Devaraj NK (2012) Live-cell imaging of cyclopropene tags with fluorogenic tetrazine cycloadditions. Angew Chem Int Ed 51:7476–7479

16. Uttamapinant C, White KA, Baruah H, Thompson S, Fernández-Suárez M, Puthenveetil S, Ting AY (2010) A fluorophore ligase for site-specific protein labeling inside living cells. Proc Natl Acad Sci 107:10914–10919

17. Puthenveetil S, Liu DS, White KA, Thompson S, Ting AY (2009) Yeast display evolution of a kinetically efficient 13-amino acid substrate for lipoic acid ligase. J Am Chem Soc 131:16430–16438

18. Hauke S, Best M, Schmidt TT, Baalmann M, Krause A, Wombacher R (2014) Two-step protein labeling utilizing Lipoic acid ligase and Sonogashira cross-coupling. Bioconjug Chem 25:1632–1637

19. Best M, Degen A, Baalmann M, Schmidt TT, Wombacher R (2015) Two-step protein labeling by using lipoic acid ligase with norbornene substrates and subsequent inverse-electron demand Diels-Alder reaction. Chembiochem 16:1158–1162

20. Hermes FAM, Cronan JE (2009) Scavenging of cytosolic octanoic acid by mutant LplA lipoate ligases allows growth of Escherichia coli strains lacking the LipB octanoyltransferase of lipoic acid synthesis. J Bacteriol 191:6796–6803

21. Wombacher R, Cornish VW (2011) Chemical tags: applications in live cell fluorescence imaging. J Biophotonics 4:391–402

22. Prescher JA, Bertozzi CR (2005) Chemistry in living systems. Nat Chem Biol 1:13–21

23. Patterson DM, Nazarova LA, Prescher JA (2014) Finding the right (bioorthogonal) chemistry. ACS Chem Biol 9:592–605

24. Saxon E, Bertozzi CR (2000) Cell surface engineering by a modified Staudinger reaction. Science 287:2007–2010

25. Agard NJ, Prescher JA, Bertozzi CR (2004) A strain-promoted [3 + 2] Azide–alkyne cycloaddition for covalent modification of biomolecules in living systems. J Am Chem Soc 126:15046–15047

26. Tornøe CW, Christensen C, Meldal M (2002) Peptidotriazoles on solid phase: [1,2,3]-triazoles by regiospecific copper(I)-catalyzed 1,3-dipolar cycloadditions of terminal alkynes to azides. J Org Chem 67:3057–3064

27. Chalker JM, Wood CSC, Davis BG (2009) A convenient catalyst for aqueous and protein Suzuki–Miyaura cross-coupling. J Am Chem Soc 131:16346–16347

28. Kodama K, Fukuzawa S, Nakayama H, Sakamoto K, Kigawa T, Yabuki T, Matsuda N, Shirouzu M, Takio K, Yokoyama S (2007) Site-specific functionalization of proteins by Organopalladium reactions. Chembiochem 8:232–238

29. Ourailidou ME, van der Meer JY, Baas BJ, Jeronimus-Stratingh M, Gottumukkala AL, Poelarends GJ, Minnaard AJ, Dekker FJ (2014) Aqueous oxidative heck reaction as a protein-labeling strategy. Chembiochem 15:209–212

30. Blackman ML, Royzen M, Fox JM (2008) Tetrazine ligation: fast bioconjugation based on inverse-electron-demand Diels–Alder reactivity. J Am Chem Soc 130:13518–13519

31. Devaraj NK, Weissleder R, Hilderbrand SA (2008) Tetrazine-based cycloadditions: application to pretargeted live cell imaging. Bioconjug Chem 19:2297–2299

32. Thalhammer F, Wallfahrer U, Sauer J (1990) Reaktivität einfacher offenkettiger und cyclischer dienophile bei Diels-Alder-reaktionen mit inversem elektronenbedarf. Tetrahedron Lett 31:6851–6854

33. Rossin R, van den Bosch SM, ten Hoeve W, Carvelli M, Versteegen RM, Lub J, Robillard MS (2013) Highly reactive trans-Cyclooctene tags with improved stability for Diels–Alder chemistry in living systems. Bioconjug Chem 24:1210–1217

34. Zeglis BM, Sevak KK, Reiner T, Mohindra P, Carlin SD, Zanzonico P, Weissleder R, Lewis JS (2013) A pretargeted PET imaging strategy based on bioorthogonal Diels–Alder click chemistry. J Nucl Med 54:1389–1396

35. Knight JC, Richter S, Wuest M, Way JD, Wuest F (2013) Synthesis and evaluation of an 18F-labelled norbornene derivative for copper-free click chemistry reactions. Org Biomol Chem 11:3817–3825

36. Schoch J, Staudt M, Samanta A, Wiessler M, Jäschke A (2012) Site-specific one-pot dual labeling of DNA by orthogonal cycloaddition chemistry. Bioconjug Chem 23:1382–1386

37. Candiano G, Bruschi M, Musante L, Santucci L, Ghiggeri GM, Carnemolla B, Orecchia P, Zardi L, Righetti PG (2004) Blue silver: a very sensitive colloidal Coomassie G-250 staining for proteome analysis. Electrophoresis 25:1327–1333

38. Zeglis BM, Emmetiere F, Pillarsetty N, Weissleder R, Lewis JS, Reiner T (2014) Building blocks for the construction of bioorthogonally reactive peptides via solid-phase peptide synthesis. ChemistryOpen 3:48–53

Chapter 24

Genetic Encoding of Unnatural Amino Acids in *C. elegans*

Lloyd Davis and Sebastian Greiss

Abstract

Site-specific incorporation of unnatural amino acids (UAAs) has greatly expanded the toolkit available to study biological phenomena in single cells. However, to address questions involving complex cellular interactions such as development, ageing, and the functions of the nervous system it is often necessary to use multicellular model organisms. The nematode *Caenorhabditis elegans* was the first organism to have its genetic code expanded. Due to its small size, ease of cultivation, and excellent UAA incorporation efficiency, *C. elegans* makes an ideal model organism to apply UAAs as tools to investigate the functioning of multicellular systems.

Here, we describe methods to generate transgenic *C. elegans* capable of UAA incorporation, as well as how to deliver unnatural amino acids and test incorporation. Furthermore, we describe methods to uncage photosensitive unnatural amino acid derivatives.

Key words Genetic code expansion, Unnatural amino acid, *C. elegans*, Multicellular organism, Biolistic bombardment, Photocaged amino acid

1 Introduction

Genetic code expansion allows enhancement of the genetic code beyond the 20 canonical amino acids. Site-specific incorporation of unnatural amino acids (UAAs) has allowed new functionalities to be added to proteins in the context of single cells [1, 2], which has facilitated structural analysis [3–6], crosslinking [7, 8], labeling [9–13], and precise spatio-temporal control of protein activation [14–17] in bacteria, yeast and in mammalian cell culture.

While a multitude of biological phenomena have been studied in the context of purified proteins or single cells, many others such as development, ageing, or the functioning of the nervous system are best studied in a multicellular context. *C. elegans* is an attractive model in which to employ UAAs introducing new functionalities. *C. elegans* are cheap and easy to maintain in large numbers. In addition, they are transparent, facilitating the easy use of photo-responsive amino acids such as photo-caged amino acids and fluorescent reporters.

Edward A. Lemke (ed.), *Noncanonical Amino Acids: Methods and Protocols*, Methods in Molecular Biology, vol. 1728, https://doi.org/10.1007/978-1-4939-7574-7_24, © Springer Science+Business Media, LLC 2018

Site-specific incorporation of UAAs is dependent upon expression of an aminoacyl tRNA sythetase/tRNA pair which is orthogonal to the host, thus does not result in cross aminoacylation of host tRNAs or aminoacylation of the exogenous tRNA with endogenous amino acids. In order to direct site-specific incorporation, a tRNA with a CUA anticodon is generally used, while the intended site of incorporation in the target gene is mutated to an amber stop codon (TAG). Thus far UAA incorporation in *C. elegans* has been achieved using the tRNA^{Pyl}/PylRS pair from *Methanosarcina mazei* [18] or the tRNA^{Leu}/LeuRS and tRNA^{Tyr}/TyrRS pairs from *Escherichia coli* [19]. These pairs differ in the UAAs they and their evolved variants are capable of incorporating, therefore the choice as to which pair to employ will depend on the UAA required. The approach described in this chapter is based on our experience using the *M. mazei* tRNA^{Pyl}/PylRS pair, and while the requirements should to the largest extent be similar, it cannot be excluded that differences may be encountered when using other synthetase/tRNA$_{CUA}$ pairs.

The incorporation of UAA in a multicellular organism requires the same genetic components as in single celled systems (Fig. 1). The use of UAA in whole animals however brings with it challenges

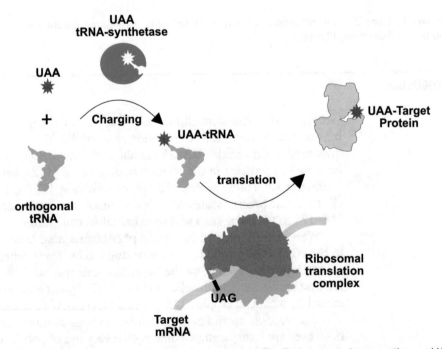

Fig. 1 Scheme for incorporation of an UAA into a target protein. The first step requires an orthogonal UAA tRNA synthetase (blue) to charge the UAA (red star) onto the cognate orthogonal tRNA (green). Once charged, the aminoacylated tRNA (UAA-tRNA) can then be recruited to the ribosomal translation complex (orange) and is incorporated in response to the amber stop codon (UAG) in the mRNA sequence of the target gene (gray) resulting in production of the target protein containing the UAA site-specifically incorporated (UAA-Target Protein)

that are either not encountered in single celled systems, or are encountered in single cells to a much lesser extent.

In this chapter, we will describe the requisite components for genetic code expansion, the generation of transgenic *C. elegans* strains for UAA incorporation [20], the UAA delivery method used, how to test for UAA incorporation, and how to uncage photocaged UAAs in *C. elegans*.

1.1 Generating Transgenic Worms

Animals contain many differentiated cell types. For each cell type, different promoters or combinations of promoters may be required to provide the requisite expression levels for incorporation. It has been shown in other systems that it is important to balance expression levels of the individual genetic components for UAA incorporation [21, 22]. We have likewise found that in *C. elegans* the ratio of aminoacyl tRNA synthetase and tRNA appears to influence UAA incorporation efficiency. In our experience the easiest way to generate transgenic lines which can efficiently incorporate UAAs is to co-transform the genetic constructs on separate plasmids, resulting in lines with varying expression ratios of the individual components. This approach usually yields lines incorporating UAA in the desired tissue.

A further complication specific to *C. elegans* is the lack of extensively characterized RNA Polymerase III (Pol III) promoters. Unlike protein coding genes, tRNA genes in eukaryotes are transcribed by Pol III from promoters internal to the tRNA. The non-eukaryotic tRNAs required for UAA incorporation in *C. elegans* do not contain internal Pol III promoters and thus need to be transcribed from upstream promoters. Two upstream Pol III promoters have been successfully used in worms: one driving expression of the stem-bulge noncoding RNA *CeN74-1* [18]; the other driving expression of *rpr-1*, the RNAse subunit of RNAseP [19]. The promoter driving expression of the spliceosomal noncoding RNA U6 has been used in mammalian cells [14] and *Drosophila melanogaster* [23]. However, U6 is expressed from around 20 different chromosomal loci in *C. elegans*, with different promoters and 3′ regions and little is known about differences in expression at each locus. In our hands both the *CeN74-1* and the *rpr-1* promoter are suitable to drive tRNA expression in a wide range of tissues. At present, no published information is available on tissue specific Pol III promoters.

In order to perform UAA incorporation in *C. elegans*, a minimum of three components are required:

1. A pyrrolysyl tRNA-synthetase driven by a relevant promoter (Fig. 2a) and also containing a selectable marker (such as resistance to Hygromycin B) (Fig. 2b).

2. A pyrrolysyl $tRNA_{CUA}$ construct expressing the $tRNA_{CUA}$ from a Pol III promoter. We generally use the *C. elegans*

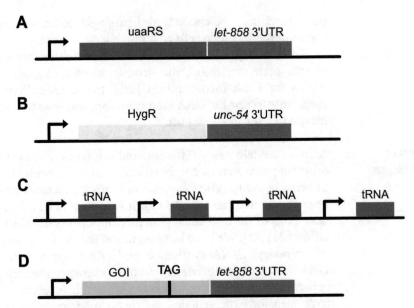

Fig. 2 The genetic components required to incorporate an UAA site-specifically in *C. elegans*. (**a**) An UAA tRNA-synthetase (uaaRS) evolved to incorporate the desired UAA must be expressed behind a relevant promoter, the let-858 3′ UTR stabilizes the mRNA in all tissues and life stages. (**b**) One of the plasmids must also contain an antibiotic selection marker to facilitate post bombardment selection, we recommend the Hygromycin B resistance gene (HygR). The HygR gene should be expressed behind a global promoter [20]. (**c**) A four copy tRNA_CUA expression cassette can be used, this can be created by iterative cloning. Each copy should be driven by a Pol III promoter such as P*rpr-1*. (**d**) The Gene of Interest (GOI) must contain the amber codon(TAG) as an in-frame mutation in place of one endogenous codon. Black arrows represent promoters, gray boxes represent terminators, colored boxes represent coding sequences

rpr-1 promoter to drive expression of the pyrrolysyl-tRNA_CUA upstream of a *C. elegans sup-7* 3′ region acting as a terminator (Fig. 2c).

3. A target gene expressed behind a suitable reporter and containing an amber stop codon at the desired site of incorporation (Fig. 2d).

We describe here a method of generating lines using biolistic bombardment and antibiotic selection [20], it is however also possible to generate transgenic lines using micro-injection. After isolating incorporating lines, the constructs can be genomically integrated using established protocols in order to remove effects of mosaic expression.

1.2 Incorporation and Testing of UAA in C. elegans

Once transgenic lines have been established, it is necessary to introduce the UAA into the organism. A significant complication in multicellular organisms that is not present in single cells is availability of UAA at non-surface tissues. In cultured cells an UAA needs only to cross the plasma membrane to access the cell. In animals, any supplemented compound must often cross several layers of tissue before reaching the intended site of expression. In order to facilitate this the UAA growth agar can be supplemented with a small volume of detergent to aid in permeabilization of the animals, allowing easier UAA uptake across membranes.

When establishing new UAAs or using new promoters to incorporate in specific tissues, it is necessary to test for incorporation and optimize incorporation conditions (e.g., UAA concentration, incubation time) with a standardized reporter construct.

Testing for UAA incorporation can be done by Western blot to check for the presence of full-length product, which would require readthrough of the amber codon. Where there is sufficient size resolution between full-length and truncated product, antibodies to the N-terminal region or to an N-terminal tag can be used. However, where such size resolution is difficult, and to ensure that only full-length product is detected, it is necessary to blot with an antibody which binds C-terminal to the amber codon or to a C-terminal peptide tag.

Alternatively, fluorescence assays can be carried out to check for readthrough of an amber stop codon. For this purpose, we have used a GFP::mCherry fusion protein in which the linker between the two fluorescent proteins contains an amber codon (Fig. 3). A C-terminal NLS domain [24] is added to concentrate the mCherry signal and provides nuclear localization as a second readout for full-length product. In addition, for this reporter, a C-terminal HA tag allows for highly sensitive detection by Western blot.

1.3 Uncaging Photocaged UAAs in C. elegans

Photocaged UAAs can allow the generation of proteins which are photoactivatable in a biological context in living animals. Due to the small, transparent nature of *C. elegans*, it provides a promising environment in which to perform such experiments. These photocaged UAAs can be uncaged globally throughout the worm or in a targeted cell-specific manner.

2 Materials

2.1 Generating Transgenic Worms

2.1.1 Chemicals and Reagents

Potassium phosphate dibasic (K_2HPO_4).

Potassium phosphate monobasic (KH_2PO_4).

Calcium chloride ($CaCl_2$).

Magnesium sulfate ($MgSO_4$).

Cholesterol.

Agar.

Peptone.

Sodium chloride (NaCl).

Disodium hydrogen phosphate (Na_2HPO_4).

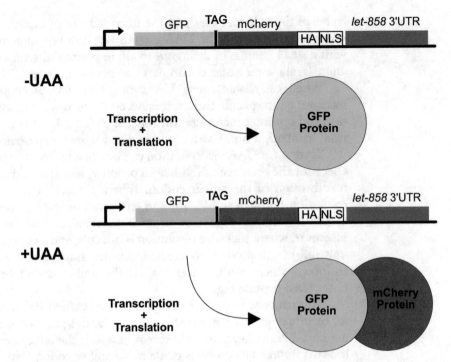

Fig. 3 Checking incorporation of an UAA by fluorescence. The GFP::mCherry reporter consists of the coding domains of each fluorescent protein fused by a short linker containing an in-frame amber stop codon (TAG). When translated in the absence of UAA (−UAA) the amber stop codon acts as a translational stop codon resulting in production of only the GFP protein. However, when expressed in the presence of UAA (+UAA), suppression of the amber codon will occur resulting in production of GFP::mCherry fusion protein that localizes to the nucleus [18]

Citrate monohydrate ($C_6H_8O_7 \cdot H_2O$).

Tri-potassium citrate monohydrate ($K_3C_6H_5O_7 \cdot H_2O$).

Disodium EDTA.

Iron sulfate heptahydrate ($FeSO_4 \cdot 7H_2O$).

Manganese chloride tetrahydrate ($MnCl_2 \cdot 4H_2O$).

Zinc sulfate heptahydrate ($ZnSO_4 \cdot 7H_2O$).

Copper sulfate pentahydrate ($CuSO_4 \cdot 5H_2O$).

Hygromycin B.

Penicillin/Streptomycin/Amphotericin B.

Nystatin.

Kanamycin.

Ethanol absolute HPLC grade.

Spermidine.

2.1.2 Preparing Media and Agar Plates

1. 1 M Phosphate buffer pH 6.0 (1 L).

    (a) Add 108.3 g potassium phosphate monobasic (KH_2PO_4) and 35.6 g potassium phosphate dibasic (K_2HPO_4) to a flask/bottle.

(b) Add water to 1 L total volume.

(c) Autoclave.

2. 1 M or 2.5 M Calcium chloride (1 L).

   (a) Add 111 g (1 M) or 277.5 g (2.5 M) calcium chloride ($CaCl_2$) to a flask/bottle.

   (b) Add water to 1 L total volume.

   (c) Autoclave.

3. 1 M Magnesium sulfate (1 L).

   (a) Add 120.4 g magnesium sulfate ($MgSO_4$) to a flask/bottle.

   (b) Add water to 1 L total volume.

   (c) Autoclave.

4. 5 mg/mL Cholesterol (1 L).

   (a) Add 5 g of cholesterol to a flask/bottle.

   (b) Add ethanol to 1 L.

   (c) Sterile filter.

5. NGM Agar (or 3× NGM Agar) (1 L).

   (a) If making normal NGM agar add 17 g agar, 2.5 g peptone and 3 g sodium chloride (NaCl) to a glassware vessel (if making 3× NGM agar add 17 g agar, 7.5 g peptone and 3 g NaCl).

   (b) Add water to 1 L total volume.

   (c) Autoclave.

   (d) Cool to 55 °C for 15 min.

   (e) Add 25 mL of 1 M phosphate buffer, 1 mL 1 M calcium chloride, 1 mL 1 M magnesium sulfate and 1 mL of 5 mg/mL cholesterol.

   (f) Mix thoroughly and dispense in petri dishes.

6. M9 Medium (1 L).

   (a) Add 3 g potassium phosphate monobasic (KH_2PO_4), 6 g disodium hydrogen phosphate (Na_2HPO_4), and 5 g of sodium chloride (NaCl) to a bottle/flask.

   (b) Add water to 1 L total volume.

   (c) Autoclave.

   (d) Add 1 mL 1 M magnesium sulfate.

7. S Basal (1 L).

   (a) Add 5.85 g sodium chloride (NaCl), 1 g potassium phosphate dibasic (K_2HPO_4) and 6 g potassium phosphate monobasic (KH_2PO_4) to a flask/bottle.

   (b) Add water to 1 L total volume.

(c) Autoclave.

(d) Add 1 mL 5 mg/mL cholesterol.

8. 1 M Potassium citrate (1 L).

(a) Add 20 g citrate monohydrate ($C_6H_8O_7 \cdot H_2O$) and 293.5 g tripotassium citrate monohydrate ($K_3C_6H_5O_7 \cdot H_2O$) to a flask/bottle.

(b) Add water to 1 L total volume.

(c) Autoclave.

9. Trace metals solution.

(a) Add 1.86 g disodium EDTA, 0.69 g iron sulfate heptahydrate ($FeSO_4 \cdot 7H_2O$), 0.2 g manganese chloride tetrahydrate ($MnCl_2 \cdot 4H_2O$), 0.29 g zinc sulfate heptahydrate ($ZnSO_4 \cdot 7H_2O$), 0.025 g copper sulfate pentahydrate ($CuSO_4 \cdot 5H_2O$) to a flask/bottle.

(b) Add water to 1 L total volume.

(c) Autoclave.

10. S Medium (1 L).

(a) To 1 L of S Basal add 10 mL of 1 M potassium citrate.

(b) Add 10 mL of trace metals solution, 3 mL 1 M calcium chloride and 3 mL 1 M magnesium sulfate.

11. 100 mg/mL Hygromycin B solution.

(a) Add 5 g of Hygromycin B to a 50 mL Falcon tube.

(b) Fill tube with 50 mL of sterile water.

(c) Sterile filter into a fresh 50 mL Falcon tube.

2.1.3 Equipment

DNA LoBind 1.5 mL tube (Eppendorf).

Petri dishes, 9 cm.

Premier polypropylene felt bag, 5 μm, size 3 (GAPS water treatment).

PDS-1000/He Biolistic Particle Delivery System with Hepta-adapter (Biorad).

0.3–3 μm gold beads (ChemPur).

Biolistic macrocarriers (InbioGold).

Biolistic rupture disks (Biorad).

Biolistic stopping screens (Biorad).

2.2 Incorporation and Testing of UAA in C. elegans

2.2.1 Chemicals and Reagents

Potassium phosphate dibasic (K_2HPO_4).

Potassium phosphate monobasic (KH_2PO_4).

Calcium chloride ($CaCl_2$).

Magnesium sulfate ($MgSO_4$).

Cholesterol.

Agar.

Peptone.

Sodium chloride (NaCl).

Disodium hydrogen phosphate (Na$_2$HPO$_4$).

Hygromycin B.

Triton-X100.

Sodium hypochlorite bleach 5% (optional).

Sodium hydroxide (1 M) (optional).

Levamisole (optional).

4× LDS Sample Buffer (Thermo-Fisher Scientific) (optional).

Dithiothreitol (optional).

2.2.2 Preparing Media and Agar Plates

All media and agar used are of same compositions as in Subheading 2.1.2.

2.2.3 Equipment

Petri dishes, 3.5 cm.

Polybead polystyrene 0.05 μm microspheres (Polysciences Inc.).

Glass slides 76 × 26 mm.

Coverslips 20 × 20 mm.

2.3 Uncaging Photocaged UAAs in C. elegans

2.3.1 Chemicals and Reagents

Potassium phosphate dibasic (K$_2$HPO$_4$).

Potassium phosphate monobasic (KH$_2$PO$_4$).

Calcium chloride (CaCl$_2$).

Magnesium sulfate (MgSO$_4$).

Cholesterol.

Agar.

Peptone.

Sodium chloride (NaCl).

Disodium hydrogen phosphate (Na$_2$HPO$_4$).

Hygromycin B.

Triton-X100.

Sodium hypochlorite bleach 5% (optional).

Sodium hydroxide (1 M) (optional).

Levamisole (optional).

2.3.2 Preparing Media and Agar Plates

All media and agar used are of same compositions as in Subheading 2.1.2.

2.3.3   Equipment

Petri dishes, 3.5 cm.

Polybead polystyrene 0.05 μm microspheres (Polysciences Inc.).

Glass slides 76 × 26 mm.

Coverslips 20 × 20 mm.

CL-1000L 365 nm ultraviolet crosslinker (UVP).

CoolLED pE-4000.

3   Methods

3.1   Generating Transgenic Worms

3.1.1   Preparing Gold Beads

1. Weigh out gold beads into an Eppendorf tube, 6 mg per bombardment.

2. Add 1 mL 70% HPLC grade ethanol to beads. Vortex for 30 min. Pellet beads by spinning briefly and remove supernatant (*see* **Note 1**).

3. Add 1 mL sterile water, vortex and remove supernatant. Repeat two more times.

4. Resuspend beads in 50% glycerol. At this point the beads can be stored in a refrigerator for a maximum of 2 weeks until required.

3.1.2   Coating Beads with DNA

1. Digest DNA to be transformed with restriction enzyme to linearize the constructs (*see* **Note 2**). After digestion, heat inactive the enzymes and mix the DNA at the ratios required for each bombardment (*see* **Note 3**). Each bombardment should have approximately 10 μg DNA in total in a 50 μL volume.

2. Add approximately 6 mg of resuspended beads in 50% glycerol to Eppendorf DNA LoBind tubes, one for each transformation. Briefly spin down and remove the supernatant.

3. Add the linearized DNA, in a total volume of 50 μL, to the beads. Then add 50 μL 2.5 M CaCl$_2$ and 20 μL 100 mM spermidine with thorough mixing. Leave with shaking for approximately 30 min (*see* **Note 4**).

4. Centrifuge briefly and remove supernatant. Add 300 μL of 70% HPLC grade ethanol, shake, spin down briefly, and remove the supernatant. Add 1 mL 100% HPLC grade ethanol, shake, spin down briefly and remove supernatant. Resuspend beads in 170 μL 100% HPLC grade ethanol. Once precipitated, the DNA-coated beads in 100% HPLC grade ethanol can be stored at −20 °C until required.

3.1.3   Preparing Plates and Food for Worms

1. Prepare 3× NGM according to the recipe above (Subheading 2.1.2). Dispense into 9 cm petri dishes. Leave to dry for 2–3 days.

2. Inoculate a starter liquid culture of HB101 bacteria with a single colony. Grow overnight at 37 °C.

3. Prepare and autoclave 6 L of bacterial medium (*see* **Note 5**). Use overnight culture of HB101 to inoculate 6 L cultures. Grow these overnight at 37 °C with shaking.

4. The rest of the starter culture can be used to seed the 3× NGM petri dishes. Use sterile technique to add approximately 1–2 mL of HB101 liquid culture to each petri dish and spread by tilting plate from side to side. About ten seeded plates per bombardment will be required (*see* **Note 6**). Leave 2–3 days for HB101 to grow and for plates to dry further.

5. After approximately 24 h remove liquid cultures from incubator and spin down to pellet bacteria. Pour off supernatant and add 15 mL of M9 medium per 1 L of spun down culture. Resuspend bacterial pellet in M9 medium. Store in a refrigerator until required.

3.1.4 Growing Worms for Bombardment

1. Chunk or pick equal numbers of worms onto four seeded 9 cm 3× NGM plates (*see* **Note 7**). Leave at 15–25 °C for worms to grow, check worms regularly until the plates are just starved and full of L1 larvae.

2. Wash worms off the plates with M9 buffer.

3. Take an autoclaved Premier polypropylene filter bag (*see* **Note 8**) and place inside an autoclaved beaker. Fill beaker with S Medium until the bottom of the felt bag is submerged in S Medium. Pipette the worms in M9 into the filter bag, aiming for the medium in the bottom of the filter bag. Allow 5 min for the L1s to travel through the filter into the medium in the beaker and then remove and discard the bag. The filtration step will remove older animals and result in a culture that is almost exclusively comprised of L1 worms.

4. Pour the worms in S Medium from the beaker into an autoclaved flask. Top up with S Medium until the total volume is approximately 400 mL. Supplement the worm liquid culture with Pen/Strep/Amphotericin B, Nystatin and Kanamycin (*see* **Note 9**). The culture can be kept at this point for several days in a shaker at 15–20 °C.

5. To initiate growth of the worms, add the concentrated HB101 from the refrigerator (*see* **Note 10**). Continue to grow the worms until there are a mix of L4 larvae and young adults.

3.1.5 Bombardment

1. Take one unseeded 9 cm plate (*see* **Note 11**) per bombardment and place in a refrigerator.

2. For each intended bombardment wash seven biolistic macrocarriers and a rupture disk in clean isopropanol and leave to dry.

3. Settle liquid culture for at least 30 min until worms form a pellet in the bottom of the flask. Remove as much of the supernatant as possible without disturbing the pellet and distribute the worms into 9 × 15 mL falcon tubes.

4. Clean the worms by settling for approximately 5 min and removing the supernatant. Add clean M9 or S Medium, resuspend the worms, allow to settle again, and remove the supernatant. Repeat until the medium is clear.

5. Once the worms are in a clear clean medium, remove as much supernatant as possible, and transfer onto the cold dry plates (see **Note 12**). Spread the worms quickly across the surface of the plate by shaking and return the plate to the fridge. Check the plates regularly until all the media is soaked in the plate, due to the low temperature the worms should not move and remain spread across the surface.

6. Begin the bombardment by inserting the seven cleaned macrocarriers inside the wells of the macrocarrier holder of the hepta-adapter. Use the seating tool to ensure they are firmly held at the bottom of the wells.

7. Spread approximately 20 µL of the DNA on gold beads in ethanol (see Subheading 3.1.2) onto each carrier. To quickly evaporate the ethanol, place the macrocarrier holder into the vacuum chamber of the PDS-1000/He biolistic particle delivery system with the gold beads facing upward and switch on the vacuum pump. Observe each carrier until all ethanol evaporates and the gold beads appear as a dry powder.

8. Turn off the vacuum pump and allow the device to vent. Remove the macrocarrier holder and place a rupture disk into the seven-way gas divider and a stopping screen onto the stopping screen holder. Fully assemble the hepta-adapter in the vacuum chamber. Ensure that the outflows of the seven-way gas divider align with the macrocarriers in the macrocarrier holder. Take a 9 cm plate of worms from the refrigerator, remove the lid of the plate, and place on the target shelf beneath the hepta-adapter.

9. Turn on the vacuum pump and allow it to continue until the vacuum dial reaches 27.5 in. of mercury. Move the vacuum switch to the hold position. Then press and hold the fire switch until you hear the rupture disk break. Change the vacuum switch to the vent position and remove the plate.

10. Allow the worms to recover at room temperature for approximately 30 min, then split the bombardment plate into ten chunks. Transfer each chunk to a different seeded 9 cm 3× NGM plate. Leave the worms overnight at 15–25 °C.

3.1.6 Post-bombardment Selection

1. When the bombarded worms begin to lay eggs, add Hygromycin B solution to the plates to a final concentration of 0.3 mg/mL.

2. After adding Hygromycin B, if worms begin to starve, a small amount of concentrated HB101 can be added to keep the worms fed. It may also be necessary to add antibiotics such as Pen/Strep/Amphotericin B to try to minimize contamination on the selection plates.

3. The worms can then be left and should be checked regularly after 1 week for populations containing transformation markers arising.

4. Transformants with the desired marker or phenotype can then be transferred to a new seeded NGM plate with Hygromycin B.

3.2 Incorporation and Testing of UAA in C. elegans

3.2.1 Preparing UAA NGM Plates

1. Prepare as for normal NGM until after the autoclaving step (Subheading 2.1.2).

2. After autoclaving, aliquot partially prepared NGM into useful portions (*see* **Note 13**). Keep warm in a hybridization oven set to 55 °C.

3. Add 1 M phosphate buffer, 1 M magnesium sulfate, 1 M calcium chloride, and 5 mg/mL cholesterol in the same proportions as for normal NGM. After adding each component return to the hybridization oven.

4. Add 10% Triton-X100 to a final concentration of 0.1% (*see* **Note 14**).

5. Add UAA stock solution (*see* **Note 15**).

6. Use a sterile serological pipette to pour molten UAA NGM into 3.5 cm petri dishes. After agar sets and if the plates will not be used on the same day, move to cold storage in a refrigerator or a cold room.

3.2.2 Growth of C. elegans on UAA NGM Plates

1. Worms can be synchronized by picking gravid adults into a drop of 1:1 5% sodium hypochlorite bleach and 1 M sodium hydroxide. When the drop is soaked into the NGM, the adults should be dead but the embryos will remain alive and after hatching will provide a population of L1s. A somewhat more approximate method is to simply allow the worms to grow to the point of starvation and monitor them until they are mostly L1s.

2. Chunk or pick the L1s onto the UAA NGM plate, then add a small drop of concentrated HB101 onto the plate to maintain them. Allow worms to grow at 15–25 °C for 24–72 h.

3. If performing this experiment for the first time with a new UAA or a new aminoacyl tRNA synthetase, it is strongly recommended to perform an incorporation test (Subheading 3.2.3).

3.2.3 Fluorescence Assay to Test for Full-Length Product

1. Grow a population of transgenic worms containing the GFP::mCherry fusion with an amber codon between the two fluorescent proteins (Fig. 3) and synchronize at L1 either by starvation or bleaching.

2. Chunk each worm line onto one UAA NGM plate and one NGM plate without UAA.

3. Add HB101 and grow for 24–72 h.

4. Prepare a glass slide by melting 3% or 8% agarose in water and placing a drop on the slide. Press down a second glass slide rapidly to flatten the agarose into a pad before it sets (*see* **Note 16**).

5. Leave a few seconds to allow the agarose to set.

6. Remove one glass slide carefully while centering the agarose pad on the other.

7. Cut away the edges of the agarose pad to leave a flat square of agarose in the center of the slide.

8. Drop 5–10 μL of 5 mM levamisole or Polybead polystyrene 0.05 μm microspheres (if using 8% agarose) onto the agarose pad (*see* **Note 17**).

9. Pick worms expressing GFP from either the UAA NGM or NGM only plate into the drop on the pad.

10. Carefully mount a coverslip onto the pad.

11. Image worms in both the green and red channels under a microscope.

12. Repeat for the other sample and compare results between UAA NGM and NGM only worms. Full-length GFP::mCherry expression should only be seen in the presence of the UAA. The sample without UAA should show only green fluorescence.

3.2.4 Preparing a Worm Lysate for Western Blot

1. Add 300–600 L1 animals onto an NGM agar plate supplemented with UAA and seeded with HB101.

2. Grow to young adult stage.

3. Wash worms off plate with M9 (supplemented with 0.01% Triton X-100).

4. Let worms settle and remove as much buffer as possible.

5. Resuspend in 50–100 μL 4× LDS sample buffer supplemented with dithiothreitol to 10 mM.

6. Freeze at −20 °C, then thaw, and lyse by boiling for 10 min at 95 °C (with shaking if possible).

7. Perform Western blot using standard protocol (*see* **Note 18**).

3.3 Uncaging Photocaged UAAs in C. elegans

3.3.1 Global Uncaging of Photocaged Lysine (PCK) in C. elegans (See Note 19)

1. Wash L4/adult worms from an NGM plate supplemented with PCK by adding approximately 200 µL of M9 media (supplemented with 0.01% Triton X-100) covering surface of the plate.

2. Remove 80 µL of worms suspended in M9, and add worms to a new unseeded NGM plate (*see* **Note 20**).

3. Repeat **step 2**.

4. Label one plate as +UV and the other as −UV.

5. Place the +UV plate in 365 nm UV crosslinker. Alternatively, LED light sources of the same wavelength can be used.

6. Illuminate with ~5 mW/cm² for 30 s to 5 min, to be determined experimentally.

7. After an appropriate time interval, depending on the phenotypical effect expected after uncaging, compare observations on the +UV and −UV sample (*see* **Note 21**).

3.3.2 Targeted Uncaging of PCK in C. elegans

1. Prepare a glass slide with an 8% agarose pad (*see* **Note 22**).

2. Add a 5–10 µL drop of Polybead polystyrene 0.05 µm microspheres to the agarose pad.

3. Pick L4/adult worms from the PCK NGM plate into the microsphere droplet.

4. Carefully mount the coverslip on top of the agarose pad (taking care to avoid lateral motion that may result in the Polybead polystyrene 0.05 µm microspheres damaging the worms).

5. Mount and image on a fluorescence microscope (*see* **Note 23**). The target cells can be illuminated at 365 nm using either ROI scanning on a confocal microscope, or by using a micropoint module integrated with the microscope. Illumination time and power parameters are determined experimentally, but it is recommended to start by stepping down from a photobleaching intensity.

6. After uncaging carefully remove the coverslip taking care to avoid damaging the worms.

7. Wash the worms onto a new seeded NGM plate and allow time for recovery.

4   Notes

1. At all steps involving centrifugation of beads care should be taken that beads do not clump together. We centrifuge for a very short duration at low rpm in a tabletop centrifuge.

2. Plasmids should be linearized prior to coating onto gold beads as this will allow them to form arrays for expression in

C. elegans. It is important that the restriction enzymes used do not cut within any important elements in the construct including coding sequence, promoters and terminators both for the incorporation components and for the selection marker. Furthermore, where multiple DNA constructs are to be bombarded together, care should be taken to ensure that the enzymes used can be heat inactivated to prevent unwanted cuts in the other constructs after mixing.

3. We recommend using an excess of all constructs that do not contain visual or antibiotic selection markers. For example, more than half of the total DNA used in our bombardments usually consists of $tRNA_{CUA}$ constructs as these are not observable by fluorescence and do not contain the Hygromycin B resistance gene. A typical bombardment mixture will consist of 2 µg synthetase plasmid (with Hygromycin B resistance), 2 µg plasmid with target gene (containing an amber stop codon to direct UAA incorporation), and 6 µg tRNA plasmid (with four copies of the tRNA expression cassette).

4. DNA precipitation ought to be complete in 5 min, but we usually incubate for at least 30 min in case of lower efficiency. We directly precipitate the DNA from the restriction enzyme buffer without prior purification.

5. Any bacterial growth medium can be used, but we find best results with rich medium such as TB or 2× TY. Using baffled flasks to increase aeration will also increase the yield of bacteria.

6. Fewer 3× NGM plates can be used if necessary, however, the fewer plates used the harder it will be to isolate independent transformants from a single bombardment. Conversely, one can also use more plates to increase confidence in the independence of the lines isolated.

7. Although *C. elegans* N2 are the most commonly used strain, the nonsense mediated decay (NMD) machinery present in these worms may result in degradation of the mRNA of the target protein. For the purpose of genetic code expansion it can therefore be advantageous to use strains without NMD. *C. elegans* employ a NMD system to identify and destroy mRNAs containing premature stop codons. In mammalian cells NMD is linked to mRNA splicing and intron-free transcripts are protected from NMD [25]. In *C. elegans* this is not the case and mRNAs containing premature stop codons are thought to be identified through their distance to the 3′ UTR of the mRNA [26, 27]. While NMD does not degrade all of the transcript, it may result in a dramatic lowering of target protein expression levels. Unlike many other animals, worms tolerate knock-outs of the NMD machinery well. The animals

are healthy, fertile and apart from slight developmental effects do not show any obvious phenotypes. If higher levels of UAA containing protein are required it can therefore be advantageous to inactivate NMD. Both genomic knock-outs and RNAi knock-down of the NMD machinery have been used for this purpose [18, 19].

8. We use premier polypropylene felt bags with a 5 μm pore size (available from GAPS water treatment). These allow L1s to pass through the walls of the bag into the surrounding medium while older worms are retained within the bag. When the bag and its contents are discarded, the remaining medium consists almost entirely of L1 worms.

9. Antibiotics and antifungals are added to prevent contamination of the liquid culture.

10. HB101 can be added until the S Medium becomes opaque. At no point during the growth of the worms should the medium be allowed to become clear.

11. Plates used at this point should not be freshly poured but should be dry in order to absorb any excess liquid when the worms are spread onto them. We find that plates over 2 weeks old work well. If older plates are not available, fresh plates can be dried quickly either in a 37 °C incubator or in a hybridization oven at 55–70 °C.

12. After settling in the 15 mL falcon tubes the settled worms should take up at least 1 mL of volume, usually we obtain approximately 1.5 mL of settled worms per tube.

13. We often keep 25 mL aliquots of NGM without adding anything after autoclaving. These aliquots can be melted in a microwave as required and the further components added.

14. Triton-X100 helps to enhance the uptake of the UAA by the worms.

15. This step involves the use of an UAA stock solution which will depend on the nature of the UAA to be used. We recommend starting with a final concentration of 5 mM when testing new UAA. For example, when making plates containing photocaged lysine (PCK), we add the required amount of PCK to 1 mL of water, we then add 20 μL of 1 M HCl to help dissolve the PCK. After adding the PCK solution to 25 mL of NGM we neutralize with 20 μL of 1 M NaOH. However, different UAAs will have different requirements, some may be directly soluble in water, while more hydrophobic amino acids may have to first be suspended in a solvent such as DMSO. It is recommended that when using a new UAA, the experimenter first check the preparations used in the literature before making a stock solution. Furthermore, if using photosensitive

amino acids all steps, including preparation of the stock solution, should be carried out in the dark or under red light. After transfer of the worms to plates containing the UAA, the time until incorporation product appears will depend both upon the amino acid and the *C. elegans* tissues in which expression is required.

16. The agarose pad should be pressed as thin and even as possible. One way to do this is to place two glass slides on a bench top in parallel at a distance of 30–50 mm apart. Fix the two slides to the bench by completely taping over them with a single layer of autoclave tape (or tape of equivalent thickness). Place a third glass slide between the two fixed to the bench in parallel to the others. Place a drop of molten agarose on the middle slide and drop a fourth slide in a perpendicular orientation onto the agarose allowing it to settle until the perpendicular slide rests upon the fixed slides. This method should result in a flat, even agarose pad of roughly the thickness of the tape used.

17. If the worms are to be used for further experiments then we recommend the use of Polybead polystyrene 0.05 μm microspheres for immobilization as levamisole can have long-lasting effects upon the behavior of the worms.

18. The amount of incorporation product can be very low, especially if the protein is expressed in only a small number of cells. Reliable detection therefore requires specific antibodies with high sensitivity and low background. Since such antibodies are not available for all proteins, we tend to add an easily detectable tag to our protein of interest. The tag is at the C-terminus and will only be present in full-length protein where translation has progressed past the amber stop codon. For this purpose, we use either an HA tag (detected with monoclonal rat anti-HA (clone 3F10) antibody from Roche), or a GFP tag (mouse monoclonal anti-GFP (clones 7.1 &13.1) from Roche). These two antibodies are very sensitive and show no background in *C. elegans* lysates even when using high sensitivity substrates such as SuperSignal West Femto Maximum Sensitivity Substrate (Thermo Fisher Scientific) together with horseradish peroxidase coupled secondary antibodies.

19. The given protocol is for a photocaged lysine (PCK), which can be uncaged at 365 nm, other photoactivatable UAAs may require different light sources depending on the wavelength of uncaging.

20. Due to absorption by the NGM agar it will not be possible to take up the full 200 μL of M9 medium.

21. We also recommend using a control which was exposed to UV illumination after being grown in the absence of UAA.

22. This should be melted very slowly to prevent drying of the agarose. It may also be preferable to use low melting point agarose.

23. We generally co-express a GFP with the target gene to label the corresponding cells and facilitate targeting. We do not observe any uncaging from the wavelength used to excite GFP when using photo-caged lysine [14], however we recommend determining the stability of the caging group experimentally for each new photocaged UAA.

References

1. Chin JW, Cropp TA, Anderson JC et al (2003) An expanded eukaryotic genetic code. Science 301:964–967. https://doi.org/10.1126/science.1084772

2. Wang L, Brock A, Herberich B, Schultz PG (2001) Expanding the genetic code of *Escherichia coli*. Science 292:498–500. https://doi.org/10.1126/science.1060077

3. Ye S, Huber T, Vogel R, Sakmar TP (2009) FTIR analysis of GPCR activation using azido probes. Nat Chem Biol 5:397–399. https://doi.org/10.1038/nchembio.167

4. Ye S, Zaitseva E, Caltabiano G et al (2010) Tracking G-protein-coupled receptor activation using genetically encoded infrared probes. Nature 464:1386–1389. https://doi.org/10.1038/nature08948

5. Li C, Wang G-F, Wang Y et al (2010) Protein (19)F NMR in *Escherichia coli*. J Am Chem Soc 132:321–327. https://doi.org/10.1021/ja907966n

6. Jackson JC, Hammill JT, Mehl RA (2007) Site-specific incorporation of a 19F-amino acid into proteins as an NMR probe for characterizing protein structure and reactivity. J Am Chem Soc 129:1160–1166. https://doi.org/10.1021/ja064661t

7. Mori H, Ito K (2006) Different modes of SecY–SecA interactions revealed by site-directed in vivo photo-cross-linking. Proc Natl Acad Sci 103:16159–16164

8. Chin JW, Martin AB, King DS et al (2002) Addition of a photocrosslinking amino acid to the genetic code of *Escherichia coli*. Proc Natl Acad Sci 99:11020–11024

9. Nikić I, Kang JH, Girona GE et al (2015) Labeling proteins on live mammalian cells using click chemistry. Nat Protocols 10:780–791. https://doi.org/10.1038/nprot.2015.045

10. Plass T, Milles S, Koehler C et al (2012) Amino acids for Diels-Alder reactions in living cells. Angew Chem Int Ed 51:4166–4170. https://doi.org/10.1002/anie.201108231

11. Lang K, Davis L, Torres-Kolbus J et al (2012) Genetically encoded norbornene directs site-specific cellular protein labelling via a rapid bioorthogonal reaction. Nat Chem 4:298–304. https://doi.org/10.1038/nchem.1250

12. Lang K, Davis L, Wallace S et al (2012) Genetic encoding of bicyclononynes and trans-cyclooctenes for site-specific protein labeling in vitro and in live mammalian cells via rapid fluorogenic Diels-Alder reactions. J Am Chem Soc 134:10317–10320. https://doi.org/10.1021/ja302832g

13. Uttamapinant C, Howe JD, Lang K et al (2015) Genetic code expansion enables live-cell and super-resolution imaging of site-specifically labeled cellular proteins. J Am Chem Soc 137:4602–4605. https://doi.org/10.1021/ja512838z

14. Gautier A, Nguyen DP, Lusic H et al (2010) Genetically encoded Photocontrol of protein localization in mammalian cells. J Am Chem Soc 132:4086–4088. https://doi.org/10.1021/ja910688s

15. Gautier A, Deiters A, Chin JW (2011) Light-activated kinases enable temporal dissection of signaling networks in living cells. J Am Chem Soc 133:2124–2127. https://doi.org/10.1021/ja1109979

16. Nguyen DP, Mahesh M, Elsasser SJ et al (2014) Genetic encoding of photocaged cysteine allows photoactivation of TEV protease in live mammalian cells. J Am Chem Soc 136:2240–2243. https://doi.org/10.1021/ja412191m

17. Luo J, Arbely E, Zhang J et al (2016) Genetically encoded optical activation of DNA recombination in human cells. Chem Commun 52:8529–8532. https://doi.org/10.1039/C6CC03934K

18. Greiss S, Chin JW (2011) Expanding the genetic code of an animal. J Am Chem Soc 133:14196–14199. https://doi.org/10.1021/ja2054034

19. Parrish AR, She X, Xiang Z et al (2012) Expanding the genetic code of *Caenorhabditis*

elegans using bacterial aminoacyl-tRNA synthetase/tRNA pairs. ACS Chem Biol 7:1292–1302. https://doi.org/10.1021/cb200542j

20. Radman I, Greiss S, Chin JW (2013) Efficient and rapid *C. elegans* transgenesis by bombardment and hygromycin B selection. PLoS One 8:e76019. https://doi.org/10.1371/journal.pone.0076019

21. Lammers C, Hahn LE, Neumann H (2014) Optimized plasmid systems for the incorporation of multiple different unnatural amino acids by evolved orthogonal ribosomes. Chembiochem 15(12):1800–1804. https://doi.org/10.1002/cbic.201402033

22. Schmied WH, Elsasser SJ, Uttamapinant C, Chin JW (2014) Efficient multisite unnatural amino acid incorporation in mammalian cells via optimized pyrrolysyl tRNA synthetase/tRNA expression and engineered eRF1. J Am Chem Soc 136:15577–15583. https://doi.org/10.1021/ja5069728

23. Bianco A, Townsley FM, Greiss S et al (2012) Expanding the genetic code of *Drosophila melanogaster*. Nat Chem Biol 8:748–750. https://doi.org/10.1038/nchembio.1043

24. Lyssenko NN, Hanna-Rose W, Schlegel RA (2007) Cognate putative nuclear localization signal effects strong nuclear localization of a GFP reporter and facilitates gene expression studies in *Caenorhabditis elegans*. BioTechniques 43:596. 598, 560

25. Kervestin S, Jacobson A (2012) NMD: a multifaceted response to premature translational termination. Nat Rev Mol Cell Biol 13:700–712. https://doi.org/10.1038/nrm3454

26. Pulak R, Anderson P (1993) mRNA surveillance by the *Caenorhabditis elegans smg* genes. Genes Dev 7:1885–1897

27. Longman D, Plasterk RHA, Johnstone IL, Cáceres JF (2007) Mechanistic insights and identification of two novel factors in the *C. elegans* NMD pathway. Genes Dev 21:1075–1085. https://doi.org/10.1101/gad.417707

INDEX

Edward A. Lemke (ed.), *Noncanonical Amino Acids: Methods and Protocols*, Methods in Molecular Biology, vol. 1728,
https://doi.org/10.1007/978-1-4939-7574-7, © Springer Science+Business Media, LLC 2018

Printed in the United States
By Bookmasters